Natural Hazards and Disaster Justice

"Natural Hazards and Disaster Justice is a timely and notable contribution to an immensely significant and generally neglected area of research. The neglect is surprising given that Injustice features so prominently in disaster preparedness, prevention, response and recovery, with huge social, economic and political consequences. This important book not only usefully describes many of the theoretical underpinnings of these consequences, but also incorporates studies from within Australia and the Indo-Pacific region to illustrate how they play out in practice. Its cogent conclusions are particularly relevant in the era of climate change, which is greatly increasing the frequency and severity of hazards and, as a consequence, amplifying disaster injustice."
—Robert Glasser, Visiting Fellow, *Australian Strategic Policy Institute, and former United Nations Special Representative of the Secretary General for Disaster Risk Reduction*

"I congratulate the editors and contributing authors for a comprehensive, insightful, diverse and provocative book. In an age of unprecedented opportunity to shape our future whilst at the same time creating unprecedented risks that threaten to destroy our existence, disaster justice must play a key role in striking the balance for a safer, prosperous, equitable and sustainable world. This book represents some of the most dynamic thinking in the relationship between disaster justice and resilience, risk reduction and climate. It provides great appeal for all public and private policy makers, strategists and tacticians to escalate disaster justice as a central ethical consideration. It is a must read for everyone."
—Mark Crossweller, Former Director General, *Emergency Management Australia, and Head of the National Resilience Taskforce*

Anna Lukasiewicz • Claudia Baldwin
Editors

Natural Hazards and Disaster Justice

Challenges for Australia and Its Neighbours

Editors
Anna Lukasiewicz
Fenner School of Environment and
Society and Institute for Integrated
Research on Disaster Risk Science
Australian National University
Canberra, ACT, Australia

Claudia Baldwin
Urban Design and Town Planning
and Sustainability Research Centre
University of the Sunshine Coast
Maroochydore, QLD, Australia

ISBN 978-981-15-0468-6   ISBN 978-981-15-0466-2 (eBook)
https://doi.org/10.1007/978-981-15-0466-2

© The Editor(s) (if applicable) and The Author(s), under exclusive licence to Springer Nature Singapore Pte Ltd. 2020
This work is subject to copyright. All rights are solely and exclusively licensed by the Publisher, whether the whole or part of the material is concerned, specifically the rights of translation, reprinting, reuse of illustrations, recitation, broadcasting, reproduction on microfilms or in any other physical way, and transmission or information storage and retrieval, electronic adaptation, computer software, or by similar or dissimilar methodology now known or hereafter developed.
The use of general descriptive names, registered names, trademarks, service marks, etc. in this publication does not imply, even in the absence of a specific statement, that such names are exempt from the relevant protective laws and regulations and therefore free for general use.
The publisher, the authors and the editors are safe to assume that the advice and information in this book are believed to be true and accurate at the date of publication. Neither the publisher nor the authors or the editors give a warranty, expressed or implied, with respect to the material contained herein or for any errors or omissions that may have been made. The publisher remains neutral with regard to jurisdictional claims in published maps and institutional affiliations.

Cover illustration: Jennifer Hart / Alamy Stock Photo

This Palgrave Macmillan imprint is published by the registered company Springer Nature Singapore Pte Ltd.
The registered company address is: 152 Beach Road, #21-01/04 Gateway East, Singapore 189721, Singapore

Preface

With the occurrence, cost and human impacts of disasters increasing in Australia and worldwide, attention inevitably focuses on who bears the burden and impacts of these events. Media, political debates and numerous post-event inquiries (142 in Australia between 2009 and 2017) have exposed a multitude of issues around fairness and equity.

Under the broad policy settings of the Sendai Framework and Australia's National Disaster Resilience Strategy, addressing the area of 'natural disasters' has shifted from a 'prepare and respond' approach to one that extends across the Preparation-Prevention-Response-Recovery (PPRR) spectrum, involves shared responsibilities and is more whole-of-society and whole-of-government.

Disaster events reveal, magnify and deepen existing injustices and can create future inequalities which lead to further injustice. Disaster justice focuses on these revealed inequities, whether they be spatially, socio-economically or politically created, seeking to better understand the uneven vulnerabilities and the policy and management options that can address them.

Disaster justice is a new and fluid area of research, drawing on multiple disciplines such as social justice, law, human geography, land use planning and social psychology. Areas such as participatory governance can provide insight into procedural justice issues (e.g. involvement in planning) that may later influence distributive justice (i.e. who gets resources or who is impacted), the latter topic further enriched by understanding of resource allocation, economics, law and social services. Disaster justice can provide fresh and useful insights into questions such as:

- Are responsibilities for disaster prevention, preparation, response and recovery as fairly distributed across society as they can be?
- In what ways are we proactive and successful in addressing both geographic and social vulnerabilities?
- How do pre-existing inequities impact on individual or community capacity to engage in aspects of disaster prevention, preparation, response and recovery?
- Do we share knowledge and opportunities for influencing disaster policy and emergency management in a just manner?
- Are existing policy and management approaches capable of delivering just processes and outcomes and of doing so in a way that the Australian public would accept as appropriate in this emotionally and politically charged policy domain?

On 19–20 November 2018, researchers from multiple Australian research institutions involved in disaster justice met in Canberra, Australia in preparation for writing this book. They discussed the meaning, importance and future of disaster justice research in Australia and its surrounds. While globally, disasters include earthquakes and terrorism, our contributors focused mainly on disasters that occur regularly in Australia, specifically floods, fires and storms. While this book focuses on natural hazards, there is nothing stopping others from applying disaster justice to manmade disasters. The book highlights issues discussed at that meeting of relevance to policy, practice and future research. It is divided into three parts: an introduction; a section on governance, including policy, planning, legal and economic approaches; and a third on vulnerability that reveals the implications of inequitable exposure to risk of disasters due to locational and social vulnerabilities but also suggestions for how to address this.

The Introduction consists of two chapters. Anna Lukasiewicz introduces the concept of disaster justice examining definitions including that developed during the workshop. She focuses on defining the characteristics of disaster justice and delineating it from other types of justice. She describes four inter-related themes that make disaster justice a unique concept; first, the Anthropocene places moral obligations on those responsible for disaster management. Second, the severity of disaster impacts is dependent not just on the force of the natural hazard but also on the social, economic and political vulnerabilities of the impacted populations. Third, the above means that governance is critical for all stages of the

PPRR spectrum; and fourth, this focus on governance emphasises the interlinkage between distributional and procedural dimensions of justice, central to which is recognition and empowerment.

Authors in this book provide real cases of issues of justice related to rapid onset natural disasters, acknowledging that slow incremental changes due to climate change (e.g. drought, sea level rise, increasing temperatures and heat waves) also require ongoing adaptation as further instances of inequity and social and environmental injustice emerge. In Australia, our most common rapid disasters are due to flooding, fire, cyclones and severe storms. Rebecca Colvin, Steven Crimp, Sophie Lewis and Mark Howden in Chap. 2 provide a well-rounded introduction to Australia's climate changed future. As global citizens and recognising that climate change and natural disasters are global problems, the book also includes learning from cases in southern and south-east Asia, our near neighbours.

Disasters 'stress test' our systems of governance, institutions, finance, livelihoods and social cohesion. Stephen Dovers' chapter leads the second section on governance, posing questions related to public policy, politics, power and responsibility. The next two authors explore the ability of planning and the legal system to adapt to Australia's flammable landscapes and whether property rights are an issue.

Jason Alexandra focuses on spatial planning, drawing on his own experiences in Victoria, Australia, a region of intensely destructive wildfires, to explore the prospects and challenges of using integrated planning for mitigating bushfire disasters. He highlights the tension between planning regimes that attempt to restrict population growth in bushfire-prone areas, and the socio-economic factors, such as housing affordability that encourage this growth.

Alan March, Leonardo Nogueira de Moraes and Janet Stanley continue the focus on planning in Victoria, as they highlight the balance of individual property rights versus collective rights and the public good. They examine a 2015 bushfire in a rural locality and consider the interplay of assumed, asserted and contested rights before, during and after a major and destructive bushfire. They end their chapter with some suggestions for new directions in the allocations of rights in terms of managing for the public good.

Tayanah O'Donnell continues the focus on private rights and public goods, albeit in the legal domain. She explores litigation involving the protection of private properties in Byron Bay (in the state of New South Wales), on the Australian East Coast. In this context, and via an analysis of case law, her chapter contemplates who the major beneficiaries of such 'justice' truly are.

Staying in the legal domain, Michael Eburn considers the legal consequences of some notable disasters and the implications of a desire to look to law for 'justice'. By examining post-disaster inquiries from recent Australian natural disasters, he persuasively argues that the adversarial legal system, which centres on proportioning blame, is not the best way to deliver 'justice' and that instead, restorative justice practices could be adopted.

In an economic analysis of post-disaster recovery, Steven Schilizzi and Masood Azeem confront the economic reality of recovery from the 2010 flood in Pakistan, in a resource-constrained environment which forces examination of equity-efficiency trade-offs, questioning feasibility of 'building back better'.

The third section delves into Vulnerability and Resilience, highlighting how disasters reveal uneven levels of exposure to risk and perpetuate socio-economic inequities in mitigating and preparing for risk, and variable levels of capacity to cope with disturbance. Authors raise how vulnerable people—the poor, those with disabilities, and some often overlooked such as youth and those who do not have recognisable disabilities, are excluded from PPRR with serious consequences.

Mohammad Shahidul Hasan Swapan, Md. Ashikuzzaman and Md. Sayed Iftekhar examine socio-economic factors that affect access to pre- and post-disaster support by the residents of informal settlements in a flood affected area of Bangladesh. Their chapter clearly links standard measures of poverty (such as income, land ownership, health conditions and membership of a female-headed household) to access to disaster support programs.

Kathryn Allan and James Mortensen then reveal the implications of lack of legal identity and formal documentation through statelessness and displacement, and the impact that can have on access to emergency relief, humanitarian aid, social benefits and freedom of movement. Their chapter focuses on 'stateless' people, such as refugees, a group whose vulnerability is well established and whose disadvantages are compounded during and after disasters.

Harriot Beazley's chapter focuses on another recognised vulnerable group—children. Using a case study centred in East Lombok, Indonesia, Beazley explores children's experiences of the impacts of earthquakes on children's lives and the life of their community. She promotes the adoption of a Child-Centred Disaster Risk Reduction (CCDRR) policy and the importance of participatory research.

Drawing attention to the broader implications of social inequity, and continuing on the theme of youth, Janet Stanley reports on research identifying the links between poor socio-economic circumstances and likelihood of arson leading to bushfires. Her chapter focuses on disadvantaged youth in the peri-urban areas of Melbourne, in the state of Victoria.

Claudia Baldwin then calls for resilient communities that can better share responsibility to minimise impact of disasters and recover quickly post-disaster. Her case study of flooding in the Brisbane suburb of Rocklea in 2011 provides an example of the unrecognised characteristics of business owners which provided barriers to access to timely information during the flood emergency. It reinforces the need to build social capital and empowerment through participatory (procedurally just) approaches.

Valerie Ingham, Mir Rabiul Islam, John Hicks and Oliver Burmeister focus on another group, that of local community leaders, whose vulnerabilities are not instantly recognisable. Their case study is set in the small, close-knit community of Blue Mountains (in the state of New South Wales) and examines the impact of the devastating 2013 fires on the role of local leaders and their participation in emergency management decision-making processes. Their chapter highlights the need for recognising, resourcing and supporting the community sector.

The themes of community resilience, and the 2013 Blue Mountains fires, are further explored in the chapter of David Schlosberg, Hannah Della Bosca and Luke Craven. Their chapter focuses on the residents' and emergency service providers' experiences of the fires using a place attachment framework. They explore how these experiences of disrupted or destroyed place attachment were connected to feelings of loss in ways that impaired both individual and community identity and wellbeing, described as an injustice.

The last two chapters in this section continue to focus on vulnerable groups, but they highlight how the knowledge and capacities of such groups can enrich disaster management. Jessica Weir, Steve Sutton and Gareth Catt use three case studies in different parts of Australia to demonstrate the benefit of respecting Aboriginal peoples' knowledge of fire management practices and the contribution they can make to contemporary bushfire risk.

As a constructive note to end the individual author contributions, Emma Calgaro, Genevieve Roberts and Michelle Villeneuve not only identify barriers to inclusion of people with disabilities but provide three case studies which provide positive examples of how to ensure universal accessibility for persons with different disabilities (including deaf people) to participate in emergency management in a non-discriminatory way. Two cases were set in New South Wales and one provided guidelines for DiDRR (i.e. Disability-inclusive Disaster Risk Reduction) in collaboration with institutional partners in Southeast Asia.

As the book editors, Anna Lukasiewicz and Claudia Baldwin summarise key issues raised in the book and identify ways forward through policy and a suggested research agenda. As the contributions in this book demonstrate, disaster justice is a rich field of research, representing many disciplines (legal, economic, planning, policy and social science), and utilising diverse methods, from very quantitative analyses (such as the chapters by Schilizzi and Azeem and Swapan et al.) to qualitative case studies. As disaster justice is still a nascent concept, these contributions engage with justice in diverse and subtle ways; however, all were chosen because they provide insights into disaster justice, and hopefully the motivation to reduce disparities revealed through disasters.

Canberra, ACT, Australia Anna Lukasiewicz
Maroochydore, QLD, Australia Claudia Baldwin

Acknowledgements

This book is the culmination of two years of effort and collaboration.

First, we would like to acknowledge the institutional support that made both the workshop and the book possible. We have received generous funding support from both the Australian National University's (ANU) Climate Change Institute and the Bushfire and Natural Hazards Cooperative Research Centre for the workshop on Disaster Justice in November 2018. We would also like to thank the Fenner School for Environment and Society, for hosting the workshop. This workshop was the first to tackle the concept of disaster justice in Australia and was a crucial foundation for this book.

We are also grateful to a host of external reviewers who generously took the time to review individual contributions, sometimes at short notice. We would like to acknowledge Rajib Shaw, Matthew Colloff, Helen Ross, Laura Stough, Gordon Walker, Sara Wright, Dave Mercer, Scira Menoni, Enid Robinson, John Bates, Ejaz Qureshi, Bruce Thom, Rosemary Lyster, Ashraful Aram, Jennifer McKay, Blythe McLennan and Julie Davidson.

Finally, we would like to thank each and every workshop participant and chapter contributor (most of whom also reviewed chapters, sometimes more than once). A special thank you is owed to Stephen Dovers who diligently reviewed the most chapters.

Thank you all, without your efforts, this book would not be possible.

Contents

Part I Introduction — 1

1 The Emerging Imperative of Disaster Justice — 3
Anna Lukasiewicz
- 1.1 Why a Book on Disaster Justice? — 3
- 1.2 What Is a Disaster? — 4
- 1.3 What Is Justice? — 5
- 1.4 What Is Disaster Justice? — 8
- 1.5 What Makes Disaster Justice Unique? — 13
- 1.6 Conclusion — 19
- References — 19

2 Implications of Climate Change for Future Disasters — 25
Rebecca Colvin, Steve Crimp, Sophie Lewis, and Mark Howden
- 2.1 Introduction — 25
- 2.2 Climate Change: The Fundamental Physical Mechanics — 27
- 2.3 Future Emissions Reductions Trajectories — 28
- 2.4 Australia's Future Under Climate Change — 28
- 2.5 Heatwaves — 29
- 2.6 Drought — 31
- 2.7 Bushfire — 34
- 2.8 Flood and Coastal Inundation — 36

2.9	Interactions in the Real World	39
2.10	Looking Towards Disaster Justice to a Climate Changed Future	40
References		41

Part II Governance   49

3 Public Policy and Disaster Justice   51
Stephen Dovers

3.1	Introduction	51
3.2	Where Are the Boundaries of 'Disaster Justice Policy'?	53
3.3	Policy or Politics?	56
3.4	What Part of the Policy Process?	58
3.5	Whose Responsibility?	63
3.6	Disaster Policy, or Something Else?	66
3.7	New Mechanisms, or the Same but Better?	67
3.8	Discussion and (In)Conclusions	67
References		70

4 Burning Bush and Disaster Justice in Victoria, Australia: Can Regional Planning Prevent Bushfires Becoming Disasters?   73
Jason Alexandra

4.1	Introduction and Overview		73
4.2	Learning to Live in a Flammable Continent		75
4.3	Welcome to the Pyrocene		77
	4.3.1	A Flammable Planet and Its Pyrocentric Civilisation	77
	4.3.2	Burning Continent: Fire Formed Australia	78
	4.3.3	Fire in Contemporary Australia	78
	4.3.4	Bushfire Disaster Mitigation and Integrated Regional Planning	79
4.4	Issues for Disaster Justice		82
	4.4.1	Land Use Planning, Vulnerability and Disaster Justice	82
	4.4.2	Reinventing Land Use Planning: A Scenario	84
4.5	Conclusions		86
References			88

5 Dimensions of Risk Justice and Resilience: Mapping Urban Planning's Role Between Individual Versus Collective Rights   93
Alan March, Leonardo Nogueira de Moraes, and Janet Stanley
   5.1   Introduction   93
   5.2   Bushfire Behaviour and the Importance of Settlement Design and Location   95
   5.3   Risk Justice: Hazards, Vulnerability and Exposure to Hazards   97
   5.4   Urban Resilience and Problematics of Disaster Justice: Rights and the Public Good   99
   5.5   Wye River and Separation Creek   101
   5.6   The Christmas Day Fire   102
   5.7   Post-Event Assessment of Risk Factors   103
   5.8   Risk Justice: Entanglements Between Rights and the Public Good   105
   5.9   More or Less Resilient? Limits and Possibilities for Urban Management   107
   5.10 Conclusions   111
   References   111

6 Climate Change Adaptation Litigation: A Pathway to Justice, but for Whom?   117
Tayanah O'Donnell
   6.1   Introduction   117
   6.2   The Legislative Context   120
   6.3   Byron Bay   122
   6.4   The Byron Litigation   123
   6.5   Justice for the Coast   126
   6.6   Conclusion   128
   References   128

7 Looking to Courts of Law for Disaster Justice   133
Michael Eburn
   7.1   Introduction   133
   7.2   Poor Outcomes Are Not Necessarily Disasters   134
   7.3   Looking to the Courts of Law to Deliver Disaster Justice   135
         7.3.1   Post-Event Inquiries   141

	7.4	A New Approach: Adopting Restorative Practices	142
		7.4.1 Sharing Responsibility	142
		7.4.2 The Use of Restorative Justice Beyond Criminal Law Is Not Unique and Is Growing	144
	7.5	Compensation	145
	7.6	Conclusion	147
	References		148

8 How to Be Fair in Prioritizing Support in the Aftermath of Disasters: Pakistan's Housing Reconstruction Challenges Following the 2010 Flood Disaster — 151
Steven Schilizzi and Muhammad Masood Azeem

8.1	Introduction	151
8.2	Analysing the Problem	155
8.3	The Impact of Equity: Efficiency Trade-Offs	160
8.4	Conclusion	163
Appendix		164
References		165

Part III Vulnerability and Resilience — 167

9 Equitable Access to Formal Disaster Management Programmes: Experience of Residents of Urban Informal Settlements in Bangladesh — 169
Mohammad Shahidul Hasan Swapan, Md. Ashikuzzaman, and Md. Sayed Iftekhar

9.1	Introduction		169
9.2	Study Area		173
9.3	Methods and Materials		175
	9.3.1	Data Collection	175
9.4	Results and Discussion		176
	9.4.1	Respondents' Characteristics	176
	9.4.2	Support to Recover from Disaster	177
	9.4.3	Relationship Between Level of Supports and Socio-Economic Conditions	178
9.5	Conclusion		181
References			181

10	**Children's Experiences of Disaster: A Case Study from Lombok, Indonesia**	**185**
	Harriot Beazley	
	10.1 Introduction	185
	10.2 Child-Centred Disaster Risk Reduction	186
	10.3 Methodology	188
	10.4 Research Findings	188
	10.4.1 Disaster Preparedness and Recovery	190
	10.4.2 Five Months On	192
	10.4.3 Psychological Responses	194
	10.4.4 Trauma Healing	196
	10.5 Post-Disaster Education and Disaster Risk Reduction	197
	10.6 Conclusion	198
	References	200
11	**How a Failure in Social Justice Is Leading to Higher Risks of Bushfire Events**	**205**
	Janet Stanley	
	11.1 Introduction	205
	11.2 The Cause of Ignition of Bushfires in Australia	206
	11.3 Who Lights Fires and Why	207
	11.4 Where Bushfires Occur	208
	11.5 Why Bushfires Occur in This Interface Area	208
	11.6 Connecting Disadvantage and Bushfires	213
	11.7 What Is the Role of Prevention?	215
	References	217
12	**Issues of Disaster Justice Confronting Local Community Leaders in Disaster Recovery**	**221**
	Valerie Ingham, Mir Rabiul Islam, John Hicks, and Oliver Burmeister	
	12.1 Introduction	221
	12.2 2 October 2013, Blue Mountains Fires	223
	12.3 The Intended Outcome of Policy Intervention	225
	12.3.1 Shared Responsibility and Resilience Policy	225
	12.3.2 Emergency Management at State and Municipal Level	226
	12.3.3 Community Sector Leadership	226
	12.3.4 Summary	227
	12.4 Approach to the Study	228

12.5	What Is Happening		229
	12.5.1	Inequality of Voice in Disaster Planning and Recovery	229
	12.5.2	Inequality of Funding vs. FACS and Office of Emergency Management Expectations	231
	12.5.3	Who Leads the Recovery?	232
	12.5.4	Community Leader Wellbeing: Burn Out, Guilt and Health Issues	233
	12.5.5	Blurring of Professional and Personal Boundaries	234
12.6	Recognising the Disaster Justice Issues		234
12.7	Recommendations		236
	12.7.1	Funding Policy to Backfill Community Leader Positions	236
	12.7.2	Structural Changes to the Composition of the Local Emergency Management Committee	236
References			237

13 Disaster, Place, and Justice: Experiencing the Disruption of Shock Events — 239
David Schlosberg, Hannah Della Bosca, and Luke Craven

13.1	Introduction	239
13.2	Place Attachment and Resilience: Identity, Emotion, and Imagined Futures	241
13.3	Methodological Context	246
13.4	The Experience of the Blue Mountains Bushfires	248
13.5	Discussion and Conclusions	253
References		256

14 Legal Identity Documenting in Disasters: Perpetuating Systems of Injustice — 261
Kathryn Allan and James Mortensen

14.1	Introduction		261
14.2	Background		262
14.3	How Is Legal Identity Currently Used in the Disaster Context?		263
	14.3.1	Pre-disaster Planning and Risk Reduction	263
	14.3.2	Response Phase: Humanitarian Aid Distribution, Healthcare, Social Benefits, Freedom of Movement	265

		14.3.3 Post-disaster Response Phase	267

		14.3.3	Post-disaster Response Phase	267
		14.3.4	Conclusion to the Review of Identity in Disasters	268
	14.4	Systemic Issues		268
		14.4.1	Entrenching Social Inequality; Women, Social Mobility, Political Vulnerability	269
		14.4.2	Vulnerabilities to Exploitation and Trafficking	270
		14.4.3	Risks to Identity Compromise	270
	14.5	Recommendations		272
	14.6	Conclusion		272
	References			275
15	**Justice, Resilience and Participatory Processes**			279
	Claudia Baldwin			
	15.1	Introduction		279
	15.2	The Concepts: Resilience, Vulnerability and Fairness		281
		15.2.1	Resilience	281
		15.2.2	Vulnerability and Fairness	284
	15.3	Case Study: Illustration of Vulnerability to Flooding in Rocklea Industrial Area		285
	15.4	Disaster Justice		288
	15.5	Role of Research in Building Community Resilience		291
	15.6	Lessons Learned: Applying Justice Principles to Build Community Resilience		292
	15.7	Conclusion		294
	References			294
16	**The Theory/Practice of Disaster Justice: Learning from Indigenous Peoples' Fire Management**			299
	Jessica K. Weir, Stephen Sutton, and Gareth Catt			
	16.1	Introduction		299
	16.2	Justice Considered and Reconsidered		302
	16.3	Aboriginal Peoples' Fire Management in Australia		307
		16.3.1	Martu Country: Gareth Catt	308
		16.3.2	Central Arnhem Land: Steve Sutton	309
		16.3.3	The Australian Capital Territory: Jessica Weir	310
	16.4	Framing Disaster Justice		311
	16.5	Conclusion		314
	References			314

17 Inclusion: Moving Beyond Resilience in the Pursuit of Transformative and Just DRR Practices for Persons with Disabilities   319
Emma Calgaro, Michelle Villeneuve, and Genevieve Roberts
- 17.1 Introduction   319
- 17.2 Disability, Disasters and Barriers to Inclusion and Greater Justice   321
- 17.3 A Systems Approach to Achieving Disaster Justice   324
- 17.4 Doing DiDRR: Demonstrating Change Through Action Using Three Projects as Grounded Examples   326
  - 17.4.1 Project 1: Get Ready! A Model for Deaf Community Leadership (2015–2016)   326
  - 17.4.2 Project 2: DiDRR in NSW: Increasing Access to DRR for People with Disabilities Through Collaboration (2015–2017)   330
  - 17.4.3 Project 3: Disability and Disasters: Empowering People and Building Resilience to Risk   333
- 17.5 Discussion: Facilitating Greater Inclusion and Disaster Justice   336
  - 17.5.1 Capacity-Building Activities and Tools That Are Rights-Based, Culturally Sensitive and Focus on Empowerment, Self-Determination and Leadership Are Essential for Success   336
  - 17.5.2 Provide Comprehensive Information in Accessible Formats That Is Delivered by Trusted Sources That Are Embedded in Their Communities   338
  - 17.5.3 Develop Capacity in the Disaster Risk Management Sector for DiDRR   338
  - 17.5.4 Create Supportive Platforms and Tools to Facilitate Inclusive Governance and Practice   339
  - 17.5.5 Sustainability and Scaling Up   340
- 17.6 Conclusion   341
- References   342

18	Future Pathways for Disaster Justice		349
	Anna Lukasiewicz and Claudia Baldwin		
	18.1	Introduction	349
	18.2	Possible Policy Approaches	350
		18.2.1 NGO and Agency Collaboration	350
		18.2.2 Integration of a Risk-Based Approach in Governance, Policy and Legislation	351
		18.2.3 Participatory Processes	352
	18.3	A Research Agenda for the Future	352
		18.3.1 Understanding Features of Vulnerability and Resilience-Building	353
		18.3.2 Investigating Existing and Perceived Rights, Responsibilities, Accountabilities, Values and Expectations	355
		18.3.3 Identifying Justice Issues Across the PPRR Spectrum	356
		18.3.4 Exploring the Interlinkages Between Procedural and Distributive Justice	357
	References		358

Index 361

Notes on Contributors

Jason Alexandra has held senior roles including as the Executive Director of the Earthwatch Institute (an international non-governmental organisation (NGO) delivering environmental research and sustainability education in over 40 countries) and as a Director of Land & Water Australia and the Port Phillip Catchment Management Authority.

Between 2008 and 2013 Jason was a senior executive at the Murray-Darling Basin Authority (MDBA) where he had responsibilities for a range of water policy, research and ecosystem management programs. These natural resources programme were delivered collaboratively between the six governments involved in the Murray Darling Initiative.

Jason left Canberra and the MDBA in early 2013 to concentrate on consulting business. As the managing director of Alexandra & Associates Pty Ltd. Jason has completed over 90 consulting projects on sustainability, natural resources management, environment and water policy.

Kathryn Allan is currently enrolled in a PhD at ANU. She is a BSocSc (Hons) graduate who has worked in international development, human rights, community organising and capacity building. She has volunteered with Amnesty International for over seven years, in both the USA and Australia.

Kathryn received the University of the Sunshine Coast (USC) Chancellor's medal when she graduated with a first class honours for a distinguished contribution to the community and high academic performance. She has worked on both international and Australian based campaigns regarding the promotion and protection of the rights of persons including: refugees, indigenous youth, LGBTQI+, children in regards to both trafficking and exploitation, women and prisoners of conscience.

Kathryn was the inaugural recipient of the IDCARE honours scholarship for her thesis entitled Examining Identity Management in Disaster Response Environments: A Child Exploitation Risk Mitigation Perspective. Her multidisciplinary thesis examined the role identity management plays in mitigating risks to children through the case study of Cyclone Winston using human factors methods and international public policy theory.

Md. Ashikuzzaman is a fulltime faculty at Khulna University, Bangladesh. He has provided his services in different Non-Governmental Organisations (NGOs) before entering the teaching profession. He has been teaching in Khulna University since 2015. Md. Ashikuzzaman studied urban planning at both undergraduate (Urban and Rural Planning Discipline, Khulna University) and graduate (Urban Engineering, University of Tokyo) level. His experiences at NGOs' level encouraged him to work on urban settings and Khulna University provided the platform. His recent published works contain public perceptions on different urban phenomenon and issues. His research interests cover planning and development management of cities in developing countries. Currently, he is working on sustainable waste management challenges in major cities of Bangladesh.

Muhammad Masood Azeem A graduate of the University of Western Australia (UWA), Dr. Masood Azeem holds a PhD in economics, MSc (Hons) in Development Economics and BSc (Hons) in Agricultural Economics. He also served as a Lecturer for more than five years at the University of Agriculture Faisalabad, Pakistan. Dr Azeem is particularly trained in the quantitative and statistical methods and he has utilised his analytical skills in areas ranging from retailing and consumer behavior, impact evaluation, performance implications of vertical relationships in the farm to fork enterprises, food insecurity, poverty and vulnerability to shocks. His ongoing work at the Centre for Agribusiness, University of New England (UNE), has a global perspective as it seeks to examine the key drivers of innovation in the food industry across Australia, Europe, Asia and Africa. Masood's research has been published in the Journal of Retailing and Consumer Services, Journal of Development Studies, Food Policy, Social Indicators Research and Journal of Asian Economics among others.

Claudia Baldwin is a Professor of Urban Design and Town Planning, and the Director of the Sustainability Research Centre at the University of the Sunshine Coast (USC). Claudia specialises in action research using participatory and visual methods related to institutional and social-environmental change especially applied to water and coastal planning and management and climate change adaptation. She was a Chief Investigator in the National Competitive Grant CSIRO (Commonwealth Scientific and Industrial Research Organisation) Coastal Cluster—Barriers to Science Uptake.

Prior to joining USC in 2006, Claudia worked for over 25 years in environmental policy and planning in government (e.g. Great Barrier Reef Marine Park Authority, Queensland Premiers Dept and Queensland Department of Natural Resources and Mines) and consulting in Australia and internationally (e.g. National Water Commission (NWC), International Centre for Water and Resource Management, and World Wide Fund for Nature (WWF)). She has considerable experience in collaboration and consultation with all levels of government, research institutions, private enterprise, not-for-profit groups and community.

Harriot Beazley is a human geographer and community development practitioner with a passion for rights-based participatory research with children and young people in Southeast Asia, especially Indonesia. Harriot is a Senior Lecturer in Human Geography and International Development at the University of the Sunshine Coast, Australia. Harriot's PhD (ANU) was focussed on the geographies and identities of street children in Indonesia. Since then she has consulted as a technical advisor to Australia's aid program AusAID, United Nations Children's Fund (UNICEF), UK's Department for Environment, Food & Rural Affairs (Defra), Save the Children, and Australia's Department of Foreign Affairs and Trade (DFAT) and has researched and published widely on a variety of child protection, social inclusion and gender issues in Southeast Asia and the Pacific. She is Commissioning Editor (Pacific) for the Routledge Journal Children's Geographies.

Oliver Burmeister is focused on ethics and on the social impact of technology. One way he achieves this is through his contributions to the field of value sensitive design, a design approach which puts ethics into practice.

Emma Calgaro is a human geographer specialising in the human dimensions of environmental change and risk responses with a regional focus on South-East Asia, Australia and the South Pacific. Being highly applied in nature, the purpose of her work is to inform both theoretical and practical debates on how to reduce vulnerability to risk and to equip communities and disaster risk reduction (DRR) and climate change adaptation (CCA) actors with the knowledge and tools needed to increase their coping capacities and resilience to future hazard events. Taking a systems approach, her current work explores two fundamental questions designed to advance greater inclusion of people with disabilities in the disaster space: (1) what does inclusion means in the context of DRR and CCA and (2) what steps—including knowledge generation and sharing, processes and practices—are needed to make disability-inclusive DRR (DiDRR) a lived reality.

Gareth Catt is a Regional Fire Management Officer for the recently established Ten Deserts Project—the largest Indigenous-led connected conservation project on earth. He previously worked as the Healthy Country Coordinator for Kanyirninpa Jukurrpa (KJ), a Martu organisation based in Newman, WA. In this, his main role was to plan and coordinate a prescribed fire program across an area of approximately 15 million hectares of the Little Sandy, Great Sandy and Gibson Deserts. This was done in collaboration with government and non-government agencies under the direction of a Martu board and community leaders. Gareth has a background in conservation and land management. Prior to arriving in Western Australia he worked for Parks and Wildlife NT in the vicinity of Alice Springs.

Rebecca Colvin is a social scientist with the Resources, Environment & Development Group at the ANU Crawford School of Public Policy. Bec's research interest is in how groups of people interact with each other—especially in settings of social and political conflict—with regard to climate and environmental issues. Much of this work has a focus on the dynamics of formalised processes for including citizens and stakeholders in decision-making, and leverages on perspectives from social psychology, especially social identity, to understand the complexities of people and process. Specific research projects have included the study of conflict about wind energy development, the psychological underpinnings of a constructive governance regime for negative emissions, the role of trust between climate researchers and policy-makers and the relationship between aggregate public opinion and conflict in environmental messaging. Before joining the Crawford School, Bec worked in knowledge exchange with the ANU Climate Change Institute and prior to that as a Lecturer in Environmental Management at The University of Queensland. Bec has a PhD from UQ (2017) on social identity and environmental conflict.

Luke Craven is a Research Fellow in the Public Service Research Group at the University of New South Wales, Canberra. Dr Craven's research focuses on developing new tools to understand and address complex policy challenges. He works with a range of public sector organisations to adapt and apply systems frameworks to support policy design, implementation and evaluation. He is known for developing the System Effects methodology, which is widely used to analyse complex causal relationships in participatory and qualitative data. He is also involved in number of collaborative projects that are developing innovative solutions to complex policy challenges, which includes work focused on food insecurity, health inequality and climate resilience. Dr Craven holds a PhD in Political Science at the University of Sydney, where he remains affiliated with the Sydney Environment Institute and the Charles Perkins Centre.

Steve Crimp is a Research Fellow with the Climate Change Institute at the Australian National University. His role in the Climate Change Institute is to examine opportunities for improved climate risk management, within primary industries, both in Australia and internationally. Steven's main interests include: the translation of climate change impact scenarios from rainfall and temperature into forms useful for decision makers, such as crop and pasture production, biodiversity, farm incomes and broader socio-economic impacts; participatory engagement with decision makers to improve the value derived from climate risk management in decision-making; the development of quantitative models and methods to derive value from seasonal climate forecasts and climate change projection information in agricultural, natural resource and biodiversity management, including the economic valuation of climate forecasts; and developing and implementing practical concepts of vulnerability, resilience and adaptive capacity across scales, including the design of nested institutional arrangements and ways of increasing the societal value of climate impacts science and re-defining biodiversity conservation objectives.

Hannah Della Bosca works as a Research Assistant at the University of Sydney, primarily on projects related to evidence-based urban policy. Hannah holds a Bachelor of Arts (Honours) and has a strong research interest in the nexus of environment, policy and place identity.

Stephen Dovers is Emeritus Professor, ANU; former Director Fenner School, ANU, 2009–2017; Fellow of the Academy of Social Sciences in Australia; Researcher, Bushfire and Natural Hazards Cooperative Research Centre;

Honorary Professorial Research Fellow, Charles Darwin University; Chair, Science Advisory Council, Mulloon Institute; Chair, Advisory Board, Sustainability Research Centre, University of the Sunshine Coast; Senior Associate, Aither. Research interests: environmental policy and institutions; emergency management and disaster policy; and climate adaptation.

Michael Eburn is the leading researcher in the area of emergency services, emergency management and the law. He is the author or co-author of three books and has made over 80 other contributions as book chapters, journal articles, professional publications and conference and professional development presentations. His blog, Australian Emergency Law, is widely read and respected throughout the sector. Michael was the chief investigator on research on law and governance in emergency management, funded by the Australian Bushfire and Natural Hazards Cooperative Research Centre (BNHCRC). Michael has been an invited expert on disaster law at events hosted by the International Federation of Red Cross and Red Crescent Societies in Kuala Lumpur and Geneva and an invited guest of the International Disaster Law project hosted by Roma Tre University and the University of Bologna in Italy. Michael an affiliate with the Disaster Development Network, University of Northumbria (UK) and the Joint Centre for Disaster Research, Massey University and GNS Science, Wellington (NZ).

John Hicks is an economist who has held a number of university positions in Australia and New Zealand and who has worked as a senior economist in the private sector. He is currently Professor of Economics at Charles Sturt University, Bathurst, Australia. His expertise encompasses labour economics and macroeconomics and his current research focuses on issues relevant to regional economic and social development.

Mark Howden is Director of the Climate Change Institute at the Australian National University. He is also an Honorary Professor at Melbourne University, a Vice Chair of the Intergovernmental Panel on Climate Change (IPCC), member of the Australian Capital Territory Climate Change Council and a member of the Australian National Climate Science Advisory Committee. He was on the US Federal Advisory Committee for the 3rd National Climate Assessment and contributes to several major national and international science and policy advisory bodies. Mark has worked on climate variability, climate change, innovation and adoption issues for over 30 years in partnership with many industry, community and policy groups via both research and science-policy roles. Issues he has addressed include agriculture and food security, the natural resource base, ecosystems and biodiversity, energy, water and urban systems. Mark has over 420 publications of different types. He helped develop both the national and international greenhouse gas inventories that are a fundamental part of the Paris Agreement and has assessed sustainable ways to reduce emissions. He has been a major contributor to the IPCC since 1991, with roles in the Second, Third, Fourth, Fifth and now Sixth Assessment Reports, sharing the 2007 Nobel Peace Prize with other IPCC participants and Al Gore.

Md. Sayed Iftekhar is an environmental and resource economist with broad interests in the interactions between human and nature. He has received training on forestry (Khulna University) and biodiversity conservation (Oxford University). He has worked on coastal zone management in Bangladesh for several years. During his pre-PhD phase, he looked at environmental issues such as sustainable forest management, ecosystem services, protected areas and environmental conservation from biological and resource management viewpoints. Later, with PhD training at University of Western Australia (UWA), he started using different economic tools such as agent based modelling, laboratory experiments and simulations to study these issues. He has also broadened my interests to combinatorial conservation auctions, fisheries quota allocation, networks and group formation, market design for environmental water buyback and intergenerational equity and risk aversion, and so on.

Valerie Ingham is a Lecturer in Emergency Management at Charles Sturt University. She is a founding member of the Disaster and Community Resilience Research Group and her research interests include community resilience, disaster recovery and the tertiary education of emergency managers and fire investigators. Her research highlights the importance of local community organisations in building community connections for disaster resilience, with a particular focus on Bangladeshi and Australian communities.

Mir Rabiul Islam is an environmental social psychologist. He received his PhD in social psychology from University of Bristol. His research interests relate to psycho-social dynamics related to race and ethnicity; human factors and environmental issues; coping, resilience and recovery from natural disasters; human rights, displaced populations and climate justice; and cross-cultural and intercultural communication.

Sophie Lewis is a senior lecturer and Australian Research Council DECRA (Discovery Early Career Researcher Award) Fellow at the University of New South Wales Canberra. She researches the role of human influences in recent extreme climate events in Australia, such as heat waves and floods. Her research also investigates changes in the severity and frequency of climate extremes in the future using climate models. These types of research areas are essential for understanding the vulnerability of human and natural systems to potential future changes in climate and weather extremes. Dr Lewis has a PhD from the Australian National University (2011) examining Australia's long-term climate history and was previously a Research Fellow at the University of Melbourne, Australian National University and the ARC Centre of Excellence for Climate System Science. She is currently serving as a Lead Author on the IPCC's 6th Assessment Report and is a Domain Editor of *WIREs Climate Change*.

Anna Lukasiewicz is a justice researcher with an interdisciplinary background in the social sciences and a keen interest in how human societies interact with nature. Since coming to ANU, Anna has focused on justice research in natural resource management and is expanding this focus to natural hazards and disaster justice. She is working on developing the Social Justice Framework for environmental decision-making. Her research includes projects on resilience to natural disasters, the social aspects of environmental ANU water stakeholder advisory groups, climate change adaptation at a catchment level and the last three decades of water reform in Australia. Anna has a PhD from Charles Sturt University in Natural Resource Management, a Masters in Social Science (International Development) from RMIT and a Bachelor of International Studies from the Flinders University of South Australia.

Alan March is an Associate Professor of Urban Planning and Director of the Bachelor of Design at the University of Melbourne. Alan is an urban planner and urban designer and has practised since 1991 in a broad range of private sector and government settings. He has had roles in statutory and strategic planning, advocacy and urban design. Alan's publications and research include examination of the practical governance mechanisms of planning and urban design, in particular the ways that planning systems can successfully manage change and transition as circumstances change. He is particularly interested in the ways that planning and design can modify disaster risks, and researches urban planning approaches to natural hazard mitigation.

He is project lead of the Bushfire Natural Hazards Cooperative Research Centre project Integrating Natural Hazard Mitigation and Urban Planning.

James Mortensen is a doctoral candidate of the National Security College at the Australian National University, having previously attained first class Honours (Religious Studies) from the University of Newcastle, Australia. His research interests include the philosophical underpinnings of security and political theory, the role of belief systems in political action and the role of technology in politics and society.

Leonardo Nogueira de Moraes is a Postdoctoral Research Fellow in Resilience and Urban Planning at the Faculty of Architecture, Building and Planning of the University of Melbourne, where he is part of the research team for the Integrated Urban Planning for Natural Hazard Mitigation project, funded by the Bushfire and Natural Hazards Cooperative Research Centre (BNHCRC). His background includes a Bachelor of Tourism (Development and Planning) degree and a Specialisation in Tourism and Hospitality Marketing Management from the University of São Paulo, Brazil. His PhD degree in Architecture and Planning (from the University of Melbourne) focused on the effects of tourism development and the implementation of protected areas to the resilience of small oceanic islands, from a social-ecological complex adaptive systems perspective. His current research on Resilience and Urban Planning targets formal and informal urban planning processes in tourism development, sustainability and resilience to disasters. From a social-ecological complex adaptive systems' perspective, he seeks to understand the roles of state regulation and self-organisation in resilience building and global-local interplays.

Tayanah O'Donnell is a lawyer and social scientist and currently Director of Future Earth Australia, the Oceania node for the global Future Earth research network. Tayanah's research is situated in the legal geography field. She has a strong interdisciplinary track record, utilising her legal skills and social research methods to contribute to current debates relating to both social and legal risks and impacts of climate change, and climate adaptation policy responses. Tayanah has published in both law and geography journals; several book chapters on property rights, law and land use planning for coastal climate change adaptation; and numerous research papers and guidance briefs. She is a regular contributor to *The*

Conversation and is highly sought after for her expertise working with local governments on the complexities of coastal management.

Genevieve Roberts is a Community Development Practitioner with extensive experience working in both the emergency preparedness sphere and the disability sector. Her work is grounded in asset-based community development and capacity building. She develops emergency preparedness as a by-product of capacity building. With a background in improvisation and wellness, she incorporates improvisation techniques and values into building capacity and empowering individuals. Examples of such techniques and values are, 'Yes and—accept all offers', and 'make each other look good'. Her approach to building capacity at individual, community and organisational levels is holistic. Working with the Country Fire Authority on the Surf Coast in Victoria, she used capacity building in areas such as the arts, environment and health to generate high levels of social engagement in individuals and deeply connected community networks. Working with The Deaf Society NSW, she used volunteerism and emergency preparedness to build community leadership, self-development and employment pathways.

Steven Schilizzi is an Associate Professor with the UWA School of Agriculture and Environment at the University of Western Australia (Faculty of Science), active in both teaching and research. He is an environmental and agricultural economist with broad international experience. He has worked on market-based instruments for environmental policy, especially conservation auctions; bio-economic modelling of agricultural and aquaculture systems; risk management; and lately, on distributional equity. In the last ten years or so, he has privileged the use of controlled lab experiments within the framework of behavioural economics, as well as interdisciplinary approaches. He has (co-)authored about 60 scientific papers and 20 book chapters and authored or edited four academic books, as well as 70 conference papers. He was recipient of GEWISOLA (*Gesellschaft für Wirtschafts- und Sozialwissenschaften des Landbaues e.V.*—Germany's Society for Economic and Social Sciences of Agriculture) and AARES (Australasian Agricultural & Resource Economics Society) best paper of the year awards and twice nominated for Excellence in Post-graduate Supervision Awards.

David Schlosberg is Professor of Environmental Politics in the Department of Government and International Relations, Payne-Scott Professor and Director of the Sydney Environment Institute at the University of Sydney. He is known internationally for his work in environmental politics, environmental movements and political theory—in particular the intersection of the three with his work on environmental justice. His other theoretical interests are in climate justice, climate adaptation and resilience and environmental movements and the practices of everyday life. Professor Schlosberg's more applied work includes public perceptions of adaptation and resilience, the health and social impacts of climate change and community-based responses to food insecurity. He is the author of *Defining Environmental Justice* (Oxford, 2007); co-author of *Climate-Challenged Society* (Oxford, 2013); and co-editor of both *The Oxford Handbook of Climate Change and Society* (Oxford 2011) and *The Oxford Handbook of Environmental Political Theory* (Oxford 2016). His latest book is *Sustainable Materialism: Environmental Movements and the Politics of Everyday Life* (Oxford 2019).

Janet Stanley is Principal Research Fellow at the Melbourne Sustainable Society Institute, Faculty of Design, University of Melbourne. Janet's work focuses on the interface between social, environmental and economic issues across policy, system design and at community levels. Janet particularly specialised in social inclusion and equity, transport, climate change, child welfare, bushfire arson, policy and planning, and evaluation. She has approximately 150 publications, including five authored and edited books.

Stephen Sutton is a PhD student at Charles Darwin University, studying the cultural and psychological drivers of good disaster risk reduction behaviour. He is also a collaborator on a number of projects focusing on improving community disaster resilience in remote northern Australia. The two sets of research have in common a focus on grass-roots disaster resilience through a better understanding of culture and its impacts upon disaster perception and motivation to prepare. Steve was Director of Bushfires NT, the Northern Territory's rural and remote fire management organisation. After witnessing countless repeated failures in disaster preparation by a broad cross-section of society, he became interested in the universal cognitive underpinnings of disaster risk reduction. He hopes his current research collaborations help build improved resilience across northern Australia and beyond. Steve's research is supported by the

Bushfire and Natural Hazards Cooperative Research Centre and an Australian Postgraduate Award.

Mohammad Shahidul Hasan Swapan is a Lecturer in Urban and Regional Planning at Curtin University, Australia. He received doctoral training on urban governance from Australia and postgraduate and bachelor training on environmental management and urban planning from New Zealand and Bangladesh, respectively. His research is currently engaged into how the community perceives environmental spaces and the antecedents of citizenship behavioural discourse. In this regard, he is looking into public open spaces and informal parks in local government areas for co-producing better environmental outcomes. Mohammad Swapan also has significant interests on urban informality and participation in South Asian cities. He has contributed to a number of socio-psychological models to explain the decreasing local engagement in metropolitan planning.

Michelle Villeneuve leads the Disability-Inclusive Community Development work stream at the Centre for Disability Research and Policy (http://sydney.edu.au/health-sciences/cdrp/work-streams/disability-development.shtml) at the Faculty of Health Sciences, University of Sydney. Michelle has over 20 years of experience working in regions of conflict and disaster to develop community-led programs and services and re-build opportunities for people with disability. In Australia, Michelle led the development of Local Guidelines for Emergency Managers Disability-Inclusive Disaster Risk Reduction in NSW as part of the Enabling Community Resilience Through Collaboration project (2015–2016) and was lead investigator on the community capacity development project PREPARE NSW (2017–2018) that created the Person-Centred Emergency Preparedness toolkit. Michelle was also a chief investigator on the first DFAT funded disability inclusive disaster risk reduction research and development project (2013–2014) in Indonesia that focused on the role and capacity of Disabled People's Organisations as policy advocates for inclusive Disaster Risk Reduction (DRR) in Indonesia.

Jessica K. Weir is a Senior Research Fellow at the Institute for Culture and Society at Western Sydney University, where she investigates how different knowledge practices inform the set of decisions faced by governments in land management policy and practice. Her work demonstrates how greater reflexivity matched with material action can lead to more just outcomes for natures and peoples. Jessica has published widely on issues such as water, climate change, Indigenous land rights, invasive plants and natural hazards. In 2011, she founded the Australian Institute for Aboriginal and Torres Strait Islander Studies' Land and Water Research Centre. Jessica holds positions as a Visiting Fellow at the Fenner School of Environment and Society, Australian National University, and an editorial board member for the Routledge Environmental Humanities series.

List of Figures

Fig. 1.1	A typology of justice that illustrates its complexity	6
Fig. 1.2	Situating disaster justice	11
Fig. 5.1	The risk triangle. (Crichton, 1999)	99
Fig. 6.1	The location of the Vaughan property, left of Manfred St Belongil Beach, Byron Bay	124
Fig. 9.1	Location of Khulna city and major informal settlements. (KCC, 2011)	174
Fig. 15.1	Inundated light industrial building recently renovated (P1)	286
Fig. 15.2	Recently rebuilt to meet Council flood standards but hazard chemical containers were tossed around by floodwaters (P2)	287
Fig. 15.3	Governance systems as mediators of community resilience	289
Fig. 17.1	Seven pillars of action to achieve DiDRR	325
Fig. 17.2	DiDRR framework. (Centre for Disability Research and Policy and Natural Hazards Research Group, 2017)	332

List of Tables

Table 3.1	Framework for policy and institutional analysis for emergencies and disasters	57
Table 3.2	A menu of policy instruments for emergencies and disasters	61
Table 3.3	Criteria for selecting policy instruments	62
Table 3.4	Linking risk, justice and disaster policy instruments: a tentative ordering	69
Table 8.1	Household types	156
Table 8.2	Number of houses afforded by households	157
Table 8.3	Subsidy rates and reconstruction outcomes	159
Table 8.4	Amartya Sen's "equality of what?"	161
Table 8.5	The role of different equity norms on final household budgets	161
Table 8.6	Percent of poorer households with a safer new house, given a 100% subsidy	162
Table 8.7	Details behind in-text Table 8.5 (The role of different equity norms on allocations and final household budgets)	164
Table 9.1	Descriptive statistics of demographic and socio-economic condition of the sample	176
Table 9.2	Distribution of respondents by types of support received	177
Table 9.3	Correlation between different types of support received by the respondents	178
Table 9.4	The relationship between the extent of support received from government organisations and the socio-economic conditions of the respondents	179
Table 9.5	The relationship between the extent of support received from non-government organisations and the socio-economic conditions of the respondents	180

Table 15.1	Three resilience concepts	282
Table 15.2	Social justice	290
Table 17.1	Barriers to the inclusion of people with disabilities in DRR processes	322
Table 17.2	Project 1: Get Ready! A model for Deaf Community Leadership (2015–2016)	326
Table 17.3	Project 2: DiDRR in NSW: Increasing access to DRR for people with disabilities through collaboration (2015–2017)	330
Table 17.4	Project 3: Disability and disasters: Empowering people and building resilience to risk	333

PART I

Introduction

CHAPTER 1

The Emerging Imperative of Disaster Justice

Anna Lukasiewicz

1.1 WHY A BOOK ON DISASTER JUSTICE?

Disasters, and their impacts, are increasing all over the world, with the Asia Pacific region being particularly susceptible. The region experiences the largest amount of natural disasters in the world; between the years 2014 and 2017 there were 55 earthquakes, 217 storms and cyclones and 236 cases of severe flooding in the Asia Pacific region, which affected 650 million people and caused 33,000 deaths (OCHA, 2019, p. 15).

Australia is not immune from these statistics. Its economic losses alone added up to $171.5 billion[1] (which includes the costs of deaths and injuries) during the period 1967–2013 and have continued to climb, mostly due to population growth, rather than a growth in the number of natural hazards, suggesting that Australia's disaster risk is driven by socio-economic trends (Handmer, Ladds, & Magee, 2018).

---

[1] Australian dollars in 2013 prices.

A. Lukasiewicz (✉)
Fenner School of Environment and Society and Institute for Integrated Research on Disaster Risk Science, Australian National University, Canberra, ACT, Australia
e-mail: anna.lukasiewicz@anu.edu.au

© The Author(s) 2020
A. Lukasiewicz, C. Baldwin (eds.), *Natural Hazards and Disaster Justice*, https://doi.org/10.1007/978-981-15-0466-2_1

1.2 What Is a Disaster?

Before launching into an explanation of Disaster Justice, a short discussion of disasters is necessary. What constitutes a disaster? Most definitions highlight the destructive nature of disasters (human lives lost, damaged infrastructure) and an acute time-delineated event (an earthquake, an epidemic), however even these lines are murky when we start to consider slow onset disasters such as droughts, cascading disasters (Pescaroli & Alexander, 2015) or when using a systemic approach (O'Connell et al., 2018). The IPCC adopted the following definition of disasters in 2012:

> *Severe alterations in the normal functioning of a community or a society due to hazardous physical events interacting with vulnerable social conditions, leading to widespread adverse human, material, economic, or environmental effects that require immediate emergency response to satisfy critical human needs and that may require external support for recovery.* (IPCC, 2012, p. 558)

This and other widely used definitions highlight the following aspects of disasters:

- A disruption or an alteration of normal functioning
- An interplay between a physical event and social condition of those affected
- Loss or impact; human, material, economic and/or environmental
- The coping capability of those impacted being affected or exceeded, leading to a need for outside intervention

There is much more to the definitions of disasters, as entire books have been devoted to this question (see Perry & Quarantelli, 2005) but the four points outlined above are particularly important in discussing climate and environmental crises and very pertinent to the following discussion of disaster justice. I have not mentioned here the delineation between a natural and a man-made disaster, because as the subsequent discussion makes clear, there is no such thing as a purely 'natural' disaster; "*natural disasters are in fact social disasters waiting to happen that may be triggered by a particular natural force*" (Pelling, quoted in Verchick, 2012, p. 34).

Another important consideration of disasters for our purpose is the expectation around their management. Lauta (2015) persuasively argues that the way we view disasters has changed. Disasters used to be seen as

'Acts of God'—divine punishment for society's shortcomings. Then they came to be seen as an unstoppable and unmanageable force of 'Nature'. Now they are mostly considered to be consequences of human decisions (such as maintaining settlements in flood, storm, or earthquake-prone areas). While the majority of disaster justice writings to date are based on this view, some point out that it is in itself a form of cultural blindness in disaster management. For example, in his criticism of this 'western' or 'North Atlantic' view of disasters, Bankoff (2018) points out that this progression suggests an 'evolution', one that is linear and absolute implying that viewing a disaster as a form of divine punishment (as some cultures still do) is somehow incorrect. This could lead to a denial of cultural knowledges and practices of affected communities, a point I will return to later in the chapter.

1.3 What Is Justice?

Justice is arguably a much harder concept to define than disasters. It has been a preoccupation of both ancient and modern philosophers and identified as the first key ethic necessary for human society. For example, it is a central virtue to Rawls, who posits justice as the basis of a social contract between citizens and their governing institutions (Rawls, 1971). John Stuart Mill called it *"the chief part, and incomparably the most sacred and binding part, of all morality"* (quoted in Riley, 2010). Justice is not just a virtue—it is the supreme virtue by which other values are measured; it is an implicit expectation of any society and something that governments are expected to provide.

Justice is an incredibly complex idea and has been pursued in isolation by many academic disciplines—philosophy, law, economics, and numerous social sciences to name but a few. Our understanding of it is thus fragmented and interdisciplinary approaches to justice research are rare. Figure 1.1[2] captures some of the different types of justice that readers of this book might come across. Starting at the bottom left corner of the figure, we can look at justice by focusing on the *subjects* of justice—as individuals (or groups, e.g. nonhumans, organisations or ecosystems). Going up to the top left, we can conversely look at justice through the lens of *space and time* (a common approach of geographers), again focusing on a specific group or multiple groups in a specific time and location. At the

---

[2] This figure is an evolution of the justice typology found in Lukasiewicz (2017a).

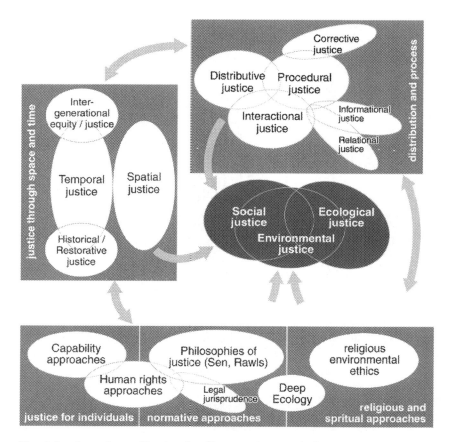

Fig. 1.1 A typology of justice that illustrates its complexity

bottom left, there are *spiritual and religious approaches* to justice sitting next to *normative approaches* such as philosophy and the law. In the middle are what I call 'composite' justice types (as they can be made up of any of the surrounding elements).

At the top, right hand side, we can look at justice as the result of *distribution* (how do we distribute rights, resources, burdens and privileges within a society?) and *process* (how do we decide this distribution?). Within procedural justice (which can be thought of as the structures, rules and methods of decision-making), some scholars separate interactional or relational justice (concerned with how stakeholders and decision-makers are treated). This is a common approach to justice, particularly practised in the discipline of social psychology and used by myself (see Lukasiewicz, 2017b) and others in this book (e.g. see Chaps. 5 and 16). The typology presented here is one representation and to some extent simplified, but serves to illustrate the complexity. For example, the principle of *recognition* (i.e. acknowledging the rights to be heard and enabling inclusion of a stakeholder) can be inserted into procedural justice or sit above the entire diagram. Another typology is a thematic approach—climate justice, water justice, food justice, and so on. A thematic typology will still incorporate the mixture of approaches presented in Fig. 1.1.

While much more can be said about the complexity of justice (see Lukasiewicz, 2017a), I will only point out a couple of issues most pertinent to disaster management.

Talking about justice immediately involves talking about authorities and institutions which are meant to administer or ensure justice. In most societies, this is the role of the legal system. Indeed, 'the law' is meant to ensure the attainment of a just society, even though it inadvertently fails due to the complexity and fluidity of justice as a concept (Sarat & Kearns, 1996). When it comes to disaster management, the most obvious authority in charge is the government; and the pursuit of justice is inextricably tied to government capability and intent, a point I will return to later. Connected to this is the concept of 'moral community', that is, those to whom justice is owed (Deutsch, 1985). Once we start talking about justice, an obvious question to ask is 'justice for whom?' This could be individuals, different groups within society; it could be nonhumans—for example, domesticated animals, wildlife or ecosystems, or future generations (both human and otherwise).

Another important point about justice is that it is often considered a specific outcome—a decision or an event, whereas in fact it should be viewed more as a process (see Patrick, 2014). Think about any resource conflict: a local community fighting for compensation following devastating flooding; or an indigenous group trying to stop the building of a mine that will affect their way of life. In any such conflict, there will be different groups of stakeholders and some form of governing authority. These groups have been interacting with each other and the resource in question outside of the specific conflict. There is a history of connections and relations that influences current actions and beliefs about what should be done. What is just (and for whom) in that specific resource conflict is not a decision that can be made outside of its specific context and any decision made will have to take into account, among other things, previous perceived injustices. It will thus not be an end (for the stakeholders at least), it will be yet another turning point shaping the progression of the stakeholders' relations.

Of course, it is impossible to talk of justice without mentioning injustice. Injustice is often described as an experience, something that is felt, rather than rationally considered (see some of the examples in Chap. 13), a violation of one's standing, beliefs or values (Gross, 2010). A very pertinent question to disaster management is what constitutes an injustice as opposed to a misfortune? Verchick argues that the key distinction is agency: in political terms, who is accountable? In moral terms, who is blameworthy? (Rumbach & Németh, 2018). Sen (2009, p. 4) puts it simply: "*a calamity would be a case of injustice only if it could have been prevented, and particularly if those who could have taken action had failed to try*". This brings us to accountability and accepting responsibility, a point elaborated on later in the chapter.

What we consider just or unjust is tied to our values, beliefs and cultural norms. Competing conceptions of what justice is or should be, not only exist within and between societies, but also continually change through time. At this point, the reader may experience exasperation and feel like giving up on any pursuit of justice, deciding it is physically impossible or at least improbable. While a singular definition of justice may be unattainable, there is much more scope for acknowledging that injustice exists and for that reason alone, the pursuit of justice is nevertheless essential. Perfect justice may not be achievable, however as long as people feel an injustice has occurred, there will be a demand for justice, and that demand will be placed at the feet of governance institutions.

1.4 What Is Disaster Justice?

Now that I have briefly introduced the two concepts of disasters and justice separately, what happens when they are put together?

Disaster justice is a new research focus in disaster management. It was coined by Robert Verchick (2012) to describe the fair treatment of all people in policies relevant to catastrophic hazards. It is slowly growing in prominence in disaster literature due to a number of factors, such as the framing of disaster management as a human rights issue (Lauta, 2015). There is not yet an agreed definition of what disaster justice is, or should be. In their special issue on disaster justice, Douglass and Miller (2018, p. 271) state:

> Our central premise is that disaster justice is a moral claim on governance, which arises from anthropogenic interventions in nature that incubate environmental crises and magnify their socially and spatially uneven impacts.

In the same issue, Verchick (2018) highlights the procedural aspects of disaster justice: "*the notion implies an inclusive decision-making process, free of discrimination, that pays attention to risks disproportionately imposed on socially disadvantaged populations*". Bankoff (2018) puts it more simply, confining disaster justice to a duty that the state has towards its citizens. Along similar lines, Williamson (2018) brings in political and legal justice—determining culpability for disasters and avoiding or accepting the responsibility for them. In writing separately on disaster justice, Shreshta, Bhattarai, Ojha, and Bajracharya (2019) highlighted three major domains for disaster justice: accountability in resource distribution, representation of different voices and recognition of different knowledge forms.

Participants in the November 2018 Disaster Justice workshop which directly preceded this book discussed the meaning and usefulness of Disaster Justice, arriving at the following description:

> *Disaster justice recognizes that disasters can expose, magnify and deepen existing injustices in society, which can then lead to further injustices. This perspective purposefully situates the disaster event in relation to past and present social choices, and also acknowledges that the disaster itself is a dynamic opportunity to investigate perceived injustice and vulnerability using different dimensions of justice. As humans live with and within nature, these matters of justice include nature and consideration of our shared future.*

While not a concise definition, this description is about how we should understand the reality of justice and injustice surrounding disasters; it acknowledges the complexity while defining the broad work of the field, which is still in its embryonic stage.

A useful starting point for this discussion is situating disaster justice in the sphere of other thematic justices. Douglass and Miller (2018) suggest that while disaster justice overlaps with both climate and environmental justice, it deserves to be considered as separate to both. Verchick (2018, p. 291) links disaster justice not only to environmental and climate justice, but to social justice as well, which he describes as "*a wider landscape devoted to concerns for human rights, economic equity, and sustainable development*". I think that this is the most apt positioning of disaster justice, at the confluence of environmental, climate and social justice. While environmental justice can be simplified as the distribution of environmental costs or negatives (such as pollution, or inequitable access to environmental resources), it also highlights the procedural dimensions of justice, such as the relative lack of access to environmental decision-making amongst vulnerable social groups (Agyeman, Bullard, & Evans, 2003), and social discrimination due to a lack of recognition (Schlosberg, 2013). While it may have started as a narrow activist movement linked to civil rights in the United States (Clark, Chhotray, & Few, 2013; Colten, 2007), it has since rapidly and thoroughly expanded into a discourse that is now applied across a wide range of environmental and climate-related experiences and conflicts in many countries (Shiva, 2015; Temper, 2019).

Disaster justice also arises out of social justice, usually understood as the allocation of goods and benefits within a society, focusing on distributive, procedural and interactional (or relational) fairness, often from the viewpoint of marginalised or disadvantaged stakeholders (Syme & Nancarrow, 2001). How a society divides resources, rights and responsibilities may be unrelated to either disasters or natural hazards, however it relates to the creation of vulnerability and marginalisation, which expose people to greater disaster impacts. Climate justice considers how climate burdens are shared and how they can be avoided (Caney, 2014). Climate justice also highlights that climate burdens (including increased frequency and intensity of natural hazards) are disproportionally distributed to the poorest whilst benefits from greenhouse gas emissions have accrued to the well-off. This is again connected to the lack of procedural justice, access, voice and recognition of marginalised groups. Climate and disaster justice coincide closely where many disasters/hazards are driven by climate variability and change.

Another important attribute in what constitutes disaster justice is that disaster justice is not just about the 'disaster' but also about what happens (both short- and long term) before the disaster and in the aftermath. Disaster management usually identifies four phases (with infinite variations): Prevention—Preparation—Response—Recovery (PPRR). Disaster justice is relevant to all four phases, as has been highlighted through the increasing literature on major natural disasters such as Hurricane Katrina. For example, Vance (2008) illustrates how Hurricane Katrina caused the collapse of the criminal justice system in New Orleans, a system that barely functioned under normal circumstances due to long-term political neglect and perpetual lack of funding. As Verchick (2012, p. 23) states: *"the heaviest burdens of disasters [is] being borne by those with the least power,—those who, for whatever social and economic reasons, are more exposed, more susceptible, and less resilient when disaster strikes*". Disasters thus expose existing 'everyday' injustices and disaster management, especially in the Preparation and Recovery phases can contribute to either the perpetuation or minimisation of existing injustices, or the creation of new ones. For example, Bullard and Wright's (2009) edited collection provides numerous examples of how prevailing class and racial inequalities have constrained efforts in the Recovery phase of Hurricane Katrina. Drake (2018) suggests that more attention needs to be drawn to the 'ambient injustices' or the 'slow violences' (such as poverty, corruption, illness, etc.) that may be tolerable in everyday life but profoundly influence the impact of a disaster (see Nixon, 2011 for an explanation of slow violence).

Figure 1.2 takes these two points and situates disaster justice at the confluence of environmental, social and climate justice, acknowledging that while it overlaps with all three, it is separate in and of itself and places it within a cycle of disaster management to highlight that justice issues are

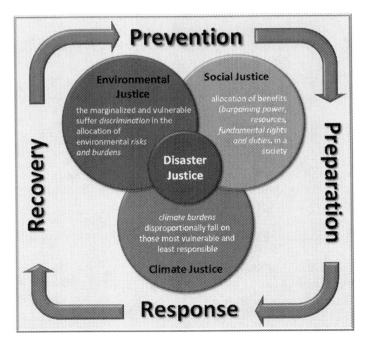

Fig. 1.2 Situating disaster justice

pertinent to all phases. The brief descriptions of environmental, social and climate justice relate to the distributive justice dimension (see Fig. 1.1) and these 'themes' of what these justices deal with are what sets them apart from each other. What unites them (and is a particular focus of disaster justice) is the procedural dimension of justice, a focus on access, voice, recognition and experiences of stakeholders in a decision-making process.

Within any thematic understanding of justice, there are multiple dimensions. The two most commonly mentioned are distributive justice and procedural justice. Distributive justice is about the intended or actual outcome of resource distribution, while procedural justice is concerned with how decisions are made and who gets to make them (Lukasiewicz, 2017b). These two justice dimensions have been discussed over many decades in the justice literature of social psychology, among others. While there is acknowledgement that they are intertwined and interlinked, they are usually discussed and analysed separately. Both the climate and environmental justice themes are especially concerned with power and discrimination, thus highlighting the procedural dimension of justice (Parthasarathy, 2018). It is no wonder then that scholarship on disaster justice so far has highlighted this interlinkage and emphasised the procedural dimension. The question of who gets to make decisions, and even who decides on decision-making processes that are used, can determine what sort of outcomes are considered. Rumbach and Németh (2018, p. 343) powerfully point out that *"decisions about who gets what, as well as the mechanisms of re-distribution, are very often left to historically powerful actors rather than the likely beneficiaries of such actions"*. Shreshta et al. (2019, p. 208) highlight 'justice as recognition', in their definition of disaster justice, referring to the recognition of knowledge systems, justice conceptions and value structures, not just the recognition of different stakeholder groups. This raises some issues for the pursuit of disaster justice that will be explored later.

In this section I've highlighted three attributes that define disaster justice: (1) its positioning at the confluence of three already established 'thematic' justices—environmental, social and climate; (2) its relevance to the four phases of the disaster management cycle, highlighting the everyday injustices that create social vulnerability; and (3) its focus on the interplay between the distributive and procedural dimensions of justice.

There is one justice type that the above points do not mention—ecological justice. Ecological justice is about what we as humans owe to non-human life and the planet more broadly (see my earlier point about 'moral community'). While environmental justice sees environmental resources and hazards as objects that are used by or impact on humans, ecological justice sees the environment as a subject to whom justice is owed. Ecological justice is a theme that runs through environmental, climate and social justice, and is growing in prominence. While it has been mentioned in disaster justice, through human-environmental complexity (Drake, 2018) and the human impacts on the environment (Parthasarathy, 2018), it is not yet a very visible thread within the disaster justice writings that have been produced to date. I contend that it should be as disaster management is slowly becoming more attentive to the needs of nonhumans during disasters (Glassey, 2018) and the power of ecosystem based adaptation to climate change is recognised (Munang et al., 2013).

1.5 What Makes Disaster Justice Unique?

Let us now examine four aspects of disaster justice that together make it a unique concept. While these will be discussed separately, they are all inter-related.

1. The Anthropocene is characterised by significant human impact on the environment, natural hazards and related crises, placing moral obligations on disaster management.

The Anthropocene, the proposed geological epoch where humanity is the dominant influence on planetary processes (Lewis & Maslin, 2015), is generally established in academia (albeit with powerful critiques, see Davis, Moulton, Sant, & Williams, 2019), especially natural sciences, social sciences and the humanities (Ziegler & Kaplan, 2019). Verchick (2016, p. 6) succinctly describes how the Anthropocene relates to disaster justice, highlighting the interplay between physical and social forces:

> *In the Anthropocene, there is no such thing as a natural disaster. Anthropogenic carbon emissions amplify the force and frequency of many environmentally triggered events. Land-use planning decisions squeeze some populous communities into the unsafe places where those escalated forces are more likely to land.*

There is wide acceptance in disaster management literature that there are no purely 'natural' disasters; the intended and unintended consequences of everyday decisions create future risk (Douglass & Miller, 2018). If the Anthropocene is in itself the result of the culmination of choices (and their unintended, indirect, diffuse effects) of past and present dominant human groups, it leaves current generations with a moral obligation to manage its consequences. However, humanity is riddled with profound inequalities, and if humanity should accept moral responsibility for the consequences of the Anthropocene, then some humans should arguably be more responsible than others. These questions fall squarely into ongoing debates of climate justice.

There is a global, planetary dimension to this aspect—all human actions, reactions and their effects, impact on other species, which is why ecological justice should be considered as an important element of disaster justice. This means that some human actions, livelihoods and lifestyles impact more than others (Althor, Watson, & Fuller, 2016). As we gain a greater understanding of how climate change is affecting the frequency and severity of natural hazards, we cannot turn away from acknowledging that those who benefit from greenhouse gas emissions are also indirectly and unintentionally the cause of increased risks from natural hazards (see Hiller, 2011). This realisation fits within the current understanding of disasters as consequences of human decisions, rather than 'Acts of God' or 'forces of nature' (Lauta, 2015). Whether human societies accept this moral obligation, and to what extent should some societies bear more responsibility than others within disaster management is of course the subject of climate justice literature (see, e.g. Dellink et al., 2009).

2. The differences in impact that natural hazards have are rooted in non-disaster related social, economic, political, cultural, spatial and temporal inequalities that produce and continue 'everyday' and 'ambient' injustices.

This aspect closely aligns with the environmental justice theme and is perhaps the most prominent feature of disaster justice. Disasters can deepen and magnify existing injustices. The poor and vulnerable in any society are less prepared for disasters; they suffer more during and are less able to recover from it (see Allen, 2007; Finger, 2014; Verchick, 2012). In other words, society's policy decisions or lack thereof, create vulnerability. This aspect of disaster justice is most firmly rooted in the classical North American tradition of environmental justice and has been strongly associated with writings on Hurricane Katrina (Hurricanes Rita, Irene and Sandy). For example, Weibgen (2015) describes how people with disabilities and the elderly were disproportionally affected by Hurricane Sandy in 2012, many being stranded in inaccessible buildings with no heat, water and electricity after evacuation instructions failed to consider their needs. Similarly, Vance (2008) points out that the support systems that poor people relied on in New Orleans were effectively shut down following Hurricane Katrina in 2005; public health clinics closed down, rental housing was destroyed or overpriced, public transportation system decimated and public schools shuttered.

The poor and vulnerable suffer more before, during and after disasters outside of the United States as well (see Cutter, 2017) and examples of disaster justice in the Asia Pacific region highlight the social and political vulnerability of disaster victims. For example, Williamson (2018) describes The Great Singapore Flood of 1954, whose impact was exacerbated by the government forcibly resettling the Bedok community of southeast Singapore into new, ill-prepared sites with poor facilities, which then flooded during the monsoon. Similarly, Parthasarathy (2018) describes how slum dwellers living along the Mithi River were blamed for the 2005 Mumbai floods, while corporate firms (who were not compliant with environmental regulations) were not blamed.

Disasters can also create new injustices. For example, they can produce 'flow on' effects that can create injustices that did not previously exist, such as post-disaster price gauging in rental accommodation (Sandel, 2009, pp. 3–5), the creation of human trafficking opportunities (Douglass & Miller, 2018) or acute food crises due to climate shocks (Wallemacq & House, 2018). If we look at one of the defining features of a disaster—a disruption of normal functioning—we can see how a disaster can lead to a breakdown of law and order, or at least to a suspension of prevailing norms of good conduct.

Disasters act as trigger points that expose existing injustice. The impact of large-scale disasters is often more dependent on the conditions of the system or group affected, rather than the scale and intensity of the hazard (Bullard & Wright, 2008). The study of a particular disaster thus becomes a study of the wider socio-political context of the area (Williamson, 2018). For example, Vance (2008) illustrates how Hurricane Katrina caused the collapse of the criminal justice system in New Orleans, painting a damning picture of a system that barely functioned under normal circumstances due to long-term political neglect and perpetual lack of funding. Disasters thus expose existing injustices and disadvantages. While this can lead to the creation of new injustices (as discussed in the paragraph above), it also highlights an opportunity for disaster management, especially in the prevention and recovery phases to contribute to minimisation or elimination of existing injustices, rather than their perpetuation or creation of new ones.

Disasters do not have to be large to have a large impact. So far, I have mentioned relatively large-scale events that overwhelm a community or some section of society. However, since it is the social condition of the affected population, more than the scale of the physical event that determines impact, relatively small events (such as frequent hailstorms) can affect vulnerable populations. This is being increasingly recognised within the disaster management community (Bull-Kamanga et al., 2003; UNISDR, 2015).

3. Disaster governance is emphasised, illuminating the political nature of disaster decisions, including issues of state accountability and citizen participation.

Disaster governance is highlighted at all levels—from the 2015 Sendai Framework for Disaster Risk Reduction, through to national strategies (such as Australia's National Strategy for Disaster Resilience) to local level plans. However what disaster justice highlights, is not just the importance of governance around the disaster but everyday governance of decisions that may lead to an increase or a decrease in vulnerability to disasters. A focus on governance underscores the role of the national government, and indeed the Sendai Framework recognises that the state has the primary responsibility for disaster risk reduction and citizens increasingly expect their governments to manage social vulnerability and shield them from physical harm (Bankoff, 2018).

Everyday government planning and development decisions can create and ameliorate risk. In many ways they are all about trade-offs between private rights and public responsibilities. A landholder in Australia might choose to live on uncleared land in a bushfire-prone area for lifestyle reasons; to what extent should governments constrict these choices? Public responsibility is also not straightforward; in their discussion of urban development, Rumbach and Németh (2018) discuss how the building of a multi-storey hotel on a steep slope increases risk in terms of a landslide (and not just to the hotel itself), but also has the potential to provide employment and thus reduce socio-economic vulnerability. Which of these public responsibilities should be prioritised? At the end of the day, these are all political decisions and as Williamson (2018, p. 325) puts it, "*[d]isaster justice (or injustice) is only possible within a political space*".

As discussed above, a calamity becomes an injustice when a responsible agent fails to act to prevent it. The Sendai Framework suggests that the national government has primary responsibility for disaster risk reduction;

however, it also promotes the idea of 'shared responsibility', that other stakeholders, including local governments and the private sectors must do their part. The implications of the Sendai Framework are about state accountability for disaster management, however to be accountable, the state must be able and willing to accept responsibility. So what happens when a government lacks the capacity to do so? This is the case of the Philippines, according to Bankoff (2018), where the government lacks resources to decisively respond to and prevent disasters and yet has enough authority to be politically judged for its failures. One could argue that a government may simply call for aid, but do others then have a responsibility to respond? Valentini (2013) makes an interesting case that the 2010 Haiti earthquake is an issue of justice (rather than charity) since Haiti's vulnerability to disasters was caused by the exploitative actions of western countries and international financial organisations, in other words: "*Haiti was made poor*" (Valentini, 2013, p. 500). Its poverty (largely determining its social vulnerability), was arguably preventable and elicits duties of justice as opposed to duties of charity.

However, what if the government is unwilling (rather than unable) to accept responsibility? Van Voorst (2014) describes the lived reality of one of Jakarta's most flood-prone neighbourhoods where government officials blame poor riverbank dwellers for living on the floodplain while ignoring inefficiencies in the city's water management and lack of affordable alternatives. Because the government does not accept responsibility for the impacts of frequent flooding on a vulnerable population and the affected do not have the political power to press their claims, there is no incentive for the government to actively reduce flooding risk.

A large part of governance, emphasised in the Sendai Framework, is the empowerment of ordinary citizens and other stakeholders in disaster management decision-making processes. This emphasis delves into questions of capabilities and power (or lack thereof) that individuals have vis-à-vis their governments. In describing the forced resettlement of the Bedok community of southeast Singapore, Williamson (2018)) describes the community's '*underlying sense of political powerlessness*', which fueled their sense of injustice.

4. The focus on governance emphasises the interlinkage between distributional and procedural dimensions of justice, central to which is recognition and empowerment.

The emphasis on the procedural dimension of justice underscores that justice is a tricky concept. Many western theories of justice, based on modern philosophy, tend to emphasise justice as freedom (Sandel, 2011). This presents two dilemmas for discussing contemporary justice issues. First, conceptualising justice as freedom invariably means the freedom to conceptualise justice in as something other than freedom. This poses a dilemma in a natural disaster situation when disaster management approaches the situation from a human rights point of view (implied in the Sendai Framework), while local communities might see the disaster in terms of 'divine justice' or God's retribution for personal wrong doing, rather than a failure of local planning authorities to appropriately plan a settlement (Bankoff, 2018).

Second, it is impossible to face a contemporary injustice situation without debating who deserves what and why. This is a conversation about virtues, which is intimately connected to cultural and religious views, as well as (inconstant) social norms and values, that are different (and continually changing) within and between different communities. There is thus *"no immutable concept of justice that defies both time and culture"* (Bankoff, 2018, p. 364). However is the existence of multiple and divergent understandings of justice a challenge for justice? More specifically, is it a challenge for disaster justice? Well, yes, given its emphasis on the procedural dimension, especially on recognition. Recognition not only refers to the recognition of different stakeholder groups but of different value systems and knowledge systems as well; whose knowledge of disasters 'counts' in disaster management (Shreshta et al., 2019), how is disaster knowledge created? (Drake, 2018) and can marginalised forms of knowledge be adequately understood and interpreted by dominant social groups? (Figueroa, 2018). Moreover, as empowerment is emphasised in disaster management, whose knowledge sets, beliefs and values get empowered?

Empowerment potentially conflicts with the government's role as the authority having primary responsibility for disaster risk reduction and emergency response as discussed by Lukasiewicz, Dovers, and Eburn (2017) in relation to Australia's National Strategy for Disaster Resilience. Even where governments accept responsibility for disaster risk reduction and are accountable, the line between private rights and the freedom of the individual to make their own choices, and the government's responsibility to act in the public good, makes empowerment a difficult concept to implement (see McLennan & Eburn, 2015 and Lukasiewicz et al., 2017 for an explanation of these tensions).

1.6 Conclusion

As the field of disaster justice is relatively new, the current discussion summarises a theoretical foundation that can be expanded upon through interdisciplinary inquiry. Despite the lack of a singular definition, disaster justice can be described as sitting on the confluence of three established justices—environmental social and climate. Disaster justice is also relevant to all phases of the PPRR spectrum, and is concerned with the interplay of distributive and procedural dimensions of justice. In this chapter, I have outlined four attributes that, taken together, delineate disaster justice into a distinct concept. These attributes include the fact that disaster management is a moral obligation due to the Anthropocene, that the political nature of disaster governance is critical in understanding disaster management, that vulnerability to disasters is rooted in everyday inequality, and that this focus on governance emphasises the importance of recognition and empowerment of those affected by disasters.

References

Agyeman, J., Bullard, R. D., & Evans, B. (Eds.). (2003). *Just Sustainabilities: Development in an Unequal World*. Cambridge, MA: MIT Press.

Allen, B. L. (2007). Environmental Justice, Local Knowledge, and After-Disaster Planning in New Orleans. *Technology in Society, 29*, 153–159.

Althor, G., Watson, J. E. M., & Fuller, R. A. (2016). *Global Mismatch Between Greenhouse Gas Emissions and the Burden of Climate Change* [Article]. *Scientific Reports, 6*, 20281. https://doi.org/10.1038/srep20281., https://www.nature.com/articles/srep20281#supplementary-information

Bankoff, G. (2018). Blame, Responsibility and Agency: 'Disaster Justice' and the State in the Philippines. *Environment and Planning E: Nature and Space, 1*(3).

Bullard, R. D., & Wright, B. (2008). Disastrous Response to Natural and Man-Made Disasters: An Environmental Justice Analysis 25 Years after Warren County. *Journal of Environmental Law, 26*, 217–253.

Bullard, R. D., & Wright, B. (Eds.). (2009). *Race, Place, and Environmental Justice After Hurricane Katrina: Struggles to Reclaim, Rebuild, and Revitalize New Orleans and the Gulf Coast*. Boulder, CO: Westview Press.

Bull-Kamanga, L., Diagne, K., Lavell, A., Leon, E., Lerise, F., MacGregor, H., … Yitambe, A. (2003). From Everyday Hazards to Disasters: The Accumulation of Risk in Urban Areas. *Environment & Urbanization, 15*, 1.

Caney, S. (2014). Two Kinds of Climate Justice: Avoiding Harm and Sharing Burdens. *The Journal of Political Philosophy, 22*(2), 125–149.

Clark, N., Chhotray, V., & Few, R. (2013). Global Justice and Disasters. *The Geographical Journal, 179*(2), 105–113.

Colten, C. E. (2007). Environmental Justice in a Landscape of Tragedy. *Technology in Society, 29*, 173–179.

Cutter, S. L. (2017). The Forgotten Casualties Redux: Women, Children, and Disaster Risk. *Global Environmental Change, 42*(42), 117–121.

Davis, J., Moulton, A. A., Van Sant, L., & Williams, B. (2019). Anthropocene, Capitalocene, ... Plantationocene?: A Manifesto for Ecological Justice in an Age of Global Crises. *Geography Compass, 13*(5), e12438.

Dellink, R., den Elzen, M., Aiking, H., Bergsma, E., Berkhout, F., Dekker, T., & Gupta, J. (2009). Sharing the Burden of Financing Adaptation to Climate Change. Global Environmental Change, 19(4), 411–421. https://doi.org/10.1016/j.gloenvcha.2009.07.009

Deutsch, M. (1985). *Distributive Justice: A Social Psychological Perspective*. New Haven and London: Yale University Press.

Douglass, M., & Miller, M. A. (2018). Disaster Justice in Asia's Urbanising Anthropocene. *Environment and Planning E: Nature and Space, 1*(3), 271–287.

Drake, P. (2018). Emergent Injustices: An Evolution of Disaster Justice in Indonesia's Mud Volcano. *Environment and Planning E: Nature and Space, 1*(3), 307–322.

Figueroa, P. M. (2018). Issues of Disaster Justice Affecting the Fukushima Nuclear Catastrophe. *Environment and Planning E: Nature and Space, 1*(3), 404–421. https://doi.org/10.1177/2514848618790894

Finger, D. (2014). 50 Years After the 'War on Poverty': Evaluating the Justice Gap in the Post-Disaster Context. *Boston College Journal of Law & Social Justice, 34*, 267–282.

Glassey, S. (2018). Did Harvey Learn from Katrina? Initial Observations of the Response to Companion Animals During Hurricane Harvey. *Animals, 8*(4), 47. https://doi.org/10.3390/ani8040047

Gross, C. (2010). *Water Under the Bridge: Fairness and Justice in Environmental Decision-Making*. PhD, Unpublished Doctoral Dissertation, Australian National University, Canberra

Handmer, J., Ladds, M., & Magee, L. (2018). Updating the Costs of Disasters in Australia. *Australian Journal of Emergency Management, 33*(2), 40–46.

Hiller, A. (2011). Climate Change and Individual Responsibility. *Monist, 94*(3), 349–368.

IPCC. (2012). *Managing the Risks of Extreme Events and Disasters to Advance Climate Change Adaptation*. In F. e. a. (Eds.), (p. 582). Cambridge, UK and New York, NY, USA. A Special Report of Working Groups I and II of the Intergovernmental Panel on Climate Change Cambridge University Press.

Lauta, K. C. (2015). *Disaster Law*. London and New York: Routledge.

Lewis, S. L., & Maslin, M. A. (2015). Defining the Anthropocene. *Nature, 519*, 171–180.

Lukasiewicz, A. (2017a). Environment and Justice: Defining the Field. In A. Lukasiewicz, S. Dovers, L. Robin, J. McKay, S. Schilizzi, & S. Graham (Eds.), *Natural Resources and Environmental Justice: Australian Perspectives* (pp. 1–11). Clayton, VIC: CSIRO Publishing.

Lukasiewicz, A. (2017b). The Social Justice Framework: Untangling the Maze of Justice Complexities. In A. Lukasiewicz & S. Dovers (Eds.), *Natural Resources and Environmental Justice: Australian Perspectives* (pp. 233–251). Clayton, VIC: CSIRO Publishing.

Lukasiewicz, A., Dovers, S., & Eburn, M. (2017). Shared Responsibility: The Who, What and How. *Environmental Hazards, 16*(4), 291–313. https://doi.org/10.1080/17477891.2017.1298510

McLennan, B., & Eburn, M. (2015). Exposing Hidden-Value Trade-Offs: Sharing Wildfire Management Responsibility Between Government and Citizens. *International Journal of Wildland Fire, 24*, 162–169.

Munang, R., Thiaw, I., Alverson, K., Mumba, M., Liu, J., & Rivington, M. (2013). Climate Change and Ecosystem-Based Adaptation: A New Pragmatic Approach to Buffering Climate Change Impacts. *Current Opinion in Environmental Sustainability, 5*(1), 67–71.

Nixon, R. (2011). *Slow Violence and the Environmentalism of the Poor*. Cambridge, MA and London: Harvard University Press.

O'Connell, D., Wise, R. M., Williams, R., Grigg, N., Meharg, S., Dunlop, M., ... Crosweller, M. (2018). *Approach, Methods and Results for Co-producing a Systems Understanding of Disaster*. Technical Report Supporting the Development of the Australian Vulnerability Profile.

OCHA. (2019). *Global Humanitarian Overview*. United Nations Office for the Coordination of Humanitarian Affairs (OCHA). Retrieved from https://www.unocha.org/sites/unocha/files/GHO2019.pdf

Parthasarathy, D. (2018). Inequality, Uncertainty, and Vulnerability: Rethinking Governance from a Disaster Justice Perspective. *Environment and Planning E: Nature and Space, 1*(3), 422–442. https://doi.org/10.1177/2514848618802554

Patrick, M. J. (2014). The Cycles and Spirals of Justice in Water-Allocation Decision Making. *Water International, 39*(1), 63–80.

Perry, R. W., & Quarantelli, E. L. (Eds.). (2005). *What Is a Disaster? New Answers to Old Questions*. Philadelphia: Xlibris.

Pescaroli, G., & Alexander, D. (2015). A Definition of Cascading Disasters and Cascading Effects: Going Beyond the 'Toppling Dominos' Metaphor. *Planet@Risk, 3*(1), 58–67.

Rawls, J. (1971). *A Theory of Justice*. London: Oxford University Press.

Riley, J. (2010). Justice as Higher Pleasure. In G. Varouxakis & P. Kelly (Eds.), *John Stuart Mill—Thought and Influence: The Saint of Rationalism*. Oxon & New York: Routledge Innovations in Political Theory.

Rumbach, A., & Németh, J. (2018). Disaster Risk Creation in the Darjeeling Himalayas: Moving Toward Justice. *Environment and Planning E: Nature and Space, 1*(3), 340–362. https://doi.org/10.1177/2514848618792821

Sandel, M. J. (2009). *Justice: What's the Right Thing to Do?* New York: Farrer, Straus and Giroux.

Sandel, M. J. (2011). Distinguished Lecture Justice: What's the Right Thing to Do? *Boston University Law Review, 91*(4), 1303–1310.

Sarat, A., & Kearns, T. R. (Eds.). (1996). *Justice and Injustice in Law and Legal Theory*. Michigan: University of Michigan Press.

Schlosberg, D. (2013). Theorising Environmental Justice: The Expanding Sphere of a Discourse. *Environmental Politics, 22*(1), 37–55.

Sen, A. (2009). *The Idea of Justice*. London, UK: Penguin Books.

Shiva, V. (2015). *Soil Not Oil: Environmental Justice in an Age of Climate Crisis*. Berkley, CA: North Atlantic Books.

Shreshta, K. K., Bhattarai, B., Ojha, H. R., & Bajracharya, A. (2019). Disaster Justice in Nepal's Earthquake Recovery. *International Journal of Disaster Risk Reduction, 33*, 207–216.

Syme, G. J., & Nancarrow, B. E. (2001). Social Justice and Environmental Management: An Introduction. *Social Justice Research, 14*(4), 343–347. https://doi.org/10.1023/A:1014628827223

Temper, L. (2019). Blocking Pipelines, Unsettling Environmental Justice: From Rights of Nature to Responsibility to Territory. *Local Environment, 24*(2), 94–112.

UNISDR. (2015). *Making Development Sustainable: The Future of Disaster Risk Management*. Geneva, Switzerland: Global Assessment Report on Disaster Risk Reduction, United Nations Office for Disaster Risk Reduction.

Valentini, L. (2013). Justice, Charity, and Disaster Relief: What, If Anything, Is Owed to Haiti, Japan, and New Zealand? *American Journal of Political Science, 57*(2), 491–503.

Vance, S. S. (2008). Justice After Disaster-What Hurricane Katrina Did to the Justice System in New Orleans. *Howard Law Journal, 51*(3), 621–649.

Verchick, R. (2012). Disaster Justice: The Geography of Human Capability. *Duke Environmental Law and Policy Forum, 23*, 23–71.

Verchick, R. (2016, November 18–19). *Diamond in the Rough: One City's Quest for Disaster Justice*. Paper Presented at the Disaster Justice in Anthropocene Asia and the Pacific, Singapore.

Verchick, R. (2018). Diamond in the Rough: Pursuing Disaster Justice in Surat, India. *Environment and Planning E: Nature and Space, 1*(3). https://doi.org/10.1177/2514848618797338

van Voorst, R. (2014). The Right to Aid: Perceptions and Practices of Justice in a Flood-Hazard Context in Jakarta, Indonesia. *The Asia Pacific Journal of Anthropology*, 15(4), 339–356. https://doi.org/10.1080/14442213.2014.916340

Wallemacq, P., & House, R. (2018). *Economic Losses, Poverty & Disasters 1998–2017*. Technical Report by UNISDR and CRED.

Weibgen, A. A. (2015). The Right to Be Rescued: Disability Justice in an Age of Disaster. *The Yale Law Journal*, 124(7), 2202–2679.

Williamson, F. (2018). The Politics of Disaster: The Great Singapore Flood of 1954. *Environment and Planning E: Nature and Space*, 1(3), 323–339. https://doi.org/10.1177/2514848618776872

Ziegler, S. S., & Kaplan, H. D. (2019). Forum on the Anthropocene. *Geographical Review*, 109(2), 249–251. https://doi.org/10.1111/gere.12336

CHAPTER 2

Implications of Climate Change for Future Disasters

Rebecca Colvin, Steve Crimp, Sophie Lewis, and Mark Howden

2.1 INTRODUCTION

The climate is changing, and human-caused emissions of greenhouse gases are accelerating the rate of change. Already we are seeing an increase in average temperatures, rising sea levels and a range of disruptions to weather patterns that are affecting the human and non-human worlds. Perhaps one of the most poignant ways in which people will experience the effects of a changing climate will be through exposure to natural

---

R. Colvin (✉)
Crawford School of Public Policy, Australian National University, Canberra, ACT, Australia
e-mail: Rebecca.colvin@anu.edu.au

S. Crimp • M. Howden
Climate Change Institute, Australian National University, Canberra, ACT, Australia
e-mail: steven.crimp@anu.edu.au; mark.howden@anu.edu.au

S. Lewis
School of Science, The University of New South Wales, Canberra, ACT, Australia
e-mail: s.lewis@adfa.edu.au

© The Author(s) 2020
A. Lukasiewicz, C. Baldwin (eds.), *Natural Hazards and Disaster Justice*, https://doi.org/10.1007/978-981-15-0466-2_2

disasters; fires, floods, drought, and severe storms. These events can—and do—cause the loss of life, loved ones, and livelihoods. Societies have navigated disasters throughout history, sometimes more successfully than others, but the unprecedented climate change we are experiencing and continuing to cause in the present time will affect our ability to predict, prepare for, and withstand future disasters. Because of climate change, what came before cannot be used as a reliable guide for what we will face in the future. In this chapter, we outline what is known about climate change—its mechanics, causes, and impacts—and describe how climate change will affect disasters in Australia into the future.

Australia is no stranger to natural disasters. In 1893, a tropical cyclone around Rockhampton in Queensland brought extreme weather and heavy rain, killing 35 people, injuring over 300, and destroying two major bridges in Brisbane city (Australian Institute for Disaster Resilience, n.d.-a). Two years later the Federation Drought began, drying reaches of the Murray–Darling River system and causing devastating effects for the agricultural industry. By 1902 the drought had broken, but the losses to the wheat, cattle, and sheep industries persisted; many farmers abandoned their farms (Australian Institute for Disaster Resilience, n.d.-c). The 1939 Black Friday bushfires in Victoria were brought on by dry conditions and extreme heat. Temperatures in the 1940s, with hot and strong winds carried bushfire through several towns, razing them to the ground along with large areas of forest, and killing over 70 people (Bureau of Meteorology, 2001b). The year of 1974 began with the Brisbane floods which killed 14 people and caused damage estimated at $200 million (over $1 billion in 2019 dollars) (Kitchener, 2011), and ended with Cyclone Tracy battering Darwin with extreme wind and rain (Bureau of Meteorology, 2001a). Cyclone Tracy killed 65 people, destroyed most built infrastructure in its path, and led to widespread evacuation.

More recently, in 2009 the Black Saturday bushfires followed an extreme heatwave in Victoria, killing 173 people and destroying over 2000 homes, resulting in over $1 billion in damages (Australian Institute for Disaster Resilience, n.d.-b). Just 10 years ago, this was one of the worst natural disasters in Australia's recorded history. But we cannot expect it will be the last. As explained in the final report of the 2009 Victorian Bushfires Royal Commission (Victorian Bushfires Royal Commission, 2009) (emphasis added):

> It would be a mistake to treat Black Saturday as a 'one-off' event. With populations at the rural–urban interface growing and the impact of climate change, the risks associated with bushfire are likely to increase.

2.2 Climate Change: The Fundamental Physical Mechanics

If we are to understand how climate change will affect future natural disasters, we must first understand the fundamental mechanics of our changing climate. Day to day we experience weather; states of rain, wind, heat, or snow that can change from hour to hour though adhere to broad patterns across the seasons. When we look at long-term patterns in weather, for example over 30 years or more, our attention turns away from weather and towards the climate. As such, when we talk about climate, we are discussing the long-term averages that describe the conditions of a place. So, it can help to think of climate as the long-term average, while weather patterns in a day-to-day sense are fluctuations around that average.

When we talk about climate change, we are discussing changes to those long-term averages. We can look at historical climate change, and view how the climate changed between warm periods and ice ages. But in its most salient contemporary use, climate change describes a very specific phenomenon; that of human-caused emissions of greenhouse gases into the atmosphere, and the resulting disruptions to our climate.

Fundamentally, human-caused climate change results from the accumulation of greenhouse gases in our atmosphere. These greenhouse gases are emitted by human activities, such as: burning coal, oil, and gas for electricity or transport; land use change; and growing livestock. These gases are long-lived so they remain in the atmosphere for upwards of thousands of years. Commonly we talk about carbon dioxide as it is one of the most prevalent greenhouse gases emitted by human activities, but there are other greenhouse gases emitted by human actions too, such as methane and nitrous oxide. Water vapour is a greenhouse gas, but because it persists in the atmosphere for only a short amount of time, it has minimal sustained impact on the climate. As is indicated by their name, greenhouse gases contribute to the 'greenhouse effect', the trapping of heat in the Earth system by the atmosphere. For life to survive on Earth, we need the greenhouse effect, but human-emitted greenhouse gases are increasing in strength, resulting in higher amounts of heat retained in the atmosphere, causing global warming. More heat in the atmosphere means more energy available to power atmospheric systems, leading to changes in the geographical distribution, frequency, and intensity of weather events. In addition, more heat in the atmosphere leads to more heat being transferred to the oceans, causing ocean warming and sea level rise as a result of thermal expansion of ocean waters and glaciers melting.

2.3 Future Emissions Reductions Trajectories

As the increased accumulation of greenhouse gases in the atmosphere is the result of human activities, global efforts have been undertaken to engender a cooperative international effort to reduce emissions. In 2015 under the Paris Agreement, all nations committed to limiting future global warming to 2 °C above pre-industrial levels, and to pursue actions in accordance with a more ambitious 1.5 °C target (Althor, Watson, & Fuller, 2016). This is based on the recognition that every half a degree matters, and there is a substantial gain to be made by limiting warming as much as is possible. Even at 2 °C, we will see substantial deleterious effects on our human and natural systems (Intergovernmental Panel on Climate Change (IPCC), 2018). At this point, our targets are to limit the extent of the negative impacts of climate change, not to prevent them.

However, the Paris Agreement is non-binding, and nations' pledged domestic emissions reductions targets are insufficient to limit warming even to 2 °C. Instead, we can expect average warming of 2.6–3.1 °C, assuming the pledges are met (Rogelj et al., 2016). What this tells us is that climate change will become ever more significant into the future. The failure to reduce emissions means that nations of large emitters—as well as private actors such as corporations—will be responsible for the consequences of climate change (Lewis, Perkins-Kirkpatrick, Althor, King, & Kemp, 2019), and those who contributed the least to the problem are most likely to be the most significantly impacted (Althor et al., 2016). This is the core premise of climate justice discussed in Chap. 1 of this book.

2.4 Australia's Future Under Climate Change

Climate change is not just a problem for the future, we are already living under conditions of human-caused global warming, and seeing the effects. In 2018, the average global temperature was 0.97 °C above the pre-industrial average; Australia was 1.14 °C above average (IPCC, 2018). Sea levels are rising by an average of 3.2 cm per decade, and within Australia the greatest rate is seen in south-east and northern Australia (Bureau of Meteorology and Commonwealth Scientific and Industry Research Organisation (CSIRO), 2018).

As a result, in 2018 Australia we saw extended drought conditions and what could be characterised as a fire year as opposed to a fire season. Despite overall low rainfall, there were rainfall events that yielded record rainfall and severe flooding (Bureau of Meteorology and CSIRO, 2018). In 2018 alone, over $1.2 billion in insurance payouts were made following extreme weather events. Meanwhile, the cost of drought to Australia exceeds $12.5 billion in 2018–2019 (Insurance Council of Australia, 2018; cited in Steffen, Dean, & Rice, 2019), and that is not accounting for the non-monetary costs such as impacts on mental health and community wellbeing (e.g. Chan, 2019). This is at 1 °C of average warming, and we are tracking towards 3 °C. For the remainder of this chapter, we explore how climate change will affect disasters across five areas: heatwaves, droughts, bushfires, and floods. We outline the intersection between climate change and disasters to lay the foundations for discussion on how justice matters for disasters, in the face of a changing climate.

2.5 HEATWAVES

Heatwaves are broadly defined as periods of prolonged heat that are anomalous for that particular location (Perkins & Alexander, 2013). Heatwaves are measured by a suite of key characteristics of minimum, maximum, or mean temperatures, such as the heatwave's intensity, duration, and spatial extent of anomalies. There is no universally applicable definition of heatwaves, as varying measures are useful in different climatic regions and for different end-users and stakeholders. Studies of Australian heatwaves typically measure excess heat (relative to local climatological values) over three or more days (Perkins-Kirkpatrick et al., 2016), providing information specific to locations. Here we will discuss only atmospheric summertime heatwaves, although note that both out-of-season and marine heatwaves increasingly impact human and natural systems in Australia (Oliver et al., 2018).

Over Australia, the frequency and intensity of observed high maximum and minimum temperatures has increased (Alexander & Arblaster, 2017). The number of heatwaves, and their duration and intensity has increased since 1950 for many parts of Australia and particularly in southern and eastern Australia (Perkins & Alexander, 2013). Heatwaves in Australia are associated with multiple physical drivers and mechanisms from within-year modes of variability (e.g. the El Niño-Southern Oscillation (ENSO)) to longer-term synoptic weather systems (Parker, Berry, Reeder, & Nicholls,

2014). Although heatwaves are naturally occurring phenomena, numerous studies have determined changes in Australian heatwave characteristics that are associated with human-caused climate change.

Linking these observed events to human-caused climate change, though, is not a simple endeavour. Studies have applied a range of methodologies to assess the role of human-caused climate change in these measured changes in Australian heatwaves, that is, studies of attribution. Examining trends in observed temperature extremes using models with and without anthropogenic greenhouse gases, Alexander and Arblaster (2009) determined that the observed changes to heatwaves were consistent with what would be expected as a response to climate change. Dittus, Karoly, Lewis, Alexander, and Donat (2016) similarly show that the observed increased area of Australia experiencing high temperatures is attributable to human-caused climate change. Further studies use an event attribution approach to examine climate change impacts on specific observed extremes such as a notable heatwave. For example, Perkins, Lewis, King, and Alexander (2014) determined that human-caused climate change increased the probability of the intensity and frequency of observed heatwaves in 2013–2014 in Australia. This record hot year of 2013 (Bureau of Meteorology, 2014) has been a target for multiple studies, which also confirm an anthropogenic signal in extreme summer temperatures (Lewis & Karoly, 2013). In summary, multiple studies have found a human-caused greenhouse gas influence in Australia's observed increases in extremes, which is consistent with changes observed across much of the globe (Perkins, 2015).

Heatwave frequency and intensity is projected to increase throughout the twenty-first century with further climate change. Using global-scale climate model information for Australia, numerous studies show that as we look towards the future, the number of cold temperature extremes substantially reduces and the number of warm temperature extremes substantially increases, depending on the greenhouse emissions scenario employed (Alexander & Arblaster, 2017). These findings are supported by targeted regional models, where the number of heatwaves and their duration is projected to increase significantly, with greater increases in the north than south of Australia (Herold, Ekström, Kala, Goldie, & Evans, 2018). Other approaches have focused on how current extreme events may change in future projections, either under different levels of global warming, in different scenarios or at different times in the future. In model simulations where warming is limited to 1.5 °C, frequency of extreme heat events in Australia, like the 2013 record summer, are substantially less likely than at 2 °C (King, Karoly, & Henley, 2017). Furthermore, what we have experienced in recent years as 'record heat' is projected to be considered mild or cool by 2035 in the majority of analysed models in high warming scenarios (Lewis, King, & Perkins-Kirkpatrick, 2017).

Extreme heat already negatively impacts human and natural systems, infrastructure, and industry in Australia. To date, heatwaves are Australia's most deadly natural hazard (Coates, Haynes, O'Brien, McAneney, & De Oliveira, 2014), affecting human health and labour capacity (Zhang et al., 2018). Heatwaves are also an established stressor of key natural ecosystems such as coral reefs (Hughes et al., 2017) and native fauna (Welbergen, Klose, Markus, & Eby, 2008). Further detrimental impacts of heatwaves on road and rail infrastructure, agriculture and aquaculture industries, and energy supply are well documented (see Perkins, 2015). While some human and natural systems have the ability to adapt to the changing characteristics of heatwaves we are seeing now, projected future changes in heatwaves likely pose further risk to natural systems (Lewis & Mallela, 2018), human health, and industry. Herold et al. (2018) specifically examined such impacts, noting that as all Australian capital cities are projected to experience at least a tripling of heatwave days each year, increases in mortality and substantial decrease in wheat production are also projected.

2.6 Drought

Drought is a human construct; depending on the definition, it may be characterised by meteorological conditions, hydrological state, or its impacts. Drought has been notoriously hard to define given its impacts are biophysical, economic, social, and environmental (Mishra & Singh, 2010; Wilhite & Glantz, 1985). For the purposes of this chapter, we refer to the broadly accepted definition of drought in Australia as '*a prolonged, abnormally dry period when the amount of available water is insufficient to meet normal use*' (Bureau of Meteorology and CSIRO, 2018). Likewise the severity of a drought can be measured in terms of rainfall deficiencies, defined as serious or severe if rainfall for the period in question is between the tenth and fifth percentiles (serious) or below the fifth percentile (severe) (Bureau of Meteorology and CSIRO, 2018). These definitions, whilst critical to triggering local, state, and/or federal action, have been found to be inadequate as they are static measures that do not account for the changing nature of extremes resulting from human-caused climate change.

As with the factors driving heatwaves, Australian rainfall variability is influenced by a range of drivers operating across timescales (King, 2014; Maher & Sherwood, 2014; Risbey, Pook, McIntosh, Wheeler, & Hendon, 2009). Within-years, the El Niño–Southern Oscillation phenomenon is a critical factor (Allan, 1988; Ashcroft, Karoly, & Gergis, 2014; Partridge, 1991; Pittock, 1978; Power, Haylock, Colman, & Wang, 2006). During El Niño, when the eastern Pacific is anomalously warm, there is an increased chance of below-average rainfall across much of Australia. In contrast, during La Niña, when the eastern Pacific is anomalously cold, there is an increased chance of above-average rainfall across much of the eastern half of the continent. On multi-year but within-decade timescales, the Interdecadal Pacific Oscillation (IPO) and Pacific Decadal Oscillation (PDO) affect Australian rainfall. Individually, these drivers account for between 20% and 40% of the year-to-year variability in rainfall, but it is their interactions that are critical in determining the length, frequency, and severity of Australian droughts (King, 2014).

Given the human-caused changes occurring in the climate system there is uncertainty about the future trends in ENSO and IPO/PDO, and as a result, drought. Recent research has indicated that the historical warming of the climate has already contributed to a southerly shift in the atmospheric system that carries cold fronts across Australia, the sub-tropical ridge (Timbal et al., 2010; Whan, Timbal, & Lindesay, 2014). The result of this is that cold fronts are steered away, resulting in rainfall declines in winter and spring and thus increasing the risk of drought conditions across southern Australia.

Australia has also experienced significant warming over the historical record, which in turn has increased evaporation and has both heightened and accelerated water stress when drought conditions occur. Research by Nicholls (2003) highlighted that whilst rainfall deficits in the Murray-Darling Basin were similar during the extreme drought years 1982, 1994, and 2002, mean maximum temperatures were in fact 1.3 °C warmer in 2002. The significant temperature increase in 2002 compared to earlier drought years resulted in much greater drought stress and larger reported crop and livestock productivity losses. In 2018, much of New South Wales received approximately half its annual rainfall, and combined with maximum temperatures 2–2.5 °C above average, resulted in the third highest evaporation values since records began (1976) and some of the lowest soil moisture values on record.

Climate change will continue in the decades ahead, with warming across the whole of Australia and rainfall declines over much of the southern parts of the country very likely (Bureau of Meteorology and CSIRO, 2018). For a global increase in mean temperature of 1.5 °C above pre-industrial levels, the likelihood of Australian drought events similar to those experienced in 2006 occurring each year increases by between 3% and 7% (King, 2014).

Drought also contributed to the enforcement of water restrictions in most major cities, to increased electricity prices, and to major bushfire events in 2003 and 2009 (van Dijk et al., 2013). Changes in future drought frequency and severity will have a significant impact on our environment, health, and wellbeing, built environment and economy. The Millennium Drought (late 1996–mid-2010) (Bureau of Meteorology, 2015) was associated with fluctuations in water bird, fish, and aquatic plant populations during the drought years in the Murray-Darling Basin (Colloff, Caley, Saintilan, Pollino, & Crossman, 2015). The drought conditions in 2017–2018 coupled with existing patterns of water utilisation in the Murray-Darling Basin have also contributed to a series of mass fish death events (Australian Academy of Science, 2019; Vertessy et al., 2019). The likely increase in the frequency of future droughts, coupled with the ongoing intensive use of water resources of the Murray-Darling Basin, is likely to result in an increase in the number and extent of mass fish death events in the future. Many ecosystems dependent upon river flows and floods are also likely to be negatively affected by future droughts, with further losses of iconic species such as the river red gums increasing in likelihood (Bond, Lake, & Arthington, 2008; Colloff et al., 2016).

A number of drought-related human health impacts have already been observed. These include physiological impacts such as heat stress, dehydration, respiratory impacts due to drought-related dust storms as well as psychological impacts such as depression, hyper-tension, and stress (Austin et al., 2018). Pre-existing health conditions such as heart disease, asthma, and high blood pressure may also be exacerbated during prolonged or extreme drought conditions, and in such times reductions in health can lead to reductions in people's adaptive capacity (Hanna & McIver, 2018).

Declines in physical health are also prevalent amongst the elderly in drought-affected rural communities in Australia. An increase in the frequency and severity of droughts coupled with an ageing Australian population may result in greater clinical or hospital presentations (Horton, Hanna, & Kelly, 2010).

2.7 BUSHFIRE

Bushfires are a part of the Australian landscape. Evidence stretching from long-past, paleo timescales to the present demonstrates that fires have been a normal and important part of the Australian landscape, and we know the impacts of bushfires varied between ice ages and warm periods. Bushfires have shaped and been shaped by the characteristically Australian sclerophyll vegetation—the bush—and the fire management practices of First Nations, and the subsequent settler-colonial European fire management regimes (Kershaw, Clark, Gill, & D'Costa, 2002; Lynch et al., 2007).

Recently, fire risk has also increased via greater amounts of more expensive infrastructure being located in highly exposed places. This risk would have occurred even in the absence of the changes to fire we are seeing as a result of global warming. Climate has always been a key driver of fire characteristics, with the palaeo record showing that the importance of fire is increased during periods when the climate is warming and especially in situations of greater climate variability (Bliege Bird, Codding, Kauhanen, & Bird, 2012; Lynch et al., 2007).

Unsurprisingly, given that the global temperature has already warmed about 1 °C above pre-industrial levels (IPCC, 2018), over the past decades there has been an increase in fire risk globally with fire seasons lengthening by about 20% across a quarter of Earth's vegetated surfaces, and with a doubling of the area prone to burning (Jolly et al., 2015). Similar increases in fire risk have been recorded for Australia with increases in the frequency of fire-prone conditions and increased length of the fire season (Clarke, Lucas, & Smith, 2013; Jolly et al., 2015). These trends are projected to continue into the future globally (Liu, Stanturf, & Goodrick, 2010; Moriondo et al., 2006) and in Australia (e.g. Cary & Banks, 2000; Clarke, Smith, & Pitman, 2011; Fox-Hughes, Harris, Lee, Grose, & Bindoff, 2014; Matthews, Sullivan, Watson, & Williams, 2012; Pitman, Narisma, & McAneney, 2007; Sullivan, McCaw, Cruz, Matthews, & Ellis, 2012) with increases in fire danger index, number of fire ignition days, rate of fire spread, and fire season length. These changes are likely to be greater for the southern parts of Australia, where rainfall declines are expected to be most pronounced, compared with the northern regions, where conditions are likely to remain the same as present, or become wetter.

The increase in risk is particularly associated with increased temperature, reduced rainfall, and decreased humidity. However, there is some indication it could also be influenced by changes in major regional drivers of climate or dominant synoptic systems including when the Pacific is in an El Niño-like state and when the Indian Ocean Dipole is positive, both of which conditions may increase in frequency with climate change (Cai, Cowan, & Raupach, 2009; Verdon, Kiem, & Franks, 2004). Additionally, many intense fires in south-east Australia are associated with strong winds channelled ahead of powerful cold fronts with the winds drawn from the hot continental interior. The frequency of these frontal systems is projected to increase by up to a factor of four by the end of this century, though whether or not this reaches the worst case scenario is dependent on the rate at which we reduce emissions—action on climate change can avoid the worst of the projected future changes (Hasson, Mills, Timbal, & Walsh, 2009).

A growing concern is that these changes in fire regimes at global scale will result in positive feedbacks through increasing emissions of greenhouse gases leading to further climate change (Bowman et al., 2013). This concern is supported by analyses that show that when global fire weather seasons are longer-than-normal or when long seasons lead to more global burnable area, net global terrestrial carbon uptake is reduced (Jolly et al., 2015; Liu et al., 2015).

Bushfires cause devastating loss of life, homes, and livelihoods, changing people and places irrevocably. Rationally, society should respond to the increasing risk of fire due to climate change. O'Neill and Handmer (2012) outline four key responses: (1) diminish the hazard through reduction of accidental and deliberate ignitions and through fuel reduction, (2) reduce the exposure of infrastructure and buildings and improve building codes to enhance fire resistance, (3) reduce the vulnerability of people via addressing individual vulnerabilities and engaging communities and other stakeholders in fire planning, management, and training, and (4) increase the adaptive capacity of institutions via insurance and fire policies such as a focus on protection of lives and critical infrastructure in periods of extreme fire danger. These changes suggest a need for increased future resources both for pro-active fire management and for post-fire recovery efforts, as well as cultural change. However, some of these responses may be affected by institutional inertia, lack of good information, and apathy, as well as by ongoing demographic trends such as reduced volunteerism which impacts on the numbers of firefighters. Additionally, extreme fire danger periods and associated heatwaves will increasingly impact on the capacity and health of firefighters and householders, reducing their ability to respond to fire risk (O'Neill & Handmer, 2012). Chapters 4, 5, 11, 12, and 13 look at justice issues around bushfires in more detail.

2.8 Flood and Coastal Inundation

Australia is characterised as a dry continent with limited freshwater resources, and our rivers are characterised by relatively low and very variable flows. In spite of being the driest inhabited continent, 80% of Australia's population can be found at or near coasts and rivers of major significance. The rate of water use in Australia is amongst the highest in the world (Department of Environment and Energy, 2011) and Australia faces challenges of a growing and urbanising population, of growing demand for water for food and fibre production, and of environmental sustainability, particularly in the face of climate change (Prosser, 2011). These challenges are not unique to Australia, but, unlike other developed nations, Australia faces the added complications of extreme rainfall variability, a widespread drying trend, and projected increases in both aridity and variability in the future. This means the distribution of rainfall across the continent and over time will change—we may see some areas that have long dry spells broken by a serious deluge not dissimilar to that which was observed in Queensland in the summer of 2019 (Thompson, 2019). Further, the concentration of human settlements along Australia's coastlines means a large proportion of the population is exposed to the impacts of rising sea levels (Abel et al., 2011). These dual drivers—extreme rainfall and sea level rise—mean floods will be central to Australia's climate changed future (Bureau of Meteorology and CSIRO, 2018).

Floods are the most costly natural disaster for Australia. The average direct annual cost of flooding between 1967 and 1999 was estimated at $314 million (Bureau of Transport Economics, 2001). Costs vary considerably between flood events (depending on flood volumes and infrastructure affected); for example, the Brisbane floods of 1974 caused $700 million damage at that time, while the damage from the 2011 floods resulted in approximately $10 billion in damages (Prosser, 2011). Chapter 15 considers justice issues around the 2011 Brisbane floods in more detail. Impacts of flooding also vary geographically with the economic cost of natural disasters for Queensland over the 2007–2016 period estimated at $11 billion per year, and floods contributing 60% of this cost. In comparison, in Western Australia in the same period natural disasters averaged $1 billion per year, with only 15% of that cost being due to floods (Deloitte Access Economics for Australian Business Roundtable for Disaster Resilience and Safer Communities, 2017).

At a national scale, statistical analysis of historical floods shows that decades of higher-than-average rainfall, such as in the 1950s–1970s, can have more extreme floods than would be expected under average rainfall conditions. During these wet decades, a once-in-a-100-year flood is likely to be twice as large as would be expected during drier times (Kiem, Franks, & Kuczera, 2003). As the climate changes due to global warming, this will become ever more a pressing issue. Both warmer ocean temperatures and ambient atmospheric conditions are likely to enhance the moisture holding capacity of the atmosphere. This in turn will result in more intense rainfall events leading to more extensive flooding. Climate modelling also supports the likely increase in the intensity of cyclones as a result of warmer oceanic and atmospheric conditions (Johnson et al., 2016). In Queensland, the Northern Territory, and Western Australia, cyclones are a major driver of floods, so the future changes in flooding are likely to be large in these states.

Meanwhile, globally sea levels have already risen by 20 cm since 1880, and this is expected to continue—and increase in the rate of rise—into the future (Bureau of Meteorology and CSIRO, 2018). Sea level rise will affect human settlements and ecosystems via a number of different processes including enhancements of astronomical tides, storm surges, and waves caused by wind, swell, local pressure systems, and coastal geographic features (McInnes et al., 2016). Around Australia, sea level rise has been most pronounced in the north-west, north, and south-east of the continent (Bureau of Meteorology and CSIRO, 2018). Sea level rise, therefore will not lead to impacts uniform in place and time, rather, coastal inundation will be a result of the combination of sea level, geography, and prevailing weather conditions. Chapter 6 considers legal issues of protecting coastal residential properties.

In the absence of effective mitigation or adaptation, it is highly likely that insurance premiums will rise as does future flood risk. Currently, not all hazards are insurable. Hazards such as bushfires, riverine flooding, and storm damage are generally covered, but events such as coastal inundation and erosion are not. More than $277 billion (in 2019 dollars) in commercial, industrial, road and rail, and residential assets are exposed to flooding and erosion hazards at a sea level rise of 1.1 m (Steffen, Mallon, Kompas, Dean, & Rice, 2019). Whilst this scenario represents the upper limit of possible sea level rise it is still plausible given the existing, inadequate, efforts to reduce emissions.

Australian coastal assets at risk from the combined impact of inundation and shoreline recession include: between 5800 and 8600 commercial buildings, with a value ranging from $58 to $81 billion (2008 replacement value); between 3700 and 6200 light industrial buildings, with a value of between $4.2 and $6.7 billion (2008 replacement value); and between 27,000 and 35,000 km of roads and rail, with a value of between $51 and $67 billion (2008 replacement value) (Department of Environment and Energy, 2011). Other national infrastructure within 200 m of the coastline include: 120 ports, five power stations, 258 police, fire and ambulance stations, 75 hospitals and health services, and 44 water and waste facilities (Department of Climate Change, 2009). All these sites critical for a well-functioning society will be threatened by floods in our climate changed future.

It is clear that due to rainfall extremes and sea level rise, both coastal flooding and riverine flooding risk will increase in the future, and this is especially critical to Australia given the concentration of our human settlements around rivers and the coasts. As the climate changes, the increased likelihood and severity of floods will amplify existing risks to Australian communities, including impacts on health and wellbeing, damage to coastal ecosystems, and disruption to individual and community livelihoods. Beyond the impacts on people, the cost of repairing damage to infrastructure is large, and is likely to result in a high cost to local economies and the Australian economy generally.

Adaptation responses to both flooding and inundation are broad but are likely to be costly. In many instances these options include infrastructure "hardening", such as development of sea walls, flood gates, or raising structures, "accommodation" which includes actions such as developing pumps and raising footpaths, and "retreat" which involves the relocation of residents from areas that are likely to become inundated or flooded (Fletcher et al., 2013). These questions, though, are value-laden, and the process by which different approaches to adapting to future floods are selected will need to be managed carefully to ensure the voice of all people is heard in the decision-making process.

2.9 Interactions in the Real World

When examining the implications of climate change on natural disasters, emphasis is often placed on changes in individual climate variables, such as rainfall. However, historically significant weather and climate events are often the result of the combined influence of extremes in multiple variables occurring simultaneously. This is often referred to as the compounding effect of climate extremes and whilst it is harder to determine the severity or frequency of such compound events, understanding the interactions and aggregated impacts provides a much more realistic estimate of future impacts.

Compound extreme events can occur in various ways (Coburn et al., 2014), for example an extreme coastal storm surge may occur in conjunction with extreme rainfall, resulting in large scale coastal inundation through a combination of surge and related overland flooding. Similarly, extreme rainfall and extreme high wind events along the New South Wales coast are often associated with the simultaneous occurrence of an intense low pressure system, cold front, and thunderstorms (Bureau of Meteorology and CSIRO, 2018). Drought can often coincide with extreme heatwave events or record high daily temperatures—an occurrence which typically results in large impacts on agriculture, human health, fire weather, and infrastructure.

Climate change can have a significant influence on the frequency, magnitude, and impact of some types of compound events. For example, the coincidence of background warming trends, background drying trends, and extreme heat and low rainfall across Tasmania during the spring, summer, and autumn of 2015–2016. October 2015 saw the third highest mean monthly maximum temperature on record for the state, record low monthly rainfall, and record high fire danger (Bureau of Meteorology and CSIRO, 2018).

As climate change continues, the combination of increases in heavy rainfall and rising sea levels means that coastal and estuarine environments may experience increased flood risk; similarly the coincidence of drought and extreme heatwave conditions is very likely in the future. Understanding the frequency and resultant severity of future compound extreme events is a significant scientific challenge, but extremely important for effective disaster risk mitigation and strategic adaption (Mora et al., 2018).

2.10 Looking Towards Disaster Justice to a Climate Changed Future

In this chapter, we have explored the way in which Australia's future disasters will be affected by a changing climate. Although Australia has a history of natural disasters—fires, floods, storms, and heatwaves—the past cannot be presumed to predict the future due to the influence of human-caused climate change. In the most general terms, climate change will take the disasters Australia already experiences, and make them more severe and less predictable. Heatwaves, the most deadly of Australia's natural disasters, will be more frequent, protracted, and hotter. Drought too will see people and the natural environment deprived of water more often and to greater extremes. Drought takes a toll on people not just through the physical impacts of the lack of water but also through substantial mental health impacts. Severe drought can destroy livelihoods and agricultural communities. We have seen the devastation of bushfire in Australia, most profoundly in the 2009 Black Saturday Victorian bushfires. Current understandings of the fire-climate relationships suggest that we may experience reinforcing feedbacks in the future, where fires exacerbate global warming, which in turn raises the risk of fire. Flooding and coastal inundation will affect many Australians in our settlements concentrated around rivers and coasts.

Even under the most optimistic scenarios for emission reductions we are still going to see the effects of climate change on natural disasters. We must consider how we can best adapt to this climate changed future (Conway et al., 2019; Howden et al., 2007; IPCC, 2014). This brings us to the question of justice, which will be examined by others in this book. Climate change is a problem caused by human choices and actions. The burning of fossil fuels and the clearing of land has created vast wealth for some people, but not all (Althor et al., 2016). And we know that the effects of climate change will hit the hardest on those who benefited the least from the economic development that brought on climate change (Lewis et al., 2019) often in complex and insidious ways (e.g. Althor, Mahood, Witt, Colvin, & Watson, 2018). The fact that we, as a species, have built for ourselves a climate changed future means that questions of disaster justice can become only more critical.

REFERENCES

Abel, N., Gorddard, R., Harman, B., Leitch, A., Langridge, J., Ryan, A., & Heyenga, S. (2011). Sea Level Rise, Coastal Development and Planned Retreat: Analytical Framework, Governance Principles and an Australian Case Study. *Environmental Science & Policy, 14*(3), 279–288. https://doi.org/10.1016/j.envsci.2010.12.002

Alexander, L. V., & Arblaster, J. M. (2009). Assessing Trends in Observed and Modelled Climate Extremes over Australia in Relation to Future Projections. *International Journal of Climatology, 29*(3), 417–435.

Alexander, L. V., & Arblaster, J. M. (2017). Historical and Projected Trends in Temperature and Precipitation Extremes in Australia in Observations and CMIP5. *Weather and Climate Extremes, 15*, 34–56.

Allan, R. (1988). El Niño-Southern Oscillation Influences in the Australasian Region. *Progress in Physical Geography, 12*, 313–348.

Althor, G., Mahood, S., Witt, B., Colvin, R. M., & Watson, J. E. M. (2018). Large-Scale Environmental Degradation Results in Inequitable Impacts to Already Impoverished Communities: A Case Study from the Floating Villages of Cambodia. *Ambio, 47*(7), 747–759. https://doi.org/10.1007/s13280-018-1022-2

Althor, G., Watson, J. E. M., & Fuller, R. A. (2016). Global Mismatch Between Greenhouse Gas Emissions and the Burden of Climate Change. *Scientific Reports, 6*, 20281. https://doi.org/10.1038/srep20281

Ashcroft, L., Karoly, D. J., & Gergis, J. (2014). Southeastern Australian Climate Variability 1860–2009: A Multivariate Analysis. *International Journal of Climatology, 34*, 1928–1944.

Austin, E. K., Handley, T., Kiem, A. S., Rich, J. L., Lewin, T. J., Askland, H. A., ... Kelly, B. J. (2018). Drought-Related Stress Among Farmers: Findings from the Australian Rural Mental Health Study. *The Medical Journal of Australia, 209*(4), 159–165.

Australian Academy of Science. (2019, February 18). *Investigation of the Causes of Mass Fish Kills in the Menindee Region NSW over the Summer of 2018–2019*. Retrieved from https://www.science.org.au/files/userfiles/support/reports-and-plans/2019/academy-science-report-mass-fish-kills-digital.pdf

Australian Institute for Disaster Resilience. (n.d.-a). *Black February Flood, 1893*. Retrieved from https://knowledge.aidr.org.au/resources/black-february-flood-1893/

Australian Institute for Disaster Resilience. (n.d.-b). *Bushfire—Black Saturday*. Retrieved from https://knowledge.aidr.org.au/resources/bushfire-black-saturday-victoria-2009/

Australian Institute for Disaster Resilience. (n.d.-c). *Federation Drought*. Retrieved from https://knowledge.aidr.org.au/resources/environment-federation-drought/

Bliege Bird, R., Codding, B. F., Kauhanen, P. G., & Bird, D. W. (2012). Aboriginal Hunting Buffers Climate-Driven Fire-Size Variability in Australia's Spinifex Grasslands. *Proceedings of the National Academy of Sciences, 109*(26), 10287–10292. https://doi.org/10.1073/pnas.1204585109

Bond, N. R., Lake, P. S., & Arthington, A. H. (2008). The Impacts of Drought on Freshwater Ecosystems: An Australian Perspective. *Hydrobiologia, 600*(1), 3–16.

Bowman, D. M., Murphy, B. P., Boer, M. M., Bradstock, R. A., Cary, G. J., Cochrane, M. A., ... Williams, R. J. (2013). Forest Fire Management, Climate Change, and the Risk of Catastrophic Carbon Losses. *Frontiers in Ecology and the Environment, 11*(2), 66–67. https://doi.org/10.1890/13.Wb.005

Bureau of Meteorology. (2001a). *Brisbane Floods, January 1974 & Cyclone Tracy, Christmas 1974*. Retrieved from http://www.austehc.unimelb.edu.au/fam/1611.html

Bureau of Meteorology. (2001b). *The Great Weather and Climate Events of the Twentieth Century*. Retrieved from http://www.austehc.unimelb.edu.au/fam/1610.html

Bureau of Meteorology. (2014). *Annual Climate Report 2013*. Retrieved from http://www.bom.gov.au/climate/annual_sum/2015/Annual-Climate-Report-2015-HR.pdf

Bureau of Meteorology. (2015). *Recent Rainfall, Drought and Southern Australia's Long-Term Rainfall Decline*. Retrieved from http://www.bom.gov.au/climate/updates/articles/a010-southern-rainfall-decline.shtml

Bureau of Meteorology and CSIRO. (2018). *State of the Climate 2018*. Retrieved from http://www.bom.gov.au/state-of-the-climate/

Bureau of Transport Economics. (2001). *Economic Costs of Natural Disasters in Australia*. Report 103. Canberra: Bureau of Transport Economics.

Cai, W., Cowan, T., & Raupach, M. (2009). Positive Indian Ocean Dipole Events Precondition Southeast Australia Bushfires. *Geophysical Research Letters, 36*(19). https://doi.org/10.1029/2009gl039902

Cary, G. J., & Banks, J. C. G. (2000). Fire Regime Sensitivity to Global Climate Change: An Australian Perspective. In J. L. Innes, M. Beniston, & M. M. Verstraete (Eds.), *Biomass Burning and Its Inter-Relationships with the Climate System* (pp. 233–246). Dordrecht: Springer Netherlands.

Chan, G. (2019). *'It Can't Get Any Worse': Why Farrer Is Turning Against the Coalition*. Retrieved from https://www.theguardian.com/australia-news/2019/apr/28/it-cant-get-any-worse-why-farrer-is-turning-against-the-coalition

Clarke, H., Lucas, C., & Smith, P. (2013). Changes in Australian Fire Weather Between 1973 and 2010. *International Journal of Climatology, 33*(4), 931–944.

Clarke, H. G., Smith, P. L., & Pitman, A. J. (2011). Regional Signatures of Future Fire Weather over Eastern Australia from Global Climate Models. *International Journal of Wildland Fire, 20*(4), 550–562. https://doi.org/10.1071/WF10070

Coates, L., Haynes, K., O'Brien, J., McAneney, J., & De Oliveira, F. D. (2014). Exploring 167 Years of Vulnerability: An Examination of Extreme Heat Events in Australia 1844–2010. *Environmental Science & Policy, 42*, 33–44.

Coburn, A. W., Bowman, G., Ruffle, S. J., Foulser-Piggott, R., Ralph, D., & Tuveson, M. (2014). *A Taxonomy of Threats for Complex Risk Management*. Cambridge: Centre for Risk Studies, University of Cambridge.

Colloff, M. J., Caley, P., Saintilan, N., Pollino, C. A., & Crossman, N. D. (2015). Long-Term Ecological Trends of Flow-Dependent Ecosystems in a Major Regulated River Basin. *Marine and Freshwater Research, 66*(11), 957–969. https://doi.org/10.1071/MF14067

Colloff, M. J., Lavorel, S., Wise, R. M., Dunlop, M., Overton, I. C., & Williams, K. J. (2016). Adaptation Services of Floodplains and Wetlands Under Transformational Climate Change. *Ecological Applications, 26*(4), 1003–1017. https://doi.org/10.1890/15-0848

Conway, D., Nicholls, R. J., Brown, S., Tebboth, M. G. L., Adger, W. N., Ahmad, B., ... Wester, P. (2019). The Need for Bottom-up Assessments of Climate Risks and Adaptation in Climate-Sensitive Regions. *Nature Climate Change, 9*(7), 503–511. https://doi.org/10.1038/s41558-019-0502-0

Deloitte Access Economics for Australian Business Roundtable for Disaster Resilience and Safer Communities. (2017). *Building Resilience to Natural Disasters in our States and Territories*. Retrieved from http://australianbusinessroundtable.com.au/assets/documents/ABR_building-resilience-in-our-states-and-territories.pdf

Department of Climate Change. (2009). Climate Change Risks to Australia's Coasts: A First Pass National Assessment. Retrieved from https://www.environment.gov.au/system/files/resources/fa553e97-2ead-47bb-ac80-c12adffea944/files/cc-risks-full-report.pdf

Department of Environment and Energy. (2011). Inland Water: Australia's Water Resources and Use in Australia State of the Environment 2011. Retrieved from https://soe.environment.gov.au/theme/inland-water/topic/australias-water-resources-and-use

van Dijk, A. I. J. M., Beck, H. E., Crosbie, R. S., de Jeu, R. A. M., Liu, Y. Y., Podger, G. M., ... Viney, N. R. (2013). The Millennium Drought in Southeast Australia (2001–2009): Natural and Human Causes and Implications for Water Resources, Ecosystems, Economy, and Society. *Water Resources Research, 49*(2), 1040–1057. https://doi.org/10.1002/wrcr.20123

Dittus, A. J., Karoly, D. J., Lewis, S. C., Alexander, L. V., & Donat, M. G. (2016). A Multiregion Model Evaluation and Attribution Study of Historical Changes in the Area Affected by Temperature and Precipitation Extremes. *Journal of Climate, 29*(23), 8285–8299.

Fletcher, C. S., Taylor, B. M., Rambaldi, A. N., Harman, B. P., Heyenga, S., Ganegodage, K. R., ... McAllister, R. R. J. (2013). *Costs and Coasts: An Empirical Assessment of Physical and Institutional Climate Adaptation Pathways*. Gold Coast: National Climate Change Adaptation Research Facility.

Fox-Hughes, P., Harris, R., Lee, G., Grose, M., & Bindoff, N. (2014). Future Fire Danger Climatology for Tasmania, Australia, Using a Dynamically Downscaled Regional Climate Model. *International Journal of Wildland Fire, 23*(3), 309–321. https://doi.org/10.1071/WF13126

Hanna, E. G., & McIver, L. J. (2018). Climate Change: A Brief Overview of the Science and Health Impacts for Australia. *Medical Journal of Australia, 208*(7), 311–315. https://doi.org/10.5694/mja17.00640

Hasson, A. E. A., Mills, G. A., Timbal, B., & Walsh, K. (2009). Assessing the Impact of Climate Change on Extreme Fire Weather Events over Southeastern Australia. *Climate Research, 39*(2), 159–172.

Herold, N., Ekström, M., Kala, J., Goldie, J., & Evans, J. (2018). Australian Climate Extremes in the 21st Century According to a Regional Climate Model Ensemble: Implications for Health and Agriculture. *Weather and Climate Extremes, 20*, 54–68.

Horton, G., Hanna, L., & Kelly, B. (2010). Drought, Drying and Climate Change: Emerging Health Issues for Ageing Australians in Rural Areas. *Australasian Journal on Ageing, 29*(1), 2–7. https://doi.org/10.1111/j.1741-6612.2010.00424.x

Howden, S. M., Soussana, J.-F., Tubiello, F. N., Chhetri, N., Dunlop, M., & Meinke, H. (2007). Adapting Agriculture to Climate Change. *Proceedings of the National Academy of Sciences, 104*(50), 19691–19696. https://doi.org/10.1073/pnas.0701890104

Hughes, T. P., Barnes, M. L., Bellwood, D. R., Cinner, J. E., Cumming, G. S., Jackson, J. B., ... Morrison, T. H. (2017). Coral Reefs in the Anthropocene. *Nature, 546*(7656), 82.

Insurance Council of Australia. (2018). *ICA Catastrophe Dataset*. Retrieved from https://docs.google.com/spreadsheets/d/1vOVUklm2RR_XU1hR6dbGMT7QFj4I0BGI_JAq4-c9mcs/edit#gid=2147027033

Intergovernmental Panel on Climate Change. (2014). *Climate Change 2014: Synthesis Report*. Contribution of Working Groups I, II and III to the Fifth Assessment Report of the Intergovernmental Panel on Climate Change, Core Writing Team, R. K. Pachauri & L. A. Meyer (Eds.). Geneva, Switzerland. Retrieved from https://www.ipcc.ch/report/ar5/syr/

Intergovernmental Panel on Climate Change. (2018). Global Warming of 1.5 °C. an IPCC Special Report on the Impacts of Global Warming of 1.5 °C above Pre-Industrial Levels and Related Global Greenhouse Gas Emission Pathways, in the Context of Strengthening the Global Response to the Threat of Climate Change, Sustainable Development, and Efforts to Eradicate Poverty. In Press:

V. Masson-Delmotte, P. Zhai, H. O. Pörtner, D. Roberts, J. Skea, P.R. Shukla, A. Pirani, W. Moufouma-Okia, C. Péan, R. Pidcock, S. Connors, J. B. R. Matthews, Y. Chen, X. Zhou, M. I. Gomis, E. Lonnoy, T. Maycock, M. Tignor, T. Waterfield (Eds.). https://www.ipcc.ch/site/assets/uploads/sites/2/2019/05/SR15_Citation.pdf

Johnson, F., White, C. J., van Dijk, A., Ekstrom, M., Evans, J. P., Jakob, D., ... Westra, S. (2016). Natural Hazards in Australia: Floods. *Climatic Change*, *139*(1), 21–35. https://doi.org/10.1007/s10584-016-1689-y

Jolly, W. M., Cochrane, M. A., Freeborn, P. H., Holden, Z. A., Brown, T. J., Williamson, G. J., & Bowman, D. M. J. S. (2015). Climate-Induced Variations in Global Wildfire Danger from 1979 to 2013. *Nature Communications*, *6*, 7537. https://doi.org/10.1038/ncomms8537. https://www.nature.com/articles/ncomms8537#supplementary-information

Kershaw, A. P., Clark, J. S., Gill, A. M., & D'Costa, D. M. (2002). A History of Fire in Australia. In R. A. Bradstock, J. E. Williams, & A. M. Gill (Eds.), *Flammable Australia: The Fire Regimes and Biodiversity of a Continent* (pp. 3–25). Cambridge: Cambridge University Press.

Kiem, A. S., Franks, S. W., & Kuczera, G. (2003). Multi-Decadal Variability of Flood Risk. *Geophysical Research Letters*, *30*, 1035.

King, A. D. (2014). Extreme Rainfall Variability in Australia: Patterns, Drivers and Predictability. *Journal of Climate*, *27*, 6035–6050.

King, A. D., Karoly, D. J., & Henley, B. J. (2017). Australian Climate Extremes at 1.5 C and 2 C of Global Warming. *Nature Climate Change*, *7*(6), 412.

Kitchener, G. (2011). *Brisbane's Flood and the Lessons of 1974.* Retrieved from https://www.bbc.com/news/world-asia-pacific-12165808

Lewis, S. C., & Karoly, D. J. (2013). Anthropogenic Contributions to Australia's Record Summer Temperatures of 2013. *Geophysical Research Letters*, *40*(14), 3705–3709.

Lewis, S. C., King, A. D., & Perkins-Kirkpatrick, S. E. (2017). Defining a New Normal for Extremes in a Warming World. *Bulletin of the American Meteorological Society*, *98*(6), 1139–1151.

Lewis, S. C., & Mallela, J. (2018). A Multifactor Analysis of the Record 2016 Great Barrier Reef Bleaching. *Bulletin of the American Meteorological Society*, *98*(6), 1139–1151.

Lewis, S. C., Perkins-Kirkpatrick, S. E., Althor, G., King, A. D., & Kemp, L. (2019). Assessing Contributions of Major Emitters' Paris-Era Decisions to Future Temperature Extremes. *Geophysical Research Letters*, *46*. https://doi.org/10.1029/2018gl081608

Liu, Y., Stanturf, J., & Goodrick, S. (2010). Trends in Global Wildfire Potential in a Changing Climate. *Forest Ecology and Management*, *259*(4), 685–697. https://doi.org/10.1016/j.foreco.2009.09.002

Liu, Y. Y., van Dijk, A. I. J. M., de Jeu, R. A. M., Canadell, J. G., McCabe, M. F., Evans, J. P., & Wang, G. (2015). Recent Reversal in Loss of Global Terrestrial Biomass. *Nature Climate Change*, 5, 470. https://doi.org/10.1038/nclimate2581. https://www.nature.com/articles/nclimate2581#supplementary-information

Lynch, A. H., Beringer, J., Kershaw, P., Marshall, A., Mooney, S., Tapper, N., … Kaars, S. V. D. (2007). Using the Paleorecord to Evaluate Climate and Fire Interactions in Australia. *Annual Review of Earth and Planetary Sciences*, 35(1), 215–239. https://doi.org/10.1146/annurev.earth.35.092006.145055

Maher, P., & Sherwood, S. C. (2014). Disentangling the Multiple Sources of Large-Scale Variability in Australian Wintertime Precipitation. *Journal of Climate*, 27, 6377–6392.

Matthews, S., Sullivan, A. L., Watson, P., & Williams, R. J. (2012). Climate Change, Fuel and Fire Behaviour in a Eucalypt Forest. *Global Change Biology*, 18(10), 3212–3223. https://doi.org/10.1111/j.1365-2486.2012.02768.x

McInnes, K. L., White, C. J., Haigh, I. D., Hemer, M. A., Hoeke, R. K., Holbrook, N. J., … Cox, R. J. C. C. (2016). Natural Hazards in Australia: Sea Level and Coastal Extremes. *Climatic Change*, 139(1), 69–83. https://doi.org/10.1007/s10584-016-1647-8

Mishra, A. K., & Singh, V. P. (2010). A Review of Drought Concepts. *Journal of Hydrology*, 391(1–2), 202–216.

Mora, C., Spirandelli, D., Franklin, E. C., Lynham, J., Kantar, M. B., Miles, W., … Hunter, C. L. (2018). Broad Threat to Humanity from Cumulative Climate Hazards Intensified by Greenhouse Gas Emissions. *Nature Climate Change*, 8(12), 1062–1071. https://doi.org/10.1038/s41558-018-0315-6

Moriondo, M., Good, P., Durao, R., Bindi, M., Giannakopoulos, C., & Corte-Real, J. (2006). Potential Impact of Climate Change on Fire Risk in the Mediterranean Area. *Climate Research*, 31(1), 85–95.

Nicholls, N. (2003). The Changing Nature of Australian Droughts. *Climatic Change*, 63(3), 3323–3336.

O'Neill, S. J., & Handmer, J. (2012). Responding to Bushfire Risk: The Need for Transformative Adaptation. *Environmental Research Letters*, 7(1), 014018. https://doi.org/10.1088/1748-9326/7/1/014018

Oliver, E. C. J., Lago, V., Hobday, A. J., Holbrook, N. J., Ling, S. D., & Mundy, C. N. (2018). Marine Heatwaves off Eastern Tasmania: Trends, Interannual Variability, and Predictability. *Progress in Oceanography*, 161, 116–130. https://doi.org/10.1016/j.pocean.2018.02.007

Parker, T. J., Berry, G. J., Reeder, M. J., & Nicholls, N. (2014). Modes of Climate Variability and Heat Waves in Victoria, Southeastern Australia. *Geophysical Research Letters*, 41(19), 6926–6934. https://doi.org/10.1002/2014gl061736

Partridge, I. J. (1991). Will It Rain?: El Niño and the Southern Oscillation. In *Information Series QI91028*. Brisbane: Queensland Department of Primary Industries.

Perkins, S., Lewis, S., King, A., & Alexander, L. (2014). Increased Simulated Risk of the Hot Australian Summer of 2012/13 Due to Anthropogenic Activity as Measured by Heatwave Intensity and Frequency. *Bulletin of the American Meteorological Society, 95*(9), S34–S37.

Perkins, S. E. (2015). A Review on the Scientific Understanding of Heatwaves—Their Measurement, Driving Mechanisms, and Changes at the Global Scale. *Atmospheric Research, 164–165*, 242–267. https://doi.org/10.1016/j.atmosres.2015.05.014

Perkins, S. E., & Alexander, L. V. (2013). On the Measurement of Heat Waves. *Journal of Climate, 26*(13), 4500–4517. https://doi.org/10.1175/jcli-d-12-00383.1

Perkins-Kirkpatrick, S. E., White, C. J., Alexander, L. V., Argüeso, D., Boschat, G., Cowan, T., ... Purich, A. (2016). Natural Hazards in Australia: Heatwaves. *Climatic Change, 139*(1), 101–114. https://doi.org/10.1007/s10584-016-1650-0

Pitman, A. J., Narisma, G. T., & McAneney, J. (2007). The Impact of Climate Change on the Risk of Forest and Grassland Fires in Australia. *Climatic Change, 84*(3), 383–401. https://doi.org/10.1007/s10584-007-9243-6

Pittock, A. B. (1978). A Critical Look at Long-Term Sun-Weather Relationships. *Reviews of Geophysics and Space Physics, 16*, 400–420.

Power, S., Haylock, M., Colman, R., & Wang, X. (2006). The Predictability of Interdecadal Changes in ENSO Activity and ENSO Teleconnections. *Journal of Climate, 19*(19), 4755–4771. https://doi.org/10.1175/JCLI3868.1

Prosser, I. (2011). *Water: Science and Solutions for Australia.* Australia: CSIRO Publishing.

Risbey, J. S., Pook, M. J., McIntosh, P. C., Wheeler, M. C., & Hendon, H. H. (2009). On the Remote Drivers of Rainfall Variability in Australia. *Monthly Weather Review, 137*(10), 3233–3253. https://doi.org/10.1175/2009MWR2861.1

Rogelj, J., den Elzen, M., Höhne, N., Fransen, T., Fekete, H., Winkler, H., ... Meinshausen, M. (2016). Paris Agreement Climate Proposals Need a Boost to Keep Warming Well below 2 °C. *Nature, 534*(7609), 631–639. https://doi.org/10.1038/nature18307

Steffen, W., Dean, A., & Rice, M. (2019). *Weather Gone Wild: Climate Change-Fuelled Extreme Weather in 2018.* Retrieved from https://www.climatecouncil.org.au/wp-content/uploads/2019/02/Climate-council-extreme-weather-report.pdf

Steffen, W., Mallon, K., Kompas, T., Dean, A., & Rice, M. (2019). *Compound Costs: How Climate Change Is Damaging Australia's Economy.* Retrieved from https://www.climatecouncil.org.au/wp-content/uploads/2019/05/Costs-of-climate-change-report.pdf

Sullivan, A., McCaw, W., Cruz, M., Matthews, S., & Ellis, P. (2012). Fuel, Fire Weather and Fire Behaviour in Australian Ecosystems. In R. Williams, A. Gill, & B. RA (Eds.), *Flammable Australia: Fire Regimes, Biodiversity and Ecosystems in a Changing World* (pp. 51–77). Collingwood, VIC: CSIRO Publishing.

Thompson, B. (2019). Queensland Farmers' Confidence Hit by Drought, Flood Double Whammy. Retrieved from https://www.afr.com/business/queensland-farmers-confidence-hit-by-drought-flood-double-whammy-20190311-h1c92m

Timbal, B., Arblaster, J. M., Braganza, K., Fernandez, E., Hendon, H., Murphy, B., ... Whan, K. (2010). *Understanding the Anthropogenic Nature of the Observed Rainfall Decline Across South Eastern Australia*. CAWCR Technical Report No. 026. Retrieved from http://www.cawcr.gov.au/publications/technicalreports/CTR_026.pdf

Verdon, D. C., Kiem, A. S., & Franks, S. W. (2004). Multi-Decadal Variability of Forest Fire Risk – Eastern Australia. *International Journal of Wildland Fire, 13*(2), 165–171. https://doi.org/10.1071/WF03034

Vertessy, R., Barma, D., Baumgartner, L., Mitrovic, S., Sheldon, F., & Bond, N. (2019, February 20). *Independent Assessment of the 2018–19 Fish Deaths in the Lower Darling*. Interim Report, with Provisional Findings and Recommendations. Retrieved from https://www.mdba.gov.au/sites/default/files/pubs/Independent-assessment-2018-19-fish-deaths-interim-report.PDF

Victorian Bushfires Royal Commission. (2009). *Final Report: Summary*. Retrieved from https://knowledge.aidr.org.au/media/4457/vbrc_summary.pdf

Welbergen, J. A., Klose, S. M., Markus, N., & Eby, P. (2008). Climate Change and the Effects of Temperature Extremes on Australian Flying-Foxes. *Proceedings of the Royal Society B: Biological Sciences, 275*(1633), 419–425. https://doi.org/10.1098/rspb.2007.1385

Whan, K., Timbal, B., & Lindesay, J. (2014). Linear and Nonlinear Statistical Analysis of the Impact of Sub-Tropical Ridge Intensity and Position on South-East Australian Rainfall. *International Journal of Climatology, 34*(2), 326–342. https://doi.org/10.1002/joc.3689

Wilhite, D. A., & Glantz, M. H. (1985). Understanding the Drought Phenomenon: The Role of Definitions. *Water International, 10*(3), 111–120. https://doi.org/10.1080/02508068508686328

Zhang, Y., Beggs, P. J., Bambrick, H., Berry, H. L., Linnenluecke, M. K., Trueck, S., ... Capon, A. G. (2018). The MJA–Lancet Countdown on Health and Climate Change: Australian Policy Inaction Threatens Lives. *Medical Journal of Australia, 209*(11), 474–474. https://doi.org/10.5694/mja18.00789

PART II

Governance

CHAPTER 3

Public Policy and Disaster Justice

Stephen Dovers

3.1 Introduction

This chapter seeks to identify where, in the policy and institutional system that shapes emergency management and disaster policy, issues of 'justice' could be or are being addressed. It is more a commentary on the intersection of public policy (as it does or could exist) in the field of emergency management and disasters, and issues of justice as they are discussed in this book, than a substantive treatment, as the former is well-covered in referenced literature. As such, it is intended to define and open up a space for research and policy discussion around disaster justice as a public policy issue, rather than an attempt to fill that space.

Because disaster impacts and lack of opportunities for engagement in disaster mitigation can arouse considerable contestation and even anger, a starting assumption here is that emergency managers and others who shape and implement disaster policy do not intentionally create injustices. Rather, they may inadvertently or unknowingly create conditions whereby disaster impacts, or opportunities for empowerment, are not evenly or 'fairly' distributed, whether by way of their own actions or inattention,

---

S. Dovers (✉)
Fenner School of Environment and Society, Australian National University, Canberra, ACT, Australia
e-mail: stephen.dovers@anu.edu.au

© The Author(s) 2020
A. Lukasiewicz, C. Baldwin (eds.), *Natural Hazards and Disaster Justice*, https://doi.org/10.1007/978-981-15-0466-2_3

lack of resources, or by the constraints imposed by the parameters of existing policy and institutional arrangements or organisational capacity. I will extend that assumption to other policy sectors and individuals and organisations within them: that, for example, land use planners, health policy makers or infrastructure builders and custodians do not seek to create inequalities of disaster impact, preparedness or recovery, but rather such are undesirable by-products of decisions made for other reasons, including lack of linkages between emergency management and other policy sectors, oversight, issues of mandate and responsibility, or resourcing. It may be that policies and institutional settings could be changed to enhance justice outcomes—or some other goal—but such policy change is a matter for elected governments, advised by senior officials and groups from outside of government, but policy and management officials cannot generally step outside the limitations of existing policy even if they believe these to be deficient. In public policy debates, such a starting point allows a non-confrontational approach to seeking redress of issues of procedural or distributional justice: constructive engagement begins with identification of an oversight and possible ways of doing things better, than it does with accusations of venality. Policy makers deserve some justice and understanding, too.

The chapter uses three sources. The first sources are the kind of justice issues identified elsewhere in this book, apparent from the disaster field. The second are frameworks and modes of analysis from the disaster policy literature (see further Cole, Dovers, Gough, & Eburn, 2018; Dovers & Handmer, 2014; Handmer & Dovers, 2013; Lukasiewicz, Dovers, & Eburn, 2017 and works referenced therein). The third is the existing institutional, policy and management system of Australian emergency management, a backdrop against which policy responses to injustice can be characterised. Apart from reference to some international situations discussed elsewhere in the book, the principles and general policy options exampled in the Australian context may be relevant elsewhere, subject to the usual caveats and cautions of comparative policy analysis and policy learning (Bennett & Howlett, 1992; Rose, 2005), that is, to seek insights and understanding rather than models for adoption.

The next section considers the scope of 'disaster justice' as a policy arena, not just in the disaster policy and emergency management sector, which is an important consideration if policy options exist in other policy sectors, which this chapter argues they do. Following that, six sections briefly explore some core (and unremarkable) perspectives from the discipline and practice of public policy and the implications of considering these in terms of disaster justice. Along the way, general arguments are grounded or informed by situations and propositions described in other chapters of this book.

3.2 Where Are the Boundaries of 'Disaster Justice Policy'?

Four issues of scope are addressed here: the need to consider all aspects of emergency management; the definition of just processes and outcomes; the importance of policy integration across sectors; and the need to include all stakeholders in considering justice. The first issue of scope is that policy attention should span across the spectrum of Prevention-Preparedness-Response-Recovery (PPRR). There has been a slow, steady evolution of emergency management and disaster policy from focusing largely on Preparedness and Response to a greater emphasis on Prevention and (better) Recovery, articulated at the highest level in the Sendai Framework (UNISDR, 2015). This is congruent with an increasing linkage between disasters, sustainable development and human development, via the 2015 Agenda on Sustainable Development (UN, 2015) and the 2015 Paris Agreement on Climate Change (UNFCCC, 2015). These linkages and larger international agendas flow into many national and sub-national level policies, and are crucial to considerations of justice: climate change will exacerbate disasters, human development is damaged by disasters and disasters cause environmental harm. Opportunities for policy remedies to avoid differential vulnerability (minimising unjust distributional impacts), and different capacities to engage in disaster risk reduction activities (procedural matters) are arguably the greatest and more systemic in the stages of Prevention and Recovery. While the more immediate (and important) stages of Preparation and Response have justice implications, especially regarding distributional justice, Prevention (mitigation) reduces the chances of disaster impacts and thus unjust outcomes, and Recovery can likewise produce settlements and socio-economic conditions where the vulnerable are less at risk, with strong distributional but also procedural justice implications.

The second issue of scope is the question of what constitutes a 'just' process or outcome, or an injustice, in a specific situation, even where loss measurement, post-event inquiry, socio-economic impact assessment and similar processes may inform the identification of uneven impacts or opportunities. This may be contested and even irresolvable, as in many cases different but arguably valid conclusions as to specific instances of (in)justice will arise due to people's different perspectives, situations or value sets. If a group of people claim injustice, then that should be investigated: any eventual policy resolution aside, recognition of their concerns is a more just response than ignoring a voiced grievance. From a pragmatic public policy perspective, here we will take propositions of injustice or prospects of improved justice seriously, but temper these against extant literature and the reality of who does what, where and how in the policy and institutional landscape. In doing so, we will consider the implications of different concepts of justice (distributive, procedural, etc.) for public policy, predicting that these will expose quite different policy questions and options. In a later section, the ultimate political (rather than policy) judgement and resolution of different positions will be briefly discussed.

The third issue of scope is that, while the core focus is on emergency management and disaster policy, the creation of injustice or opportunities to improve justice in other policy sectors must also be considered (planning, health services, communications, housing, etc.). This recognises that it is in other policy sectors where vulnerabilities to hazards are often created, maintained or overlooked through inattention: disaster policy and emergency management seeks to manage or reduce, and eventually attends the consequences of, those vulnerabilities. Policy options to enhance just outcomes may be found more in policy sectors other than emergency management and disaster policy, but possibly harder to formulate and implement amidst other mandates and imperatives. In particular, the issue may be a lack of horizontal policy integration, whereby natural hazard and disaster considerations are not taken into account sufficiently in these other policy sectors. This may be via lack of mechanisms to allow or enforce such integration, or by lack of use of available mechanisms.

The final issue of scope is justice for whom? Issues of fairness and justice are mostly discussed with a primary focus on 'victims' of disasters, or those who may become so, through being vulnerable to the impacts of hazards. However, from a public policy perspective, a view that is both wider and of finer resolution is required. Justice for those impacted, or vulnerable to impacts, is indeed central, but injustices may be borne by responders, both paid and volunteer, emergency service organisation leaders and political figures, and firms. In this book, Ingham et al. (Chap. 12) identify the arguably unjust or at least untenable positions that community leaders may find themselves in a post-disaster event environment, subject to high expectations, with little support and scarce resources, on top of their own disrupted lives and normal demands on their time. Their study evidences the potential for a wider and finer resolution of the question 'justice for whom?' and generates both understanding of the situation and practical policy responses (in their case, back-filling leaders' positions and reconstituting Local Emergency Management Committee membership). Within sub-groups, access to justice, or the imposition of injustices, may vary: for example those who pay for flood insurance and thus increase their resilience, versus those who pay but later discover a clause differentiating types of flooding that restricts their claim for damages, versus those who cannot afford insurance and call on the public purse post-event, versus those who are capable of affording insurance but do not on the expectation that the state will assist them nonetheless. Are those in the final category being unjust against the other two, and on the wider public?

Returning to justice afforded to emergency managers themselves, in Australia the prime lesson-drawing and blame-attributing mechanism—post-event inquiries—focus heavily on emergency management agencies and their professional officers, and very little on firms, volunteers, households or individuals (Cole et al., 2018). With 'shared responsibility' and similar policy principles stated internationally and nationally, requiring contributions across the PPRR spectrum from public, private and community sector actors, this is a strange lack of attention. As discussed by Eburn in this book (Chap. 7), many post-event inquiries are adversarial and blame-seeking rather than restorative, and may be seen as not delivering justice to professional emergency managers and responders.

Issues of scope summarised, the following sections explore some implications for disaster justice issues of six basic (and often interrelated) lenses of public policy: (1) whether justice-relevant decisions are a matter for public policy or more value-laden politics; (2) if the former, where in policy process they can be attended to; (3) which sector, agency or individual has the major or ultimate responsibility or power over the matter at hand; (4) in addition to the legal or political mandate, what actors outside of government contribute to justice or injustices; (5) whether it is a matter of disaster policy or emergency management practice, or a matter for another policy sector; and (6) whether enhancement of existing powers or policies can provide remedy, or whether the issue requires the creation of new mechanisms.

3.3 Policy or Politics?

While some may view policy as a rational, neutral practice, or believe that it should be, the mechanisms of public policy implement social and political goals that have been argued and negotiated more broadly across society. 'Political decisions' are often viewed poorly, but in the best sense are where sufficiently supported social goals, or indeed unpopular but necessary decisions, are defined by elected governments and then passed through the public policy system of a jurisdiction for implementation (element 1, Table 3.1). Davis et al. (1993, p. 257) put it well (original emphasis):

> Politics is *the* essential ingredient for producing workable policies, which are more publicly accountable and politically justifiable … While some are uncomfortable with the notion that politics can enhance rational decision-making, preferring to see politics as expediency, it is integral to the process of securing defensible outcomes. We are unable to combine values, interests and resources in ways which are not political.

Policy is imbued with values, including different interpretations of the public interest, as manifestations of political decisions. A justice issue (or occurrence of injustice) might be a matter for public policy and public administration, and can be attended to by public officials with non-government partners within existing policy settings, or might rather be a value question requiring a political judgement and decision to rectify. Often, attention to injustices require a political decision to change policy goals or resourcing levels for 'policy' to then proceed to address the situation. Consider the alternate scenario presented by Alexandra in this book (Chap. 4), of a very different manner of land use planning in fire-prone peri-urban and rural areas. Land use planning is, constitutionally, a matter for states and territories, and while all planning regimes consider natural hazards, giving prime consideration to that above all over policy goals has never occurred. While existing settlements and dwellings could not conceivably be altered in the near term or substantially, unless as dwellings age they are prohibited from being rebuilt, it is entirely possible that a jurisdiction faced with increasing exposure to bushfire (aka wildfire) could decide to manage future risk by only permitting new

Table 3.1 Framework for policy and institutional analysis for emergencies and disasters

1. Problem framing	1. Social debate of problems.
	2. Monitoring, research.
	3. Identification of direct and indirect causes, including vulnerability and resilience.
	4. Assessment of other policies.
	5. Assessment of risk and uncertainty.
	6. Definition of policy and institutional problems.
2. Policy framing and strategic policy choice	7. Choice of general policy style/styles.
	8. Identification of policy principles.
	9. Definition of policy goals.
	10. Policy statement.
3. Policy design and implementation	11. Policy instrument choice.
	12. Implementation planning.
	13. Provision of resources (multiple forms).
	14. Communication and information strategies.
	15. Enforcement and compliance provisions.
	16. Establishment of monitoring mechanisms.
4. Policy monitoring and learning	17. Monitoring and routine data capture.
	18. Learning from events.
	19. Evaluation.
	20. Adaptation, cessation, problem redefined.
Cross-cutting policy principles	Whole-of-government coordination.
	Transparency and accountability.
	Public participation.
Institutional design imperatives	Coordination of actors and organizations.
	Use of legal systems and instruments.
	Clarity of responsibilities.
	Purposefulness and persistence over time.
	Information richness and sensitivity.
	Flexibility and adaptability.

Source: Adapted from Handmer and Dovers (2013, p. 64)

This is *not a model*, but a heuristic checklist of elements that often apply and may be crucial in a given situation. In both policy analysis and reality, the implicit linearity below will generally not hold: focus may be on a small number of elements; an exercise or attention may commence at a number of points; reiteration or feedback loops will often occur, causing reconsideration of an earlier element settlement in defined areas in the manner Alexandra describes. That could be expressed in a new land use and dwelling approval regime enabled by state legislation, which would then be implemented by the appropriate state and local government agencies. However, such a policy shift would be considered by some as a radical shift in the face of actual and assumed rights, historical patterns of land use, and social expectations, and would require a 'brave' political decision from a state government, with certain strong opposition in public debate. Many would see such constraint on peoples' choice of where and how to live as deeply unjust, even if it saved some from exposure to risk of property loss and death. If land already zoned to allow future development was affected, legal recourse would likely be pursued. Public policy and public administration could implement such a regime, but initiating it would be a deeply political matter.

Consider another example of the justice dimension of political decisions. The patchy adoption of flood insurance (not insuring, as well as inadequate policies) and thus the uneven ability to recover losses, repeatedly arises as an issue following significant floods in Australia and elsewhere. It would be entirely possible for a government to define a policy whereby those who are uninsured in instances of known risk would not be eligible for recovery aid (perhaps with a pro-justice subsidy or exemption for poorer households). That would incentivise self-reliance and use an available, well-known fiscal policy instrument (insurance). Now consider a political leader confronting affected households and whole communities saying "bad luck, you were told to insure and you didn't", and being repaid with community anger and electoral loss. However economically logical and effective the policy, in the face of human suffering, a back down from that policy position might not be only 'political' in the negative sense, but 'political' also as an expression of human sympathy and sense of justice for fellow community members.

3.4 What Part of the Policy Process?

Behind the call or need for a 'policy response' there lies significant and important detail, as the public policy process is complex, comprising many parts and the required response may be available or not in various parts. So, at which 'stage' of the policy process, or via which element or combination of elements of a policy process is a particular justice issue (1) caused by, or (2) can be attended by: policy formulation, policy design, policy implementation, policy monitoring and evaluation? Those four 'stages' are typical of simple policy models, but Table 3.1 (from Handmer & Dovers, 2013) provides greater detail of specific and general elements of the disaster and emergency policy process. Usually a justice issue, or a suite of issues, will need to be attended to in multiple parts of the policy process, or it may already be catered for but could be more effectively implemented. An issue may require broad social and political debate (element 1, important regarding public participation and thus procedural justice) and/or further investigation (elements 2–5), before a problem suitable for policy attention and action (element 6) can be defined. Whether those preliminary elements are required or not (the situation and problem may be evident already), policy formulation (elements 7–10) is a substantial task, well before the detail of policy design and implementation (elements 11–16). Monitoring of the success or failure—it is usually somewhere between—of a policy programme should be an ongoing effort, with clear goals and metrics defined and appropriate data capture that allows rigorous evaluation. Within those broad categories of elements of the policy process, some will demand detailed analytical and design effort, such as choosing the policy instrument/s to be deployed (a tax incentive, a stricter regulation, public education campaigns, etc.) or the statutory settings to enable implementation (see shortly below in the case of bushfire). Policy instrument choice involves the application of a range of selection criteria, including justice-relevant ones such as distributional impacts and communicability (Table 3.1).

Importantly, public policy is almost always an iterative (indeed messy) rather than linear process, with attention starting at different 'stages' or with different elements, and in line with different stakeholders' values and motivations, and moving back and forth between problem identification, policy design, implementation and policy monitoring as experience and understanding evolve such as when new information becomes available.

The study by March et al. in this book (Chap. 5) can illustrate the complexities and multiple choices outlined above. The Wye River-Separation Creek bushfires exposed causes and effects that share similar characteristics to other fires: a risky environment where existing settlement patterns, dwelling construction and household behaviours led to significant loss when combined with severe fire weather. Should society wish to prevent future harm, what are the policy questions? One is whether the lessons should inform future planning and development control via a strong, statutory and regulatory regime that tightly controls future development in a shift from historical and current patterns (similar to the Alexandra scenario discussion above): a deeply political policy direction that might well be contested by some stakeholders (Table 3.1 element 1, with subsequent action through the parliamentary and policy processes). Less controversial would be a focus on more targeted communication to encourage actions to increase the fire resistance of dwellings, or funding programmes to enable better Prevention and Preparation, or a fiscal incentive such as property rate rebates to achieve the same. These options are the detail beneath element 11 in Table 3.1—policy instrument choice—where the means to the ends of policy are chosen, detail exposed further in Table 3.2. We know the social/policy outcome that is desired (better management of households and surrounds to minimise/reduce fire hazard) and have to select the policy tool to achieve that. Instrument choice is fundamental to public policy, and has justice implications that we now explore briefly.

Table 3.2 provides a menu of policy instruments—the tools available to governments and their partners to achieve agreed social and policy goals—and Table 3.3 presents a set of criteria against which instruments can be compared (from Handmer & Dovers, 2013, tailored to the disaster field but not dissimilar to many other iterations in the public policy literature). There are many instrument options, and many criteria, and choice is not a deterministic, easy matter but the two lists can guide discussion and choice.

The strong, planning regime option identified above in a case such as Wye River (Chap. 5) would call upon a legislated and enforced land use planning regime (options 6 and 7 in Table 3.2). The less confronting but complementary options to encourage self-defence involve options 3–4 and 8–10 (typically, a policy response will involve use of more than one policy instrument). The considerations regularly taken into account by policy makers (the criteria in Table 3.3), whether satisfactory or not, already include justice-relevant factors, even if not explicitly stated.

Table 3.2 A menu of policy instruments for emergencies and disasters

Class	Selected, major instruments
1. Research and monitoring	Policy initiatives designed to increase knowledge in a general or specific sense, re hazards, vulnerability, success of policy initiatives, community awareness, and so on, e.g. by research grants or inquiries.
2. Improving communication and information flow	Aiding information flow between research and policy; of policy imperatives to research; between agencies, firms and individuals, through a wide range of mechanisms such as indicator systems, community-based monitoring, and so on.
3. Training and education	General public education, education targeting sub-sets of community; formal curricula in schools, universities; specific skill development and training.
4. Consultative	External mediation over conflicts; negotiation; facilitated planning procedures; dispute resolution; inclusive policy processes.
5. Intergovernmental agreements	Intergovernmental agreements/policies, memoranda of understanding, and so on, between countries or within countries, for cooperation, joint response, information sharing, and so on.
6. Legal	1. Statute law: Statutes or regulations under existing law to: Create institutional arrangements; establish statutory objects and agency responsibilities; guarantee public rights in policy processes; prohibit certain activities; zone land and control development; define and enforce standards; create penalties. 2. Common law: Applications of doctrines such as negligence or nuisance to prevent or punish.
7. Planning and assessment procedures	Incorporation of emergency and disasters in land use planning, social and environmental impact assessment; risk assessment.
8. Self-regulation	Incorporation of disaster/emergency considerations within industry or firm codes of practice or ethics, professional standards, recommended procedures.
9. Community participation	Community-based risk assessment and management; public participation in higher level policy formulation; freedom of information laws; rights to comment on development proposals; community monitoring of hazards; joint government-community implementation of programmes.
10. Market and economic	Taxes/charges; use charges; subsidies; rebates; penalties; performance; competitive tendering.
11. Institutional change	New or revised institutional system or organizational features, to enable implementation of other instruments.
12. Adjustment of other policies	The assessment and if necessary alteration of incentives, goals or processes in other policy settings that increase vulnerability or decrease resilience, or which block desired policy change.
13. Do nothing	Inaction is usually seen as a policy failure, but may be justified after analysis.

Source: Adapted from Handmer and Dovers (2013, pp. 126–7)

Table 3.3 Criteria for selecting policy instruments

Criteria	Question, relative to other instruments
1. Dependability	How certain is it that the instrument will achieve policy goals?
2. Timeliness	Can the instrument be developed and applied in time?
3. Cost and efficiency	What is the cost and efficiency of the instrument, relative to achieving goals?
4. Systemic potential	Does the instrument address underlying causes, or only address symptoms?
5. Information and monitoring requirements	Is needed information available to design and implement the instrument?
6. Distributional impacts	Will implementation of the instrument have inequitable impacts across the affected population; can these be managed?
7. Political and institutional feasibility	Is use of the instrument feasible in terms of political support and institutional capacity?
8. Ability to be enforced or avoided	Can implementation/uptake of the instrument be ensured or enforced?
9. Communicability	Can the desirability of using the instrument be communicated to those implementing it or impacted by it?
10. Flexibility	Is the instrument capable of being adapted as knowledge or circumstances change?

Source: Adapted from Handmer and Dovers (2013, p. 129)

Consider the following policy instrument choice criteria and the case of bushfire, where one criterion may not rule out an option but will focus attention on a justice issue and how that can be attended:

- Criterion 6 (distributional impacts): A policy instrument may be deemed effective and efficient, but impose financial or other burdens unfairly on a segment of the community, such as the cost of required fire-safety retrofitting for poorer households (distributional justice). That may be a reason to not select an instrument, or to lessen the distributional impact through associated grants or transfer payments.
- Criteria 5 (information) and 9 (communicability): Access to information and therefore the ability to comply with or gain benefit from measures such as building code changes, flood risk overlays, financial payment programmes or similar may be uneven across a community (procedural and informational justice). This instructs careful design to ensure policy changes and opportunities are made known and accessible to specific groups in the community (e.g. Chap. 17).

- Criterion 8 (compliance): Although frequently considered regarding the public finance implications, compliance also applies to households and firms, particularly with regulatory instruments, where the costs of compliance and reporting (a subset of distributional impacts) may be either undesirably high overall or for a particular segment of the community. An example would be monitoring fire preparedness measures at the household level to qualify for rate or insurance rebate incentives. Again, this may rule out an instrument, or suggest measures to alleviate these burdens.
- Finally, criterion 3 (cost and efficiency). These have justice implications in so far as gross cost and relative efficiency determines the quantum of resources required. It may well be possible to significantly reduce one category of risk (say, from bushfire) but only at the expense of other risks (say, flood) from a given budget, or only in some places and not others, with implications for distributional justice. Such difficult choices of priority and spending are universal in public policy, and every choice to address one problem, for some people, limits the ability to do something else for others. All policy choices involve trade-offs and potential creation of actual or perceived injustice.

A stronger justice orientation in disaster and emergency management could be achieved by stating criteria such as those above more clearly in overarching policy and decision making procedures, and emphasising their importance. That would empower decision makers implementing policy to act with a clear mandate and greater confidence in seeking justice outcomes.

Utilising standard frameworks from public policy such as Tables 3.1, 3.2, and 3.3 does not make formulating policy responses easier, but encourages more explicit consideration of multiple factors, including those with justice implications.

3.5 Whose Responsibility?

In deploying instruments to achieve a social and policy goal, a key question is what individual or organisation has the mandate and capacity to initiate, develop and implement the policy. Two considerations here, within government and outside of government: (1) who has that legal and political authority, generally being a government or part thereof, or delegated by that 'responsible authority' to someone else; or (2) someone outside of government but necessarily involved in the formulation, implementation and/or monitoring of a policy initiative.

Regarding (1), with respect to a justice-related decision (the policy goal with distributional impacts or a procedural decision on who is allowed to have input into a decision), at which level of government, (national, state/provincial, local) or within the power of which government agency at a particular level of government does primary responsibility lie? Mistaking who is responsible will dilute otherwise valid arguments for redress of injustice. While policy involves partners outside of government, the state has a foundational role and is generally the primary agent in provision of resources and underpinnings of statutory, financial, informational and administrative competence. Clarity over who could or should have responsibility is important and needs to consider jurisdiction, portfolio, source of authority and legal mandate. Certain responsibilities and roles can be defined by legislation and government agency mandates, including who may participate in policy processes, such as through submissions or identified avenues for engagement. These can be clear, and also can be politically and legally challenged in cases where justice issues such as protection of vulnerable groups or inclusion of diverse voices are perceived not to have been fulfilled.

In regards to (2), the PPRR spectrum and the modern policy principle of shared responsibility (or, collaborative approaches) requires actions by non-government actors, including households, community groups, NGOs and private firms. Calgaro et al. (Chap. 17) show the potential power of collaborative engagement between government and NGOs to implement and achieve policy goals that are directly relevant to justice considerations, enhancing the safety of the disabled in disaster risk reduction. For a policy maker, however, the allocation of roles to non-government actors is at once highly desirable for both practical and ideological reasons, and yet problematic for reliable acquittal of 'responsibilities'.

While the primary expectation for most aspects of PPRR has been placed on governments (e.g. Victorian Government, 2010), the policy goal of 'shared responsibility' instructs that seeking to locate problems or opportunities regarding disaster justice must consider non-government actors (households, firms, community groups, industry associations, professions, NGOs) as well (Lukasiewicz et al., 2017). While broader consideration of policy partnerships and poly-centric or network governance are relevant, the focus here is the justice implications that individuals and organisations undertake a PPPR function even though not required legally or politically nor funded to do so. There are three justice-related aspects to this: (1) seeking to impose a responsibility on an entity that may not have the resources to fulfil that expectation might be unfair (unjust); (2) from a pragmatic public policy perspective, to rely on an unenforceable expectation involves some risk of policy failure and thus unintended outcomes, including justice impacts; and (3) voluntary engagement in a policy process raises issues of representation and the validity of inputs to policy.

Some roles in disaster response by non-government actors are well-established and predictable, and often the subject of clear agreements with governments (for example the prominent role of Red Cross and Red Crescent). Other roles may be less prominent or sustained but highly effective, targeting a specific need: Calgaro et al. (Chap. 17) describe a positive case. Other roles and contributions are locally regular, critical but tenuous (e.g. Ingham et al., Chap. 12). Others again are dynamic, unpredictable and present a tension of both problem and opportunity for disaster management. A high profile example is 'spontaneous volunteers' post-disaster (on this and related volunteer issues, see McLennan, Whittaker, & Handmer, 2016)—such waves of help are unpredictable, can be crucial, but can create risks to both those affected and responders if undertaken without due care, and questioning or controlling the right of people to assist their own community has justice implications. More broadly, there is the procedural justice question of a community's right to identify risks and assets and fashion their own strategies versus more top-down, professionally driven and agency-driven approaches: the values and thus priorities (and losses) from a community perspective may be overlooked by outsiders, as discussed by Schlosberg et al. (Chap. 13).

The most unpredictable but crucial non-government actors are individuals and households, who are central to the 'shared responsibility' model, but who are diverse in their adoption of risk reduction actions and in many cases free from specific obligations. The household that loses their well-prepared property due to ignition spread from unprepared, fuel-rich neighbouring houses may feel a sense of injustice. If coercive policy measures were in place to force pre-fire season preparation, some households would doubtless view this imposition as unjust. This is a space replete with highly varied personal, political and theoretical positions on the role of individuals, communities and the state that are central but beyond scope here (Lukasiewicz et al., 2017; McLennan & Eburn, 2015). The fact that post-event inquiries, at least in Australia, are largely silent on the role of individuals, households and even volunteers (Cole et al., 2018) means an important potential impetus for policy debate on these actors is lacking.

3.6 Disaster Policy, or Something Else?

Importantly, issues of justice may not be caused within or be addressable by 'disaster policy', but in fact more likely can only be rectified in other policy domains such as land use planning, communications, infrastructure or health services. Bluntly, emergency management and disaster policy is as much about reactively addressing the impacts of vulnerabilities created or ignored by other policy sectors, as it is about proactively reducing vulnerability or enhancing resilience. This invites consideration of policy integration (aka mainstreaming) where disaster-relevant measures are implemented within those other sectors and portfolios. This is a large topic and a constant but often unachieved aim of disaster and emergency professionals and agencies whose primary mandate is Preparedness and Response, but that is beyond scope here. However, some indications of the justice-relevant aspects of this emerge in this book and can serve as indications.

The role of land use planning (Chaps. 4 and 5; March & Dovers, 2017) is one of the more prominent, where vulnerability is created, or at least claimed to be so, by planning decisions or oversights that place households in disaster-prone areas. While significant measures are included in many planning schemes, such as identified flood or fire risk zones and standards for buildings, there are tensions between risk reduction and the many other imperatives that drive planning. Housing affordability and consumer preference for bushland proximity, as well as sheer economic development pressure, raise questions of justice, or perceived injustice, if housing and development choices are constrained or made more expensive by imposed risk reduction measures. Chapter 11 delves behind the usually simplistically treated issue of arson to expose evidently unjust precursors to many offenders' ignitions. As with many policy problems, policy can and should attend the symptom (policing, education, punishment) but could also consider the causes related to alienation and disaffection that lie well outside fire management and policing (criterion 4, Table 3.3). This would require a decision by society and government to widen policy scope.

3.7 New Mechanisms, or the Same but Better?

A central question from a public policy perspective is whether addressing issues of justice (assuming this is found to be needed) requires (1) wholly or substantially new policy framing, policy instruments or institutional settings; or (2) revisions to the implementation of existing policy mechanisms (i.e. more or differently applied of the same). The likelihood of both success and the speed of reform is almost always greater in the case of the latter—management actions are changed more quickly than policy details, which are more easily and quickly changed than entire policies, and creation of a new institutional setting is an even longer-term prospect. An example from this book, addressing procedural justice, is Ingham et al.'s (Chap. 12) recommendation for adjusting membership of local emergency management committees, easily done, likely effective, and through an existing structure.

That example raises the question of how quickly a policy change can be implemented to address a justice or other issues (see further Handmer & Dovers, 2013, pp. 158–9). Institutional settings and major legislative instruments take significantly longer to change, for good reason, than organisational detail or subsidiary regulations. Whole policy programmes are not renewed quickly, but implementation procedures or guidelines under them may allow more rapid adjustments. If a positive outcome can be achieved by adjustment within broader policy and institutional settings, then a more rapid redress of justice concerns may be possible (or, longer term, substantial institutional change may indeed be required).

3.8 Discussion and (In)Conclusions

The above discussion of six policy issues and their disaster justice implications opens a space about how to address (in)justice within the pragmatic and complex disaster policy environment. It also shows how many long-standing disaster policy and emergency management matters have clear, and sometimes less clear, justice implications. That should not surprise, as disasters lead to death, destruction and loss unevenly, so of course injustices occur. Two questions arise: does or can consideration of the disaster-justice nexus contribute to (1) justice research, and (2) disaster policy?

On (1), the high stakes, context-dependency, uncertainty and dynamic nature of disasters does make for a fertile field of inquiry to explore issues of justice. Other contributions in this book demonstrate this in many detailed instances. On (2), issues of unequal access to decision making, vulnerability and impacts are standard topics in disaster policy and emergency management, but a justice research lens can enhance thinking in two ways. One is that the very concepts of justice and injustice sharpen the importance of inequalities: as Calgaro et al. (Chap. 17) identify, these become matters of human rights. The second is that the language of justice research—distributional, procedural, informational, relational, and so on—provides a more comprehensive set of descriptors and thus deeper perspective than usual in the disaster field, compared to simply 'equality', and this can increase the resolution of understanding (and, therefore, potentially the effectiveness and targeting of policy responses).

In Chap. 8, Schilizzi and Azeem in their analysis of post-flood housing in Pakistan demonstrate the potential for justice principles, via different and valid approaches to equality, to aid policy analysis. While their analysis does not provide a singular policy answer, it does lead to a more explicit understanding of the logic and implications of different approaches to a particular disaster policy question. That kind of approach has significant potential.

To conclude this chapter, Table 3.4 presents a tentative matrix filled with selected, illustrative examples that ask if issues of justice emerge as a result of risk borne unevenly or unfairly, then (1) what are the risks, (2) who bears that risk, (3) at what 'stage' on the PPRR spectrum, (4) what is the justice issue, and (5) what are the public policy options applicable to redressing the issue?

Table 3.4 invites further detail and discussion, but indicates a way to structure a discussion between justice research and disaster policy in a potentially constructive way among the respective concerns and core concepts of disaster policy makers, emergency managers, justice researchers and practitioners, as well as of those vulnerable to and impacted by disasters. Disasters, policy and justice have commonalities: all three are complex, unpredictable and contested. This chapter and this book suggest that a conversation between the two fields of research and practice can be better structured and fruitful.

Table 3.4 Linking risk, justice and disaster policy instruments: a tentative ordering

Example risk	Risk bearer	PPRR[a]	Justice type[b]	Selected policy instruments
Unpreparedness due to lack of knowledge or information.	1. Individuals, households, firms, local communities. 2. responders	PP	Procedural, interactional.	Educative, informational.
Increased exposure becomes apparent due to changed scientific understanding of climate extremes.	Individuals, households, firms.	PPR	All?	Ratchet up all policy instruments.
Material damage from impact of event: Life, property, livelihood.	Individuals, households, firms, local communities.	R(esponse)	Distributional	Insurance, transfer payments, grants, public works.
Longer term, ongoing losses of income or opportunity.	Individuals, households, firms, local communities.	R(ecovery)	Distributional	Insurance, transfer payments.
Loss of income, injury, and so on, by responders.	Professional and volunteer responders; spontaneous volunteers.	R(esponse)	Distributional	Workplace conditions, compensation.
Lack of access to mitigation options (eg insurance, works) due to socio-economic status; own property.	Individuals, households, firms.	PP	Distributional	Transfer payments, grants for works, subsidised insurance.
Ditto: Rental property.	Ditto.	PP	Distributional	Regulation, incentives, and so on: Landlords.
Reputational, psychological or material damage from blame (media or from inquiries).	Professional responders (rarely volunteer).	R(ecovery)	Procedural, interactional	Change media reporting norms; restorative versus adversarial inquiries; citizen attitudinal change.
Political damage (blame) post-event (media or from inquiries).	Political leaders.	R(ecovery)	Procedural	Ditto.

[a]PPRR = Prevention (of impacts), Preparedness (for onset of hazard), Response (when hazard event occurs), Recovery (from impacts). There are several variants of this schema; the PPRR terminology is used here consistent with Australian practice

[b]Using the terms Distributive, Procedural and Interactional (including Interpersonal and Informational) Justice, but not Spatial and Temporal (which here can be considered within Procedural) (Graham & Barnett, 2017)

References

Bennett, C. J., & Howlett, M. (1992). The Lessons of Learning: Reconciling Theories of Policy Learning and Policy Change. *Policy Sciences, 25*, 275–294.

Cole, L., Dovers, S., Gough, M., & Eburn, M. (2018). Can Major Post-Event Inquiries and Reviews Contribute to Lessons Management? *Australian Journal of Emergency Management, 33*(2), 34–39.

Davis, G., Wanna, J., Warhurst, J., & Weller, P. (1993). *Public Policy in Australia*. Sydney: Allen & Unwin.

Dovers, S., & Handmer, J. (2014). Disaster Policy and Climate Change: How Much More of the Same? In A. Ismail-Zadeh, J. Urrutia-Fucugaughi, A. Kijko, K. Takeuchi, & I. Zaliapin (Eds.), *Extreme Natural Hazards, Disaster Risks and Societal Implications* (pp. 348–358). Cambridge: Cambridge University Press.

Graham, S., & Barnett, J. (2017). Accounting for Justice in Local Government Responses to Sea-Level Rise. In A. Lukasiewicz et al. (Eds.), *Natural Resources and Environmental Justice: Australian Perspectives* (pp. 91–103). Melbourne: CSIRO.

Handmer, J., & Dovers, S. (2013). *Handbook of Disaster Policies and Institutions: Improving Emergency Management and Climate Adaptation* (2nd ed.). London: Routledge.

Lukasiewicz, A., Dovers, S., & Eburn, M. (2017). Shared Responsibility: The Who, What and How. *Environmental Hazards, 16*, 291–313.

March, A., & Dovers, S. (2017). Mainstreaming Urban Planning for Disaster Risk Reduction. In K. Vella & N. Sipe (Eds.), *Australian Handbook of Urban and Regional Planning* (pp. 231–246). Taylor & Francis.

McLennan, B., & Eburn, M. (2015). Exposing Hidden-Value Trade-Offs: Sharing Wildfire Management Responsibility Between Government and Citizens. *International Journal of Wildland Fire, 24*, 162–169.

McLennan, B., Whittaker, J., & Handmer, J. (2016). The Changing Landscape of Disaster Volunteering: Opportunities, Responses and Gaps in Australia. *Natural Hazards, 84*, 2031–2048.

Rose, R. (2005). *Learning from Comparative Public Policy: A Practical Guide*. London: Routledge.

(UN) United Nations. (2015). *Transforming Our World: The 2030 Agenda for Sustainable Development*. A/RES/70/1. New York: United Nations.

(UNFCCC) United Nations Framework Convention on Climate Change. (2015). Report of the Conference of the Parties on Its Twenty-First Session, Held in Paris from 30 November to 13 December 2015. Addendum. Part Two: Action Taken by the Conference of the Parties at Its Twenty-First Session. In *Conference of the Parties 21*. Paris, France: United Nations.

(UNISDR) United Nations Office of Disaster Risk Reduction. (2015). Sendai Framework for Disaster Risk Reduction 2015–2030. In *Third UN World Conference*. Sendai, Japan: United Nations.

Victorian Government. (2010). *2009 Victorian Bushfires Royal Commission*. Final Report. Retrieved from www.royalcommission.vic.gov.au/Commission-Reports/Final-Report.html

CHAPTER 4

Burning Bush and Disaster Justice in Victoria, Australia: Can Regional Planning Prevent Bushfires Becoming Disasters?

Jason Alexandra

4.1 INTRODUCTION AND OVERVIEW

In recent decades catastrophic wildfires have occurred in many countries such as USA, Russia, Greece, and Australia—where the scale, intensity and destructive force are consistent with predictions of fire behaviour in a warming climate (Moritz et al., 2014).

Along with climate change, social inequity and settlement patterns amplify bushfire impacts, highlighting the dimensions of community vulnerability and resilience. However, the majority of "*bushfire research has traditionally focused on the physical properties of fire hazards and disasters, with relatively little consideration of how cultural, economic, political and social factors shape vulnerability*" (Whittaker, Handmer, & Mercer, 2012, p. 161).

One of these factors, spatial or land use planning—including its strategic and procedural elements—is often a key determinant of people's exposure to bushfires (Kornakova, March, & Gleeson, 2018). Spatial

---

J. Alexandra (✉)
Alexandra & Associates Pty Ltd, Eltham, VIC, Australia

RMIT School of Global, Urban and Social Studies, Melbourne, VIC, Australia

© The Author(s) 2020
A. Lukasiewicz, C. Baldwin (eds.), *Natural Hazards and Disaster Justice*, https://doi.org/10.1007/978-981-15-0466-2_4

planning as a form of hazard mitigation contrasts with disaster response and recovery phases, occurring well in advance of disaster events (Bond & Mercer, 2014). It has roles in minimising fire impacts by determining spatial relations and configurations of where human activities and environmental processes interact. Therefore, planning has important roles in disaster mitigation and disaster justice by reducing exposure to, and impact of fires, including for vulnerable members of the community.

This chapter explores the intersection of land use planning, disaster justice and wildfires (known as bushfires in Australia). It explores bushfires and land use planning in Victoria—a state that periodically suffers catastrophic fires, sustaining over 50% of Australia's economic damage from fire, in an area approximately 3% of the continent (Buxton, Haynes, Mercer, & Butt, 2011). The central question examined is: What can be learnt from Victoria's experience of land use planning in a highly flammable landscape for disaster mitigation and disaster justice?

Given the intensification of the impacts of wildfires occurring globally (Moritz et al., 2014), Victoria's experience may provide valuable lessons. Victoria sits in one of the world's most intensely flammable regions due to a combination of climatic conditions that produce high fuel loads during moister months (autumn, winter, spring) that are baked dry in long hot summers. During summers a recurrent weather pattern brings:

> *seasonal winds, associated with cold fronts that draft scorching, unstable air from the interior across whatever flame lies on the land. At such times the region becomes a colossal fire flume that fans flames which for scale and savagery have no equal elsewhere on Earth.* (Pyne, 2009, p. 1)

This chapter is informed by personal experience of two catastrophic fires, reports, policy documents and peer-reviewed literature. It is structured in five parts:

- Learning to live in a flammable continent
- Welcome to the 'Pyrocene'
- Bushfire disaster mitigation and integrated regional planning
- Land use planning and disaster justice
- Conclusions

4.2 Learning to Live in a Flammable Continent

For most of my life, I've lived in the fertile, productive hills on Melbourne's periphery where I've experienced two significant bushfires—Ash Wednesday (February 1983) and Black Saturday (February 2009). Both were catastrophic in terms of loss of lives and property, with the latter defined as Australia's worse civilian disaster (Attiwill & Adams, 2013).

Black Saturday fires burnt over 430,000 hectares, 173 people died and over 5000 were injured, 2133 houses and 61 commercial premises were destroyed (Teague, McLeod, & Pascoe, 2010). This horrific toll could have been substantially higher if the cold front and associated wind change had come later in the day. The change altered the fire's direction, away from Melbourne's northern fringes (Manne, 2009). Here drought affected Eucalyptus trees would have carried the fire into heavily populated suburban interface areas, including along the Diamond Creek and Yarra corridors where bushland and suburbs mingle. The wind change shifted the fire's trajectory to the northeast where it ravaged peri-urban communities around the Kinglake Ranges approximately 50 kilometres from central Melbourne.

On Ash Wednesday 1983, bushfires killed 75 people, and destroyed over 2500 dwellings, mostly on the outskirts of greater metropolitan Melbourne (Hughes & Mercer, 2009). Again, fatalities may have been significantly higher except for a previous, unrelated bushfire on February 1 that burnt the northern slopes of Mount Macedon. In a disorganised, spontaneous mass evacuation, hundreds of people desperately fled by car into this burnt forest, traversing a narrow mountain pass away from the fire's destructive front, driven by ferocious southwesterly winds. Many may have perished on this road had the previous fire not provided for safe passage as a result of fortunate circumstance rather than planning.

The Ash Wednesday fires that swept through the Macedon Ranges destroyed my house, business and virtually all worldly possessions. However I do not regard this as disastrous, because I had the skills, resources and social connections to recover. The experience was a significant learning exercise, triggering a life-long interest in our socio-ecological relationships with bushfires.

I spent the night of the fires as a volunteer on a Country Fire Authority (CFA) fire truck, returning at dawn to the burnt out villages. Initially it appeared the fire had burnt indiscriminately, randomly taking houses and leaving others. However, my closer observations revealed that surviving houses were typically surrounded by gardens of exotic plants. Deciduous trees like oaks, birches and elms had protected numerous houses, while those amongst eucalypts had burnt, regardless of age, design or construction materials. In short, landscape relationships mattered. Location in the landscape, relative to surrounding vegetation types had determined their fate (at Mt. Macedon few fire-fighting resources were deployed through the night).

The formal post bushfire inquiry found that spatial relationships mattered, recommending that land use planning regimes should become central in mitigating future disasters (Norman, Weir, Sullivan, & Lavis, 2014).

On Black Saturday (2009) I was in Gippsland's mountainous foothills helping prepare a friend's property against the anticipated fires. As market gardeners they had their diesel powered pumps, abundant water, irrigation systems and fields of irrigated vegetable as defence, but nothing was certain because the property adjoined mountain ash forests, that burn infrequently but furiously (McCarthy, Malcolm Gill, & Lindenmayer, 1999). Fortunately, a wind change spared us testing our survival plans.

Many were not so fortunate. The Black Saturday fires devastated peri-urban districts to Melbourne's north that were strikingly similar in terms of landscape types, settlement patterns and demographics to the Macedon and Yarra ranges that burnt on Ash Wednesday 26 years earlier.

These similarities highlight the pressing need for intergenerational policy learning, and the use of historical and cultural knowledge to guide planning reforms that are science based, effective and politically acceptable (Buxton et al., 2011; Kornakova et al., 2018). Catastrophic fires are episodic occurrences. We need to be learning about the nature of the flammable continent and how to live well in it—but how best to do this?

Learning from major episodic bushfires like Black Friday (1939), Ash Wednesday (1983) and Black Saturday (2009) requires active intergenerational remembering (Griffiths, 2016), comprehensive research and monitoring and ways of ensuring that knowledge gained from infrequent, episodic events is institutionalised and acted on (Eburn & Dovers, 2015).

Disastrous fires catalyse significant post-fire inquiries—like Royal Commissions—yet, institutionalising policy learning remains challenging with deep concerns that numerous recommendations are neither implemented nor evaluated (Eburn & Dovers, 2015). This concern extends to those recommendations on greater use of the integrated planning system to reduce exposure to fire risks.

Land use planning can mitigate the fire's destructive impacts through settlement patterns that reduce exposure. Many of these destructive impacts are a consequence of cultural, economic and socio-political factors that shape differential vulnerabilities including policy and planning decisions that determine land use configurations (Buxton et al., 2011; Moritz et al., 2014).

Catastrophic fires raise questions about how well we are learning to live in this flammable continent, and how should fire and, more generally, ecological literacy inform public policies, settlement patterns, construction, farming and gardening practices in our fire prone landscapes (Campbell, Alexandra, & Curtis, 2017)?

4.3 Welcome to the Pyrocene

4.3.1 A Flammable Planet and Its Pyrocentric Civilisation

The prominent scholar of fire, Stephen Pyne (2018) argues we are pyrophilic creatures, immersed in pyrocentric civilisations whose 'new combustion regime based on fossil biomass' has resulted in a pyric transition that has not only shaped our technologies, mobility and habitation patterns, but also our values and relationships with landscapes. This escalating 'firepower through industrial combustion underwrites the Anthropocene' so completely that we could re-name it the Pyrocene (Pyne, 2018, p. 1).

Regardless of the epoch's name, climate change has profound implications for the scale, intensity and impacts of fires, intensifying wildfire hazards/risks in many parts of the world (Moritz et al., 2014; Pyne, 2018). However, it is dangerous to simplistically and deterministically reduce the future to climate (Hulme, 2011) when fires are intensifying due to compounding climate, land use and ecosystem changes, for which humans are causal agents (Moritz et al., 2014). It is industrial firepower—including motor vehicles—that makes the rural-urban interface possible, powering the spread of populations amongst high-amenity, high-fuel and high-risk environments (Pyne, 2018). This is occurring in an era when many of the world's forests and woodlands are on the cusp of fire-driven 'tipping points' resulting in changes in vegetation types, fire and ecosystem dynamics (Adams, 2013).

Given these compounding change drivers, linear projections based on static views of 'natural' systems and 'natural' fire regimes have limited utility. Instead an acceptance of a new dynamism and an embrace of uncertainty is needed in our models of socio-ecological systems and their capacity to shift to radically altered states (Alexandra, 2012). A world of multiple transformations needs non-static, non-linear paradigms (Folke et al., 2002; Folke, Hahn, Olsson, & Norberg, 2005). Yet, incorporating climate change driven dynamism, with its dynamic drivers and feedbacks, into the policies and programmes of public sector organisations remains problematic (Alexandra, 2017a; Keenan & Nitschke, 2016).

4.3.2 Burning Continent: Fire Formed Australia

Victoria's pyric transition has been powered through state sponsored development of brown coal (lignite) reserves, like those in the Latrobe Valley (Alexandra, 2017b). These gigantic coal deposits contain compelling evidence that landscape fires predated human occupation of the southern continent. Fossilised remains of Gondwanna rainforests—including Antarctic Beech (Nothofagus) and Kauri (Agathis)—are overlain by more recent coal containing charcoal from forest fires and fossils of sclerophyll vegetation that became dominant during the Middle Miocene's drying climate (11.6–16 million years ago) (Holdgate et al., 2007, 2009).

Indigenous burning practices have a long and proud tradition (45,000 to 60,000 years), and are accepted as formative in shaping Australia's extensive grasslands and open woodlands (Bowman, 2008; Gammage, 2011). British squatters found these grazing landscapes so desirable for sheep and cattle they forcibly occupied land beyond formally declared boundaries of British settlements (Kerkhove, 2015; Watson, 2014). During the ensuing wars of occupation, destructive fires were skillfully used by indigenous 'guerrillas' to destroy squatters' settlements and disrupt their enterprises (Kerkhove, 2015).

4.3.3 Fire in Contemporary Australia

Generations of rural settlers have experienced the destructive power of bushfires (Watson, 2014) so fear of destructive wildfires may be both pragmatic and culturally engrained. Most districts in Victoria formed volunteer fire brigades with the state's CFA evolving into one of the world's largest volunteer fire-fighting organisations with over 53,000 volunteer members (CFA, 2019). While effective at controlling smaller and moderate fires, their capabilities are frequently overpowered by major conflagrations during extreme conditions (Teague et al., 2010; Whittaker et al., 2012).

Fire frequency and intensity influences many ecosystem processes, and in northern Australia, deliberate use of fire as a savanna management tool is increasingly used for carbon sequestration, continuity of cultural practices and biodiversity conservation (Russell-Smith et al., 2015). In southern Australia, burning practices inspired by indigenous cultural practices, while rare, are increasingly occurring and publically celebrated (ABC, 2017; SBS, 2017). In contrast, cool season fuel reduction burning on public land is established practice in southern Australia (Keenan & Nitschke, 2016). While many are advocating for expanding its use (Attiwill & Adams, 2013), the practice remains controversial.

The fire prone nature of Australia makes bushfires a pervasive risk in rural, regional and peri-urban areas, which often house a larger percentage of disadvantaged people compared with inner city locations due in part to the pressures of housing affordability (Foster, Towers, Whittaker, Handmer, & Lowe, 2013; Whittaker et al., 2012). Wildfire risks are particularly acute in the peri-urban and rural-urban interface, with estimates that over 1.6 million dwellings are located in or within 700 m of bushland in and around capital cities in Australia (Gill & Stephens, 2009).

Ash Wednesday and Black Saturday attest to the fact that peri-urban wildfires can be catastrophic in terms of loss of human lives and property damage, yet settlements continue to expand in areas that amplify risks (Butt, 2013; Buxton et al., 2011). Despite deep concerns about the potential for more disastrous bushfires, there is a failure to achieve a sustained consensus about how to apply integrated land use planning to bushfire hazard mitigation (Kornakova et al., 2018).

Without doubt, many of Australia's landscapes are fire formed. However, in the divisive politics of blame following major bushfires, discourses about the 'naturalness' of bushfires have been used in attempts to absolve governments of responsibility for fire prevention and disaster mitigation (Whittaker & Mercer, 2010). Such conceptual gymnastic would not be required if inquisitorial-style post-disaster inquiries were reformed to focus on learning key lessons rather than finding blame (Eburn & Dovers, 2015).

4.3.4 Bushfire Disaster Mitigation and Integrated Regional Planning

Sixteen major inquiries since 1939 have strongly recommended more use of spatial planning for bushfire mitigation, particularly after Ash Wednesday in 1983 (Norman et al., 2014). However, this has proven difficult to implement due to the realpolitik of policies that involves complexity and interagency dynamics (Kornakova et al., 2018). Land use planning strategies are recognised as critical to disaster mitigation. By influencing the types and locations of development, spatial planning provides significant opportunities for risk reduction (March & Henry, 2007) including through land use allocation, infrastructure and construction standards (Bond & Mercer, 2014). However, realising this potential requires planning regimes with professionals capable of informed judgement and policy integration across land use, housing, economic development and emergency management, that involves multiple agencies and land use types (such as farmland, parks, urban areas) (Buxton et al., 2011; March & Rijal, 2015).

The critical importance of not increasing populations exposed to bushfire risk in peri-urban areas is widely recognised. However, the planning system has facilitated this with expanding peri-urban populations a sustained phenomenon over many decades (Buxton et al., 2011; McKenzie, 1997).

Actual planning practice in Victoria, over many decades, has been mostly pro-development incrementalism, resulting in low density, peri-urban sprawl, across much of the state, particularly around outer suburban areas and regional cities. It has enabled thousands of rural residential subdivisions, adding many thousands of dwellings in high fire risk locations, including in those shires impacted disastrously on Black Saturday (Bond & Mercer, 2014; Butt, 2013).

Black Saturday demonstrated that the lack of effective peri-urban planning frameworks have been sustained policy failures, with lethal consequences (Buxton et al., 2011). Entrenched patterns of small lot, peri-urbanisation persisted after several inquiries into Ash Wednesday (1983) recommended tighter planning and development controls (Norman et al., 2014). Since then, statutory planning regimes have given explicit approval to thousands of additional peri- urban dwellings, dispersing large numbers of people into highly flammable country, often in or adjoining bushland, increasing their exposure to extreme bushfire risks. More people do not just mean more exposure to hazards, they also mean more ignition sources—cooking fires, cool burns, machinery—grinders, welders, slashers—power lines (Syphard & Keeley, 2015) and arson (Cozens & Christensen, 2011). The electricity network, strung up like spider webs, tends to fail under the most extreme fire conditions of high temperatures and high winds (Mitchell, 2013) with faulty power lines igniting numerous major fires when dangerous weather produces the most destructive fires (Teague et al., 2010).

For several decades debates have raged about permitting, restricting or stopping further dispersed peri-urban housing developments into high-risk areas. Stopping dispersed housing would have large costs in terms of restrictions on personal liberty, lowering property values and possibly reducing employment in non-urban areas. However, these debates are often conflated with debates about housing affordability, but low density semi-rural development is often on titles between 2 and 10 hectares and is separate to questions of land release for higher density suburban development where dwelling densities are often between 15 and 20 units per hectare and where hundreds of thousands of new dwellings have been added in defined, and carefully planned, urban growth corridors (Buxton & Scheurer, 2007).

There are no simple ways of determining how the compromise between individual choice and collective responsibilities should be made. However, land development policies enabling incremental, small lot semi-rural subdivision is also expensive, with the Royal Commission estimating that Black Saturday alone cost the Victorian economy over $4.4 billion (The Age, 2010), not to mention the personal tragedies, losses, hardship and suffering.

The failure to adopt more restrictive planning policies can be explained by a series of political upheavals during the pro-development, pro-deregulation Kennett government in the 1990s when reforms radically altered land use planning practice and resulted in loss of professional capacity from planning and natural resources agencies (Attiwill & Adams, 2013; Buxton, 2007). Buxton et al. (2011) document professional, commercial and institutional constraints to developing strategic planning frameworks capable of wise locational decisions warning that continuing to increase populations in high fire risk areas will result in further deaths.

History provides several useful examples of wise locational decisions. The Stretton Royal Commission into the Black Friday fires of 1939 found that many of the 71 people killed had worked in sawmilling settlements within the state forests. Acting on Stretton's recommendations sawmills were moved to towns reducing workers' exposure to bushfire risks. Secondly, in the 1970s, the conservative Victorian government bought *"hundreds of undeveloped blocks of land in the Dandenong Ranges to the east of Melbourne on the grounds that dwellings should never be allowed to be constructed in such a potentially dangerous, bushfire-prone area"* (Bond &

Mercer, 2014 p. 8). In contrast, following Black Saturday, the Victorian Government offered a voluntary buy back of blocks in fire prone areas, but this has had limited and patchy uptake and with growth in peri-urban populations, many people now live in similarly dangerous bushland areas.

Attempts to minimise risks by introducing planning instruments that zone land as bushfire-prone have proven problematic. Attempts to specifically delineate fire risk through defined zoning and planning policies has been fiercely contested by the property development lobby and in 2013, after *"intense lobbying by the Urban Development Institute of Australia, some 200 000 properties in Victoria were summarily 'removed' from bushfire-prone zones on the grounds that the risks had been overstated"* (Bond & Mercer, 2014, p. 20).

This is a justice issue because this kind of zoning provides publically available information to prospective purchasers about areas of land prone to bushfire risk and if retained may have limited the number or types of dwellings constructed, and therefore future exposure to bushfire risks.

In summary, integrated planning—from regional to site scale—offers tangible prospects for bushfire disaster mitigation but requires institutional capacities and political will (Kornakova et al., 2018). More than a technocratic planning challenge, it is fundamentally political because it involves balancing individual freedoms with collective responsibilities in ways that are fair and just. The central questions are about the state's roles and rights to impose constraints on individuals, their freedom and assumed property rights in the interest of disaster prevention (Hughes & Mercer, 2009).

4.4 Issues for Disaster Justice

4.4.1 Land Use Planning, Vulnerability and Disaster Justice

The demographics of peri-urban areas are often a mixture of tree changers, farmers, renters and people attracted to country living but close to urban employment, education and cultural opportunities. Socio-economic analysis typically indicates higher levels of social and economic disadvantage with more poverty, lower average incomes and educational attainment, meaning many households may not undertake risk reduction measures like retrofitting houses or purchasing insurance (Foster et al., 2013). Housing affordability often attracts people to locations beyond the suburbs, but often limited public transport and a dependency on car-based commuting contributes to poverty traps (Foster et al., 2013).

Socio-economic and cultural factors of peri-urban communities—including economic disadvantage and cultural and linguistic diversity—contribute to their vulnerability (Foster et al., 2013). The Black Saturday Royal Commission classed half of those who died as 'vulnerable' because they had a disability or chronic illness or were under 12 or over 70 years old (Teague et al., 2010). Others classed as vulnerable include people from non-English speaking backgrounds, those with mobility difficulties and the economically disadvantaged (Norman et al., 2014). These and other factors contribute to vulnerability, which can be defined as exposure to specific hazards (e.g. bushfires) in combination with a person's personal and collective resources (material, social, financial etc.) and capabilities (education, experience etc.) (Whittaker et al., 2012).

Spatial planning can reduce vulnerable people's exposure to hazards through determining locations of activities and the type and nature of built infrastructure (including roads, housing, schools and facilities) and is therefore centrally important to disaster mitigation and disaster justice. Norman et al. (2014) identify opportunities for mitigating bushfire risk whilst also attending to equity considerations arguing planning policies need to take into account socio-spatial issues, including ensuring equitable access to economic development opportunities for vulnerable communities through provision of social and physical infrastructure and services. However, the planning system alone is ill equipped to deal with many dimensions of social disadvantage such as housing affordability, and many people, including renters, will move to existing housing in fire prone areas due in part to its affordability compared with established urban areas.

In all Australian jurisdictions, governments create and implement strategic planning regimes that are expected to anticipate risks, including bushfire (Bond & Mercer, 2014; Hughes & Mercer, 2009). These planning regimes determine permissible uses of land, and specify development types, locations and through granting of development consents, specify site-specific controls, all of which (at least in theory) accord with strategic, longer-term planning agendas (Kornakova et al., 2018; O'Donnell & Alexandra, 2018). For example, Victoria's Wildfire Management Overlay aims to determine "*where development will be permitted and how it must occur to minimise risks to life and property*" focusing mostly at the site scale, however this "*regulatory power can only be exercised for new developments or redevelopments of land*" and does not cover existing dwellings (Whittaker et al., 2012, p. 166). While each jurisdiction has specific planning mechanisms intended to reduce exposure to bushfire risks, these have proven difficult to implement because of vested interest in land development and culturally entrenched private property rights particularly in areas of high population growth near metropolitan or regional centres (Bond & Mercer, 2014; Hughes & Mercer, 2009).

Bushfire disaster mitigation requires more comprehensive and expansive use of planning strategies and instruments from the site to the regional scale (see Bond & Mercer, 2014; Buxton et al., 2011; Hughes & Mercer, 2009). Opportunities include planning strategies that determine settlement patterns through planning and development controls, specifying settlement locations through regional planning policies that include 'no-go' zones, and retreat and resettlement strategies for high-risk locations determined through modelling, mapping and zoning based on comprehensive fire risk assessments (Norman et al., 2014). However, the Victorian experience of introducing more risk-adverse planning controls after the 2009 fires were overturned as being too restrictive (Bond & Mercer, 2014) due to lobbying and the realpolitik of restrictive planning regimes (Kornakova et al., 2018). This example illustrates that use of zoning and restrictive land use planning are often contested and inevitably involve compromises because they are part of a dynamic system of governance that attempts to give effect to policies that sometimes have complex and at times competing goals (for example, housing affordability, fire risk reduction and biodiversity conservation).

4.4.2 Reinventing Land Use Planning: A Scenario

Imagine an alternative scenario to historical patterns of ad hoc peri-urbanisation. Imagine if Victoria vigorously implemented planning reforms after Ash Wednesday to reduce future exposure to bushfire hazards. Imagine that instead of either stopping peri-urban housing growth or allowing it to sprawl, the planning system had mandated that all new housing be formed into fire safe villages/clusters in appropriate locales, protected by water bodies, irrigated gardens and fire breaks. Imagine if under a reinvigorated planning regime all regions were comprehensively modelled, mapped and zoned for fire risks to determine appropriate, low risk and defendable locations for settlements. Any site and subdivision planning, would need to be subservient to broader, landscape scale siting factors which are important determinants of bushfire intensity and therefore risk (Bond & Mercer, 2014). Imagine if local councils had been required to end 'cookie cutter' small block, rural subdivisions but population growth was accommodated via cluster settlements in suitable, safer locations. Imagine that each cluster complies with robust standards for bushfire protection—including fire defence infrastructure of water supplies, ring mains, independent fire-fighting equipment, perimeter access roads and a community fire bunker in the event of worst-case fires—and that higher standards in building design and construction had been used to reduce fire risks on all houses built since the 1980s. Imagine these clusters were protected not only by their location but also by land management techniques like grazing, clustering of houses near dams, use of low flammability vegetation for fuel breaks and the separation of housing from existing bushland areas.

Developing such a scenario for peri-urban areas draws on decades of detailed science that, for example, offers information on improved building standards and land management options that can reduce fire risks (Syphard, Massada, Butsic, & Keeley, 2013). The impacts of wildfire can also be modified by the location, spatial distribution, density and flammability of buildings (Spyratos, Bourgeron, & Ghil, 2007). Fuel reduction burning in cooler seasons can be used to reduce fuel loads (Attiwill & Adams, 2013), but remains controversial. An alternative is the use of fuel reduction zones where ground fuel levels are controlled by grazing or mowing and shaded by wide spaced canopy trees that can assist in reducing the spread of fires (Agee et al., 2000; Agee & Skinner, 2005). Bands of deciduous trees can reduce wind speed and slow fire spread, because unlike pines or eucalyptus species, these have lower flammability, and steam rather than exploding during fires (Alexandra et al., 2017). European studies confirm that deciduous broad leaf species modify fire behaviour, disrupting the rates of fire spread (Fernandes, 2009; Moreira, Vaz, Catry, & Silva, 2009). This has far reaching implication for species used in gardens and for revegetation in fire prone areas (Murray, Martin, Brown, Krix, & Phillips, 2018; see also Campbell et al., 2017 for a review of four decades of Australian revegetation).

This scenario is offered because land use planning does not simply involve restricting or approving proposed developments, but includes enabling, facilitating and creating more sustainable settlements. By skilfully combining creativity, foresight and the knowing and governing of landscapes with integrated planning, we can develop settlement patterns more suited to Victoria's Pyrocene conditions.

4.5 Conclusions

The intensity and magnitude of bushfire risks—property damage, trauma and loss of life, pressures on emergency services, and extensive post catastrophe recovery costs—are being amplified due to climate change (Alexandra, 2012). This is compounded by settlement patterns, especially in peri-urban areas, where demographics result in higher levels of vulnerability (Foster et al., 2013). That the potential for catastrophic loss of lives exists in Victoria during episodic droughts is a historical fact. These conditions are predicted to increase in frequency, duration and intensity (Alexandra, 2012). However, proactive planning policies intended to mitigate bushfire disasters remain tentative and contested (Bond & Mercer, 2014). Several factors contribute to this: (1) institutional inertia and perceived and actual rights bestowed on private property; (2) a failure to recognise and use the full suite of planning provisions legally available; (3) the episodic nature of severe bushfire events that enable backsliding on reforms in intervening periods; and (4) an uncertainty and/or unwillingness to formally incorporate bushfire risk assessments, including climate change amplification of risks, into policies, exacerbated by divisive debates about climate change.

These are disaster justice issues because of the differential exposure to bushfire impacts that fall on the more vulnerable members of the community. Despite the substantive challenges of integrated spatial planning that reduces this exposure, there remains an expectation that governments will take responsibility for the citizenry while still allowing some freedom (Hughes & Mercer, 2009). However, it remains unclear, how best to balance responsibility sharing between governments and citizens (Holmes, 2010). The locus of responsibilities between citizen centric and governmental approaches shifts depending on policy fashions, circumstances and the time elapsed since the last major fire (Bond & Mercer, 2014; Hughes & Mercer, 2009). However, the central challenge remains of determining how personal freedoms and property rights should be balanced or traded off against collective responsibilities for ensuring safety in ways that are respectful of personal choices, sound public administration and democratic governance. Achieving greater coherence in the land use planning system requires negotiating deep, unconstructive and polarising controversies, about private property rights versus collective rights to determine settlement patterns that achieve an acceptable level of exposure to significant bushfire hazards. However, while never simple, negotiating these challenges is a regular requirement of governments as they attempt to juggle competing societal demands and expectations of safety.

Such a negotiation requires adequate information provision and education about bushfires in ways that empower informed decisions. However, the most vulnerable are often least engaged, or otherwise occupied in their survival. Any democratisation of fire knowledge requires dissemination widely through culturally and linguistically diverse networks and communities, in ways meaningful to them. Therefore approaches to delivering warnings about bushfire risk could be improved, including attaching warning to land titles, building permits and rental deeds to inform prospective purchasers or renters of houses about predicted bushfire risks.

This chapter emphasises policy learning as an adaptive process, and that adaptive policies rely on institutions with capabilities for incorporating the lessons derived from experience and reflection. The challenges of learning about living well in a flammable continent raise questions about how governance is improved and how governments are held to account, over the longer term, other than by episodic post-fire commissions of inquiry (Eburn & Dovers, 2015). In liberal democracies accountability processes require the citizenry to hold governments to account and provide oversight and evaluation of institutional competence (Dean, 2018).

Finally, this chapter argues it is not useful to appeal to naturalism to explain catastrophic fires. In the Anthropocene, neat definitional boundaries between 'human' and 'natural' binaries are breaking down. Arguing that fuel loads and fire behaviour are *'matters for nature'* does not help (Adams, 2013) nor do claims about the naturalness of bushfire (Whittaker & Mercer, 2010). If we accept that this epoch is the Anthropocene (or the Pyrocene), then appeals to naturalism have neither a sound basis, nor utility.

Accepting coupled socio-ecological systems in which humans are a dominant influence, obliges us to learn about and work with systems, in terms of developing capabilities to set and achieve goals, including minimising the disastrous impacts of bushfires, including on those most vulnerable. Without naturalism as a source of benchmarks we need to be setting forward-looking objectives for landscapes—for carbon, for biodiversity, disaster resilience and for fairness and equity etc.—and developing the planning, knowledge and governing capabilities to achieve these. Disaster justice requires functioning learning institutions. Otherwise without the capacity for institutionalised learning about pyric landscapes, unjust disaster impacts will continue. Disaster justice depends on a capacity for social learning and policy evolution that will enable us to learn to live well in a flammable continent and adapt to the Pryocene.

References

ABC. (2017). Cultural Burning Being Revived. Retrieved January 6, 2019, from https://www.abc.net.au/news/2017-06-19/cultural-burning-being-revived-by-aboriginal-people/8630038

Adams, M. A. (2013). Mega-Fires, Tipping Points and Ecosystem Services: Managing Forests and Woodlands in an Uncertain Future. *Forest Ecology and Management, 294*, 250–261. https://doi.org/10.1016/j.foreco.2012.11.039

Agee, J. K., & Skinner, C. N. (2005). Basic Principles of Forest Fuel Reduction Treatments. *Forest Ecology and Management, 211*(1), 83–96.

Agee, J. K., Bahro, B., Finney, M. A., Omi, P. N., Sapsis, D. B., Skinner, C. N., ... Weatherspoon, C. P. (2000). The Use of Shaded Fuel Breaks in Landscape Fire Management. *Forest Ecology and Management, 127*(1), 55–66.

Alexandra, J. (2017a). Risks, Uncertainty and Climate Confusion in the Murray-Darling Basin Reforms. *Water Economics & Policy, 3*(3), 1–21. https://doi.org/10.1142/S2382624X16500387

Alexandra, J. (2017b). Water and Coal—Transforming and Redefining 'Natural' Resources in Australia's Latrobe Region. *Australasian Journal of Regional Studies, 23*(3).

Alexandra, J. (2012). Australia's Landscapes in a Changing Climate—Caution, Hope, Inspiration, and Transformation. *Crop & Pasture Science, 63*, 215–217. https://doi.org/10.1071/CP11189

Alexandra, J., Norman, B., Steffen, W., & Maher, W. (2017). *Planning and Implementing Living Infrastructure in the Australian Capital Territory Canberra Urban and Regional Futures.* Canberra: University of Canberra.

Attiwill, P. M., & Adams, M. A. (2013). Mega-Fires, Inquiries and Politics in the Eucalypt Forests of Victoria, South-Eastern Australia. *Forest Ecology and Management, 294*, 45–53. https://doi.org/10.1016/j.foreco.2012.09.015

Bond, T., & Mercer, D. (2014). Subdivision Policy and Planning for Bushfire Defence: A Natural Hazard Mitigation Strategy for Residential Peri-Urban Regions in Victoria, Australia. *Geographical Research, 52*, 6–22. https://doi.org/10.1111/1745-5871.12040

Bowman, D. M. J. S. (2008). The Impact of Aboriginal Landscape Burning on the Australian Biota. *New Phytologist, 140*, 385–410. https://doi.org/10.1111/j.1469-8137.1998.00289.x

Butt, A. (2013). Exploring Peri-Urbanisation and Agricultural Systems in the Melbourne Region. *Geographical Research, 51*, 204–218. https://doi.org/10.1111/1745-5871.12005

Buxton, M., & Scheurer, J. (2007). Density and Outer Urban Development in Melbourne. *Urban Policy and Research, 25*(1), 91–111.

Buxton, M. (2007). Victoria's Kennett Government: Its Impact on Urban and Regional Planning. *Urban Policy and Research, 19*, 367–372. https://doi.org/10.1080/08111140108727884

Buxton, M., Haynes, R., Mercer, D., & Butt, A. (2011). Vulnerability to Bushfire Risk at Melbourne's Urban Fringe: The Failure of Regulatory Land Use Planning. *Geographical Research, 49*, 1–12. https://doi.org/10.1111/j.1745-5871.2010.00670.x

Campbell, A., Alexandra, J., & Curtis, D. (2017). Reflections on Four Decades of Land Restoration in Australia. *The Rangeland Journal, 39*(6), 405–416. https://doi.org/10.1071/RJ17056

CFA (Country Fire Authority). (2019). Retrieved January 9, 2019, from https://www.cfa.vic.gov.au/about/history

Cozens, P., & Christensen, W. (2011). Environmental Criminology and the Potential for Reducing Opportunities for Bushfire Arson. *Crime Prevention and Community Safety, 13*, 119–133. https://doi.org/10.1057/cpcs.2010.24

Dean, R. J. (2018). Counter-Governance: Citizen Participation Beyond Collaboration. *Politics and Governance, 6*(1), 180–188. https://doi.org/10.17645/pag.v6i1.1221

Eburn, M., & Dovers, S. (2015). Learning Lessons from Disasters: Alternatives to Royal Commissions and Other Quasi-Judicial Inquiries. *Australian Journal of Public Administration, 74*, 495–508. https://doi.org/10.1111/1467-8500.12115

Fernandes, P. (2009). Combining Forest Structure Data and Fuel Modelling to Assess Fire Hazard in Portugal. *Annals of Forest Science, 66*(4), 1–9.

Folke, C., Carpenter, S., Elmqvist, T., Gunderson, L., Holling, C. S., & Walker, B. (2002). Resilience and Sustainable Development: Building Adaptive Capacity in a World of Transformations. *Ambio, 31*, 437. https://doi.org/10.1639/0044-7447(2002)031[0437:RASDBA]2.0.CO;2

Folke, C., Hahn, T., Olsson, P., & Norberg, J. (2005). Adaptive Governance of Social-Ecological Systems. *Annual Review of Environment and Resources, 30*(1), 441–473.

Foster, H., Towers, B., Whittaker, J., Handmer, J., & Lowe, T. (2013). Peri-Urban Melbourne in 2021: Changes and Implications for the Victorian Emergency Management Sector. *Australian Journal of Emergency Management, 28*, 6.

Gammage, B. (2011). *The Biggest Estate on Earth – How Aborigines Made Australia*. Sydney, NSW: Allen and Unwin.

Gill, A. M., & Stephens, S. L. (2009). Scientific and Social Challenges for the Management of Fire-Prone Wildland–Urban Interfaces. *Environmental Research Letters, 4*, 034014. https://doi.org/10.1088/1748-9326/4/3/034014

Griffiths, T. (2016). An Unnatural Disaster? *History Australia, 6*, 35.1–35.7. https://doi.org/10.2104/ha090035

Holdgate, G. R., Cartwright, I., Blackburn, D. T., Wallace, M. W., Gallagher, S. J., Wagstaff, B. E., & Chung, L. (2007). The Middle Miocene Yallourn Coal Seam—The Last Coal in Australia. *International Journal of Coal Geology, 70*(1), 95–115.

Holdgate, G. R., McGowran, B., Fromhold, T., Wagstaff, B. E., Gallagher, S. J., Wallace, M. W., ... Whitelaw, M. (2009). Eocene–Miocene Carbon-Isotope and Floral Record from Brown Coal Seams in the Gippsland Basin of Southeast Australia. *Global and Planetary Change*, 65, 89–103. https://doi.org/10.1016/j.gloplacha.2008.11.001

Holmes, A. (2010). A Reflection on the Bushfire Royal Commission—Blame, Accountability and Responsibility. *Australian Journal of Public Administration*, 69, 387–391. https://doi.org/10.1111/j.1467-8500.2010.00701.x

Hughes, R., & Mercer, D. (2009). Planning to Reduce Risk: The Wildfire Management Overlay in Victoria, Australia. *Geographical Research.*, 47, 124–141. https://doi.org/10.1111/j.1745-5871.2008.00556.x

Hulme, M. (2011). Reducing the Future to Climate: A Story of Climate Determinism and Reductionism. *Osiris*, 26, 245–266. https://doi.org/10.1086/661274

Keenan, R. J., & Nitschke, C. (2016). Forest Management Options for Adaptation to Climate Change: A Case Study of Tall, Wet Eucalypt Forests in Victoria's Central Highlands Region. *Australian Forestry*, 79, 96–107. https://doi.org/10.1080/00049158.2015.1130095

Kerkhove, R. C. (2015). Aboriginal 'Resistance War' Tactics—'The Black War' of Southern Queensland. *Cosmopolitan Civil Societies: An Interdisciplinary Journal.*, 6, 38–62. https://doi.org/10.5130/ccs.v6i3.4218

Kornakova, M., March, A., & Gleeson, B. (2018). Institutional Adjustments and Strategic Planning Action: The Case of Victorian Wildfire Planning. *Planning Practice & Research*, 33(2), 120–136. https://doi.org/10.1080/02697459.2017.1358505

Manne, R. (2009). Why Weren't We Warned: The Victorian Bushfires and the Royal Commission. *The Monthly.* Retrieved from https://www.themonthly.com.au/monthly-essays-robert-manne-why-we-weren-t-warned-victorian-bushfires-and-royal-commission-1780

March, A., & Henry, S. (2007). A Better Future from Imagining the Worst: Land Use Planning and Training Responses to Natural Disaster [Online]. *Australian Journal of Emergency Management*, 22(3), 17–22. Retrieved May 22, 2019, from https://search.informit.com.au/documentSummary;dn=943610337205268;res=IELHSS. ISSN:1324-1540.

March, A., & Rijal, Y. (2015). Reducing Bushfire Risk by Planning and Design: A Professional Focus. *Planning Practice & Research*, 30(1), 33–53. https://doi.org/10.1080/02697459.2014.937138

McCarthy, M. A., Malcolm Gill, A., & Lindenmayer, D. B. (1999). Fire Regimes in Mountain Ash Forest: Evidence from Forest Age Structure, Extinction Models and Wildlife Habitat. *Forest Ecology and Management*, 124, 193–203. https://doi.org/10.1016/S0378-1127(99)00066-3

McKenzie, F. (1997). Growth Management or Encouragement? A Critical Review of Land Use Policies Affecting Australia's Major Exurban Regions. *Urban Policy and Research, 15*(2), 83–99. https://doi.org/10.1080/08111149708551508

Mitchell, J. W. (2013). Power Line Failures and Catastrophic Wildfires Under Extreme Weather Conditions. *Engineering Failure Analysis, 35*, 726–735. https://doi.org/10.1016/j.engfailanal.2013.07.006

Moreira, F., Vaz, P., Catry, F., & Silva, J. (2009). Regional Variations in Wildfire Susceptibility of Land-Cover Types in Portugal: Implications for Landscape Management to Minimize Fire Hazard. *International Journal of Wildland Fire, 18*, 563–574.

Moritz, M. A., Batllori, E., Bradstock, R. A., Gill, A. M., Handmer, J., Hessburg, P. F., …, Syphard, A. D. (2014). Learning to Coexist with Wildfire. *Nature, 515*, 58 EP.

Murray, B., Martin, L., Brown, C., Krix, D., & Phillips, M. (2018). Selecting Low-Flammability Plants as Green Firebreaks Within Sustainable Urban Garden Design. *Fire, 1*, 15. https://doi.org/10.3390/fire1010015

Norman, B., Weir, J., Sullivan, K., & Lavis, J. (2014). *Planning and Bushfire Risk in a Changing Climate*. Australia: Bushfire CRC.

O'Donnell, T., & Alexandra, J. (2018). *Regulatory Boundaries and Climate Adapted Futures*. CURF Research Report, University of Canberra.

Pyne, S. (2009). Peeling Back the Bark Blog of the Forest History Society. Retrieved from https://fhsarchives.wordpress.com/2009/02/10/historian-stephen-j-pyne-on-the-australian-fires/

Pyne, S. (2018). Big Fire; or, Introducing the Pyrocene. *Fire, 1*, 1. https://doi.org/10.3390/fire1010001

Russell-Smith, J., Yates, C. P., Edwards, A. C., Whitehead, P. J., Murphy, B. P., & Lawes, M. J. (2015). Deriving Multiple Benefits from Carbon Market-Based Savanna Fire Management: An Australian Example. *PLoS One, 10*, e0143426. https://doi.org/10.1371/journal.pone.0143426

SBS. (2017). Retrieved January 6, 2019, from https://www.sbs.com.au/news/traditional-burning-reviving-indigenous-cultural-burns-for-bushfire-management

Spyratos, V., Bourgeron, P. S., & Ghil, M. (2007). Development at the Wildland–Urban Interface and the Mitigation of Forest-Fire Risk. *Proceedings of the National Academy of Sciences, 104*(36), 14272–14276.

Syphard, A. D., & Keeley, J. E. (2015). Location, Timing and Extent of Wildfire Vary by Cause of Ignition. *International Journal of Wildland Fire, 24*, 37–47. https://doi.org/10.1071/WF14024

Syphard, A. D., Massada, A. B., Butsic, V., & Keeley, J. E. (2013). Land Use Planning and Wildfire: Development Policies Influence Future Probability of Housing Loss. *PLoS One, 8*, e71708. https://doi.org/10.1371/journal.pone.0071708

Teague, B., McLeod, R., & Pascoe, S. (2010). *Victorian Bushfires Royal Commission Final Report*. Parliament of Victoria.

The Age. (2010). Black Saturday Cost $4.4 Billion. December 30, 2018, from https://www.theage.com.au/national/victoria/black-saturday-cost-44-billion-20100801-11116.html

Watson, D. (2014). *The Bush*. Penguin Group Australia.

Whittaker, J., Handmer, J., & Mercer, D. (2012). Vulnerability to Bushfires in Rural Australia: A Case Study from East Gippsland, Victoria. *Journal of Rural Studies, 28*, 161–173. https://doi.org/10.1016/j.jrurstud.2011.11.002

Whittaker, J., & Mercer, D. (2010). The Victorian Bushfires of 2002–03 and the Politics of Blame: A Discourse Analysis. *Australian Geographer, 35*, 259–287. https://doi.org/10.1080/0004918042000311313

CHAPTER 5

Dimensions of Risk Justice and Resilience: Mapping Urban Planning's Role Between Individual Versus Collective Rights

Alan March, Leonardo Nogueira de Moraes, and Janet Stanley

5.1 Introduction

Bushfires, otherwise known internationally as wildfires, pose considerable risks at the interfaces between large areas of vegetation and urban areas (Gill & Stephens, 2009). The state of Victoria, Australia is understood to be one of the most bushfire-prone regions internationally, as a result of its topography, large tracts of flammable vegetation, and the potential for extremely hot and dry weather, sometimes combined with long dry periods or even severe droughts. Significant fires have occurred over time

A. March (✉) • L. N. de Moraes
Melbourne School of Design, Urban Planning Program,
The University of Melbourne, Parkville, VIC, Australia
e-mail: alanpm@unimelb.edu.au

J. Stanley
Melbourne Sustainable Society Institute, School of Design,
The University of Melbourne, Parkville, VIC, Australia
e-mail: janet.stanley@unimelb.edu.au

© The Author(s) 2020
A. Lukasiewicz, C. Baldwin (eds.), *Natural Hazards and Disaster Justice*, https://doi.org/10.1007/978-981-15-0466-2_5

since European settlement, including one of the worst recorded bushfire seasons in the period 2008–2009 when the most severe and deadly fires occurred on 'Black Saturday'—7 February 2009 (Teague, McLeod, & Pascoe, 2009). The consequences of the Black Saturday fires include 173 deaths, 2056 houses destroyed and extensive loss of other property, livestock and other long-term impacts (Teague et al., 2009).

Grasslands, forests or other vegetation are often proximate to homes, workplaces and other parts of human settlements. Fires can be started by lightning, intentionally or by accident (Padilla & Vega-García, 2011). When bushfires exceed human capacities to prevent, control or withstand their negative effects at a large scale, they are considered disasters (Mileti, 1999). The risks associated with bushfires include three main aspects. First is the likelihood of a fire based on the probability of ignition and subsequent spread to, and interaction with, inhabited areas. Secondly, risks are a function of the likelihood of negative consequences of a bushfire having impacts on a human settlement, such as loss of property, injury or deaths and loss of business capacity (Stephenson, 2010). Third is the risk to the natural environment, an issue not under direct consideration in this chapter. The morphology of human settlements in these interface areas is a key factor in the frequency and intensity of wildfires, and the types of risks faced by people, their properties and the wider environment. In addition, the impacts of climate change is increasing the frequency of extreme weather associated with dangerous bushfires, particularly in Australia's south-east, southern and south-west regions (Steffen, Dean, Rice, & Mullins, 2019).

This chapter applies a justice, rights and public good lens to the complex of tensions between individual rights and the public good relating to bushfire risks as they are currently managed by urban planning and building codes. It examines the case of the Wye River–Jamieson Track bushfire that occurred on Christmas Day 2015 in Victoria, Australia. Analysis of this event reveals the complex interplay of assumed, asserted and contested rights that play out before, during and after major and destructive bushfire events, and their implications for justice. The chapter goes on to consider the various roles that a reconceived approach to risk justice might play within the structural frames influencing government responses to major bushfire risks and actual events. These include acknowledgement and treatment of ensuing risks as complex manifestations of ongoing decisions over time. This occurs in the context of an existing emphasis on individual property rights eroding collective rights and the public good. Thus, setting minimum standards of overall risk management to be achieved in all settlements might be necessary, with the need to actively integrate overall actions across building, property management, planning, emergency response, public education and government actions, to achieve risk justice. It concludes with suggestions for new directions in the allocations of rights in terms of the management of the public good, specifically pointing to lessons relevant to the case examined.

5.2 Bushfire Behaviour and the Importance of Settlement Design and Location

The specific ways that fires progress in areas of vegetation (fire behaviour) and the ways they interact with human systems such as buildings and other structures are core drivers of immediate fire risk from a human perspective. Overall, three categories of factors influence the behaviour of a fire in a given situation. First is topography, the gradient or 'hilliness' of terrain being highly influential on fire speed and the level of intensity at which fuels burn. A general rule of thumb is that a fire will double its rate of spread (ROS) for every $10°$ of uphill slope (CFA, 2013). Second, atmospheric conditions are highly influential. The humidity, wind and other factors such as wind direction, gustiness and other factors impact on the speed, direction, size and intensity of fires. A very large, hot fire can generate its own dangerous weather, including severe winds that drive the fire rendering it extremely hard to extinguish (Badlan, Sharples, Evans, & McRae, 2017). These factors play a role in the likelihood of initial ignition as well the direction and behaviour of fires as they progress (CFA, 2013). If winds are strong a fire will be driven forward and its oxygen supply will be intensified, in addition to embers also being carried beyond the main fire (Ramsay & Rudolph, 2003). Thirdly, vegetation and other fuel sources are influential. The particular composition of fuels, the moisture content, their size and shape (i.e. fine and easily combustible but short lived versus heavy fuels that may be slow to ignite but once on fire, will continue burning for a long period) influence fire speed, the width of a fire front, overall intensity, and the existence and type of embers (CFA, 2013; Ramsay & Rudolph, 2003).

Structures burn in bushfires when there is fuel, heat and oxygen at levels that is sufficient to ignite and subsequently maintain a fire (Cohen, 2000). During a fire, the processes of radiation and convection heating may preheat the house for ignition, subsequently bringing about conditions that are more likely to cause a structure such as a dwelling to burn. Accordingly, if direct flame contact, radiant heat or ember attacks occur, a structure that is susceptible may ignite (Mikkola, 2008).

Overall, ember attacks are now understood to be the main mechanism causing house destruction in bushfires (Blanchi, Leonard, & Leicester, 2006; Cohen, 2008). Winds are capable of carrying embers for a number of kilometres, although even embers that are carried only a short distance may be destructive if structures are located close to their source. Overall, embers can pose risk before, during and after the passage of an actual fire front (Mikkola, 2008).

Building materials, structural design, site location and vegetation management can make a building more or less vulnerable to ignition during a wildfire (Blanchi et al., 2006; Cohen, 2008; Price & Bradstock, 2013). Fire-driven winds can also be a mechanism of fire attack due to the damage that it can cause to structures, impacting upon people sheltering there, and also allowing embers to enter a structure, setting it on fire (CFA, 2013; Ramsay & Rudolph, 2003).

Proximity to large areas of flammable vegetation is the most predictable indicator of fire risk (Blanchi et al., 2014). Heat transfer reduces rapidly as distance from a fire increases as does the potential for direct flame contact (Standards Australia, 2018). As noted, complexities such as the particular combination of fuels, slope, wind direction, the character of structures, access and fire weather behaviours such as strong updrafts also play a part in particular settlements' risk profiles. Further, built-up areas themselves may also contain significant amounts of flammable materials. This can include vegetation in parks and gardens, dwellings themselves and commercial buildings, outbuildings and their contents such as wood-piles, rubbish, and fuel stores for heating or cooking. The arrangement of potential fuel sources can have significant influence upon the potential for fires to ignite and continue progress within settlements. It is common in many low density settlements to include a lot of vegetation between structures that can facilitate ongoing fire progress.

Human factors also impact on vulnerability and the ability to prepare for or mitigate fire risks. In general, communities and persons who have experienced fires previously, and who have arranged or joined with others who have resources and systems for taking considered action, are better equipped in dealing with wildfire. Alternatively, places that include transient populations or significant numbers who are vulnerable are usually less capable of managing risks (Gonzalez-Mathiesen & March, 2018).

5.3 Risk Justice: Hazards, Vulnerability and Exposure to Hazards

When broad ideas of justice are applied to risk and disasters in human settlements, powerful dilemmas and paradoxes emerge. Justice can be understood simplistically as people and communities *"getting what they are due"* (Woodruff, 2019, p. 5). However, here lies considerable complexity such as: how do we determine what people should get, whether people got (or will get) what they deserve, or whether governance and justice systems need to regulate and modify processes? (Audi, 1998, p. 395). In this chapter we focus mainly upon procedural and distributive conceptions of justice, and refer to foundational definitions to allow generative application to the substantive focus of the chapter and its case study.

Procedural justice is generally understood as the fairness (perceived or 'real') of the means and justifications used to make decisions (Thibaut & Walker, 1975). Many aspects comprise this view of procedural justice: the creation and application of consistent processes, suppression of bias, use of accurate information, application of correction mechanisms, representation of participants' concerns, and these are based on prevailing moral and ethical standards (Leventhal, Karuza, & Fry, 1980). In contrast, distributive justice is concerned with the reality and perceptions related to the fairness of outcomes and the distribution of costs and benefits (Folger, 1987). Lukasiewicz and Baldwin (2017) summarise this as including:

> *the principles of need (that which is necessary for survival), equity (where distribution is proportional to one's inputs) or equality (where all stakeholders have equal access to the resource), self-interest (which is practised by governments as agents acting on behalf of their jurisdictions), and efficiency (wise use of resources and avoidance of wastage).* (Lukasiewicz & Baldwin, 2017, p. 1044)

We argue that both distributive and procedural aspects are fundamental to achieve risk justice whether they have been realised as actual consequences (outcomes) in an actual event, or not. However, when disasters occur, the outcomes are often profoundly unfair (Alexander, 1999) in ways that cut across most intuitive and theoretical ideals of justice, as well as perceptions of fairness. This can be tempered with an understanding that justice does not only exist along the balancing of rights and responsibilities through distributive and/or procedural mechanisms (Bulkeley,

Edwards, & Fuller, 2014). Rather, it is also a matter of understanding the impacts of pre-established and historical conditions (Agyeman, Schlosberg, Craven, & Matthews, 2016; Caney, 2010).

The overall consequences of disasters commonly impact certain groups more than others (Stanley, 2017). Women, children and those with a disability are often injured or die more frequently (Bolin & Kurtz, 2018; SAMSHSA, 2017); poorer and less educated communities often suffer more from initial impacts and longer term post-event consequences at a higher rate (Dash, Peacock, & Morrow, 1997; Stanley, 2014) and those with uncertain tenure and in circumstances of forced migration, homelessness or informality commonly suffer injury, death and economic hardship disproportionately (Elliott & Pais, 2006).

This chapter concentrates on the rights of the average person and their responsibility to protect the rights (in this case not to be subject to fire) of others. The corollary of these questions is to question the role of urban planning in terms of the real or potential consequences of disaster events in existing urban environments (Schlosberg & Collins, 2014), even while broad commentary such as that provided by the Sendai Framework (2015) would suggest the centrality of planning. We suggest first that an understanding of risks must inform any development of solutions.

A subset of risks (explained further below) is hazards. A hazard such as a bushfire is a *'potential or existing condition that may cause harm to people, or damage to property or the environment'* (Australian Institute for Disaster Resilience, 2015, p. 32). Risk can be understood as the likelihood of unwanted consequences occurring as a result of an event, such as a bushfire destroying property or causing fatalities (Australian Institute for Disaster Resilience, 2015, p. 113). Another way of considering risks in a way that more directly takes us to potential actions that could improve risk profiles in human settlements is to consider Crichton's risk triangle. Shown in Fig. 5.1, this suggests that the risks faced in a given place are a function of the levels of exposure, the hazard itself and the vulnerability of a community (Crichton, 1999).

If the precepts suggested by the risk triangle are applied to a human settlement, it means that a complex of existing, potential risks and consequences exist. Vulnerability is the:

> *extent to which a community, structure, service or geographic area is likely to be damaged or disrupted by the impact of a particular hazard, on account of their nature, construction and proximity to hazardous terrain or a disaster-prone area.* (Australian Institute for Disaster Resilience, 2015)

Fig. 5.1 The risk triangle. (Crichton, 1999)

Exposure refers to the '*elements within a given area that have been, or could be, subject to the impact of a particular hazard*' (Australian Institute for Disaster Resilience, 2015, p. 32) driving particular risk profiles to emerge for a given place, for a given hazard such as bushfire.

Using the categories above, the main elements of human settlements that drive bushfire risks in terms of the hazard are proximity to vegetation and wider landscapes that support and propagate bushfires (March & Rijal, 2015). As mentioned, longer potential fire runs, challenging topography and heavier fuel loads associated with greater amounts of flammable vegetation increase risks. Vulnerability is a function of physical layout and road systems, design and materials used in structures, the capabilities of human response systems such as evacuation and firefighting, and the characteristics of occupants themselves in terms of decision making, evacuation, firefighting, and overall susceptibility to negative effects such as smoke, heat and stress (CFA, 2013). In terms of bushfire, exposure is understood as the proximity of humans and structures to the impact elements of bushfire such as heat, ember, flame, smoke and fire-driven winds (Blanchi et al., 2014).

5.4 Urban Resilience and Problematics of Disaster Justice: Rights and the Public Good

As set out above, challenges to bushfire risk management and resilience exist in urban areas due to the variety of factors that influence risk profiles, but also includes the imperfect scope of urban planning and its temporal 'position' oriented to development control as a locus of impact. We suggest that directions for improvement to urban planning's functions in terms of risk justice can be provided by examining the dilemma of rights versus the public good (putting aside other factors that interact with the public good, such as markets and environmental preservation) (Dahl, 1982), measured against achievement or otherwise of resilience. In this chapter we recommend that improvements to resilience, for all groups of people, are based upon actions that positively improve elements of the risk profile, recognising that comprehensive treatment is ideal. We take urban resilience to be:

> *the ability of an urban system-and all its constituent socio-ecological and socio-technical networks across temporal and spatial scales-to maintain or rapidly return to desired functions in the face of a disturbance, to adapt to change, and to quickly transform systems that limit current or future adaptive capacity.*
> (Meerow, Newell, & Stults, 2016, p. 39)

Urban planning systems embody fundamental assumptions and ongoing reassertions of rights, balanced and sometimes traded off against the public good[1] (March, 2003) cutting across procedural and outcome-oriented focuses (Alexander, 2002). Rights-centric approaches seek to establish inviolable standards of process or distributive justice outcomes, linking in with approaches to justice as outlined above. However they are derived, rights are socially established standards and potentially contestable value judgements that are maintained for individuals, groups or a societal belief, such as 'a fair go for all'. It is noted, however, that rights need to be re-asserted and re-interpreted on an ongoing basis as circumstances change. The public good, while having utilitarian origins as consideration of overall human well-being, welfare, happiness or satisfaction of wants and desires is generally subsumed to understandings of wider interests and the goals of government in pursuing general community goals (Campbell & Marshall, 2002).

For urban planners, the concept of public good has provided, theoretically at least, legitimation for spatial and land use planning and development control that modifies market processes; as a foundational principle for the profession; as a normative base for professional judgement and

---

[1] We take the position that while rights are a foundational component of the public good, there are situations where the exercise of individual rights can undermine achievement of the public good.

expertise; and, as an ethical norm for practice (Alexander, 2002). Debates regarding the tensions and overarching tests of rights and the public good remain contested, cutting across norm-based, moral and substantive tests founded on the thinking of philosophers and sociologists such as Dworkin (1977), Habermas (1984), Sen (2000) and Rawls (1971). For the purposes of this chapter, in keeping with Alexander (2002), consideration of the various competing models has been put aside in favour of an empirical examination of rights and the public good in the practical mechanisms of land use and development.

In an operational sense, land in Australia is typically used, held and regulated as a commodity which carries with it a "bundle" of rights—including the property of others (Neave, Rossiter, & Stone, 1994). Ownership of land is *"not so much possession, as the rights that accrue as its result"* (Mattei, 2000, p. 77). Accordingly, while an owner has rights over his or her land, these are not absolute and generally common law has established that the exercise of one person's rights must not excessively over-ride another's (March, 2003). While generally supporting many aspects of this rights-based approach, urban planning introduces another principle, that higher order impositions may be applied to land, based on the protection or achievement of public good goals. Familiar examples include mechanisms ensuring minimum participation rights, health standards, heritage controls, efficiency, facilitating education provision and utilities, to name a few. Importantly, these concepts intersect with many of the principles underpinning distributive and procedural justice, as discussed below.

5.5 Wye River and Separation Creek

Wye River and Separation Creek are small adjacent settlements located on the south coast of Victoria, Australia on a scenic coastline. The towns are approximately 2.5 hours' drive from the centre of Melbourne, Victoria's capital city. They have attractive beaches, proximity to forests and access to a range of recreational activities central to their popularity as tourist destinations. As a result of their mainly summer-oriented tourism, there is considerable population variation seasonally. The permanent populations of Wye River and Separation Creek are 144 and 24, respectively (ABS Australian Bureau of Statistics, 2006). The settlements (pre-fire) included 469 and 37 dwellings in each settlement, respectively. Holiday homes, unoccupied during the census count, amount to approximately 70% of housing. The estimated population during summer holiday periods, in contrast, is in the order of 1600 due to the large capacity allowed by caravan parks and camping grounds in the towns.

The proximity of the two towns is such that their low density 'urban edges' partially merge over the ridgeline that separates them. The central and original settlement of each town focuses upon the lower altitude watercourse after which each is named, sloping up to the ridgelines surrounding each. As a result of the vegetation reserves around each town, their historical development and the hilly topography, only one road provides access—the iconic Great Ocean Road. The road is located directly on the coastline and is highly trafficked, particularly during tourism peak periods. The road is also a slow speed carriageway with many narrow components and tight turns that include breathtaking coastal views but is also limited in its capacity.

5.6 The Christmas Day Fire

On 25 December 2015 a bushfire caused significant damage to Wye River and Separation Creek. The fire was not rated as extreme on the Fire Danger Index (FDI) (notionally with a highest rating of 100 that has been exceeded considerably in recent years) at 49. Nonetheless a total of 166 houses were destroyed or significantly damaged in the two towns (Leonard, Opie, Blanchi, Newnham, & Holland, 2016).

The events leading up to the fire interacting with the towns provide significant information that contextualises the fire's progress and consequences. Prior to the fire, December 2015 had warmer weather than usual. Average temperatures were the sixth-highest on record for Australia as a whole, and highest on record for Victoria, Tasmania and South Australia (Bureau of Meteorology, 2016). Fires were initiated in the Cape Otway Ranges by lightning strikes on 19 December 2015 near the town of Lorne (Emergency Management Victoria, 2015). Due to the combination of heavy vegetation, challenging topography, poor road access and a long period of hot weather, the fire was difficult to contain.

On the morning of 24 December, the fire was approximately four kilometres north-west of Separation Creek. The Country Fire Authority (CFA) issued warnings to a number of towns in the general area. The police and CFA had already door-knocked and held a number of evacuation information meetings from 23 December. While evacuation was voluntary, it was suggested that it would be appropriate for people to leave, and many did so soon after receiving this information. The emergency siren was sounded by Wye River Fire Brigade Captain Roy Moriarty at 11:30 am on the 25th. At this time winds were estimated to be 33 km/h with humidity reaching its lowest of 17% at 4:30 pm (Hillard, 2016, pp. 4–6). The fire is estimated to have arrived at the outskirts of Wye River at approximately 3:00 pm. From 11:30 am, after the siren sounded, the few remaining residents and visitors gathered their possessions and left in vehicles in an orderly way (Hillard, 2016, p. 5).

The mechanisms of fire progress and interaction with the towns is summarised by the CSIRO (Commonwealth Scientific and Industrial Research Organisation)-CFA as:

> *The bushfire burned towards Wye River in native bushland to the north-west. Even though the winds responsible for driving the fire towards the township on the day of the fire came from the north or north-east, surveys did not reveal any evidence of a fire front arriving from this direction and interacting directly with structures. It appears that terrain and coastal wind effects around the township meant that spotting ahead of the approaching fire fronts ignited and burnt back upslope towards the approaching fire, reaching the perimeters of the townships before the approaching fire front. Fire-impacted areas of Wye River and Separation Creek were typically the steepest areas, which reach gradients approaching 30 degrees.* (Leonard et al., 2016, p. 23)

The response activities of the emergency services were orderly, well-resourced and orchestrated. They included considerable water bombing (18 aircraft including a high capacity Chinook), local crews and approximately 50 Department of Environment, Land, Water and Planning (DELWP) slip-ons (four wheel drive fire response trucks). A total of five fire-tankers operated in lower areas, including one located at the fire station, two at the surf club and two protecting the pub (Hillard, 2016, pp. 4–5).

5.7　Post-Event Assessment of Risk Factors

The post-event study undertaken by CSIRO and the CFA found that the fire spotted embers in a south-east direction into the town, and then, drawn by low level convection winds of the higher level fire, ember initiated fires progressed upslope away from the coast in a generally north-west direction through the town. The forward spotting was described as a mass event. This challenged the attempts to predict the path and progress of a fire front per se (Leonard et al., 2016: 24). No deaths occurred, while

116 houses were destroyed in the fire with many others damaged of the total 506 dwellings in the twin settlements. The fire itself continued to burn in highly inaccessible terrain and difficult weather conditions until 21 January 2016, and burnt approximately 2500 hectares in total (Country Fire Authority, 2016). Drawing on a range of sources, including key studies undertaken by the CSIRO and the Country Fire Authority, the remainder of this paper documents and critically considers the drivers of risk in the Wye River-Separation Creek 2015 fire.

Using the risk triangle (Crichton, 1999) as an organising framework, it can be seen that interactions between the fire hazard, high levels of exposure and vulnerabilities of structures led to the outcomes described previously. Wye River and Separation Creek's proximity to large tracts of forested and challenging topographical land means that the bushfire hazard is characterised by the potential for fires to approach the towns with considerable intensity from a number of directions. This means that there is potential for high levels of heat radiation, flame contact, ember attack and fire-driven wind. Further, the predominance of high levels of leaf litter and other highly flammable elements throughout the settlements meant that the fire could propagate after embers found flammable receiving environments well ahead of the fire front within the settlement itself.

The difficulties of conducting fuel reduction burns in the challenging surrounding terrain, combined with ecological considerations, also means that fire intensity is unlikely to be mediated over time by human interventions. Directly related to this is the challenge of the settlements containing a range of vulnerable elements exposed to the damaging effects of the fires. These elements were primarily dwellings which were constructed between, under and directly adjacent to vegetation. Further, the dwellings were predominantly built upon slopes that facilitated the progress of the fire in, around and under the structures themselves, bringing about high levels of direct exposure. A complicating factor that became apparent during the post-event evaluation was that houses, when ignited, become additional new hazards themselves, emitting extreme heat, long flame lengths and embers. Proximate structures are highly exposed to this new threat, putting them at extreme risk.

The settlements' vulnerability can be understood as a function of multiple factors. In terms of physical aspects, the majority of structures were built prior to introduction of bushfire design and planning codes, and many extensions and ad hoc modifications have been undertaken by owner builders. Overall, most structures were not designed and constructed to a standard capable of withstanding the impacts of a bushfire of any intensity. Further, the ongoing use and maintenance of many structures and their surrounds increased vulnerability to fire. These factors included unenclosed underfloor spaces allowing ember ingress, use of underfloor spaces for storage of flammables, use of flammable retaining walls adjacent to structures, addition of flammable cladding after formal completion of building, flammable furniture or other materials on decks and verandahs, location of vegetation adjacent to windows, and tree strike during fires allowing ember entry into structures. As mentioned above the close proximity of many houses side by side on narrow lots meant that burning houses themselves became hazards and generated heat and flame outputs well in excess of the resistance characteristics of adjacent houses.

Two main human vulnerability aspects also contributed to the losses suffered. The voluntary evacuation of all residents was undertaken from the perspective of prioritising human life, as expressed in a range of integrated policy documents (CFA, 2013; Victoria, 2018). Additionally, firefighters were not able to enter the majority of the areas at risk due to the narrow, steep and dangerous road network. This, in effect meant that the only form of active response available was water drops from aircraft. While evidence suggests that the water drops undertaken were effective, the detailed monitoring and firefighting that ground crew are sometimes capable of, to prevent escalation of small house fires immediately after ignition, was not possible for the majority of structures during this fire.

5.8 RISK JUSTICE: ENTANGLEMENTS BETWEEN RIGHTS AND THE PUBLIC GOOD

The complex of elements that embody risk profiles such as those summarised above are only partly under the influence of urban planning and related systems before and after disaster events such as the 2015 fires. Further, they are a combination of spatial, physical, social, environment and economic characteristics. Urban planning and building codes exist in a series of moments of 'emergence' during which attempts are made to deal with past and current circumstances, with a view to achieving particular goals into the future (Albrechts, 2010; Hillier, 2010; Hopkins, 2001). This transitionary process, sometimes described as becoming (Albrechts, 2011) highlights the dynamic nature of planning, and indeed of justice.

Fundamental challenges to justice are embodied in conceptions and practices of risk acceptance and risk reduction, of justice itself and in the role of urban planning. In the dynamic processes of change management and the challenges and opportunities that exist in a given place, risks, just as they are socially constructed during human processes of production and consumption (Beck, 1986, p. 5) are borne by individuals or groups whether or not they are aware of this (Beck, 2008). Similarly, procedural and distributive rights are not universal, but are constructs in a given time and place—even when they are assumed or presented as fundamental and enduring (Campbell & Marshall, 2002). Urban planning and building systems in Australia are enmeshed with property rights systems, even while they seek to modify and improve upon these by reference to the public good (Alexander, 2002). We argue here that parallels can be drawn between rights systems and risk 'ownership' rationales for collective action.

As illustrated above, the bushfire risks faced on a given piece of land result from a complex of interactions between the characteristics of the hazard, levels and types of exposure, and the vulnerability (or otherwise) of the people and structures on that land. The rights associated with land ownership are not absolute but normally include the rights of transfer, meaning that land and buildings can be bought and sold, along with the right (within limits) to build and use land within the bounds of relevant rules such as urban planning and building regulations. This means that in effect, bushfire risks associated with a piece of land are able to be bought, sold or rented between different parties. This provokes a number of interrelated questions: should this information be determined in advance; who would determine it; should it be public or at least declared to prospective owners; are all tenure types deserving of the same levels of knowledge, and so forth?

The imposition of urban planning and building rules that seek to improve the standards of construction, firefighting water and access, materials and siting are, in effect, restricting the absolute rights of land owners on their land, in favour of achieving a minimum standard of fire resistance in any newly constructed dwelling on that land. The assumption is that this provides the greatest chance of protecting human life, in addition to retaining the structure. This might also be understood as establishing a minimum standard of substantive individual rights for any occupant in the present or future to expect that a structure can deal with the risks that can be reasonably expected to be faced in a given location at some point of the building's life.

However, risks do not accrue just to individual parties on isolated sites at a single point in time, but rather exist as a series of dynamic, temporal and interconnected relationships—invoking questions of the need for and type of government intervention (May, 1991). A number of dimensions of risk need to be considered. One individual determining that they are prepared to accept residual risk on land does not necessarily mean that another owner or occupant of that land and building at some point in the future is able to fully understand or appreciate the risks associated with that site—especially if they are not made transparent. The risks borne on one site that result in a structure burning may cause adjacent structures to burn, transferring the consequences of untreated risks across property boundaries. Further, the risks potentially faced by firefighters, and the burden placed on government, insurance companies and reconstruction authorities is effectively transferred from one individual exercising his or her individual rights, onto the wider community, eroding others' rights and ultimately the public good.

Parties that own land that was built upon at a low standard of bushfire risk treatment enjoy the same procedural rights in receiving firefighting, reconstruction and post disaster relief as those who have built more recently to comply with the bushfire code. However, the risks they and their dwelling face are greater, and the potential for their structure to transfer fire impacts to the rest of the community is also higher. At the same time, however, the burden upon that individual if they wish to improve their property, to build it to a high bushfire standard is such that, perversely, they may avoid improving the bushfire resistance of their structure due to adjacent vulnerable buildings (March & Kornakova, 2017). The current regulatory arrangements are such that only if substantial changes are made to a structure that it must be brought up to modern bushfire risks standards—meaning that many owners avoid significant change for fear of incurring major costs.

5.9 More or Less Resilient? Limits and Possibilities for Urban Management

We argue here that considering the public good as part of risk justice requires particular principles to be more fully developed and applied. These principles will require in some ways a re-assertion and re-ordering of some elements of current assumptions of private property rights and the public good. Using the case study presented as a basis, we suggest that risk justice as it relates to bushfire and urban planning mechanisms can be understood in a number of inter-related ways that *"contribute to improve overall adaptation to change to transform systems that limit current or future adaptive capacity"* (Meerow et al., 2016, p. 39). As it stands, most aspects of the recovery and rebuilding process tend to reproduce the underlying risk factors outlined above, with the exception of building to higher Bushfire Attack Levels in keeping with the building code. For example, lot design and proximity of structures (i.e. houses near to each other) that can transfer fire to other structures are not modified in reconstruction processes. Road layouts and overarching settlement patterns remain the same. Risk factors outside of planning and building codes are typically reproduced: flammable storage under houses; use of flammable retaining walls; and use of flammable gas bottles for domestic supply.

In parallel with approaches to resilience, risk justice cannot be seen as a static 'destination' that can be reached through passive regulatory approaches alone, even while minimum design standards do have an important role. For each party and community to get a just outcome in an existing settlement, such as Wye River and Separation Creek, will require a suite of actions over a long period of time. Further, these may require interventions that actively modify elements of risk, such as modifications to road systems or lot boundaries to improve risk profiles for individuals and for overall public good outcomes. Thus, to summarise the complexities:

1. Risks in a given place are the manifestations of a series of ongoing decisions by a range of parties over a range of time periods. Individual acts (or the absence of these) can have significant impacts on individual and others' rights, and on the public good in the present and future. This suggests a need to actively assess and manage these implications. The question then becomes by whom and through what authority.
2. Conceptions of property rights as automatically allowing autonomous use of land erode the maintenance of wider risk justice rights and the achievement of public good outcomes such as attaining minimum standards of acceptable risks and associated actions to achieve these.
3. Developing a public good conception of risk justice suggests that extraordinary actions might be required in certain locations where risks are unusually high, such as Wye River and Separation Creek.

4. Achievement of risk justice outcomes would require that overall settlement policies address minimum standards and their achievement at a range of spatial and temporal scales. These would encompass the integration of risk across a range of actions, tensions of growth, change and decline, as well as competing goals across social, economic and environmental concerns.

Following from the above, and applying these principles to the case study of Wye River and Separation Creek, the following issues emerge:

1. The silence of the processes for sale and transfer of land and property in a way that is essentially 'silent' regarding risks and the treatment of these erodes the ability for community members to understand and appreciate risk levels: as a characteristic of individual sites and the overall community. This includes appreciation of the impacts individual risk profiles might have on neighbours, firefighters and the wider community. Accordingly, a bushfire star rating or similar approach should be part of the information provided during all land and property transfers in bushfire-prone areas.
2. Current regulations and standards do not extend to dealing in sophisticated ways with ongoing maintenance, retrofitting, human vulnerability and its changeability over time, or tradeoffs with environmental implications of risk treatment processes.
3. A minimum standard of risk acceptance for all relevant parties needs to be established and governments (including urban planning agencies) must seek to actively achieve this universally to achieve rights justice. While a pure property rights-based approach may argue that individuals may make their own choices to live in a risky location, a risk justice approach would consider the entirety of the community and the risks faced, including those who may occupy land in the future, and the overall costs and benefits on the community and broader population, including tax payers. If risks cannot be managed sufficiently, then a settlement may need significant modification or even to be compulsorily acquired in part or entirety.
4. Current urban planning bushfire risk assessments and treatment occur only at the individual site level. A risk justice approach would also assess risk and its treatments at the overall community level,

ensuring that all interactions and transfers of risks and benefits are appropriately managed in a just and integrated manner over time. For example, it is plausible to imagine it will be decided in the case study area that it is not possible to reduce the extensive current fuel loads in between structures due to practical and environmental constraints. This would suggest that the vulnerability of already exposed houses needs to be reduced to account for this, at a wider community level. Appropriate retrofitting, choices of materials and ongoing development control processes need to reflect this.

5. Urban planning needs to actively integrate with other actors and processes relevant to risk justice and management, while considering and dealing with competing demands. This includes service providers, infrastructure providers, land acquisition agencies and landscape management bodies.

6. Current approaches to participation in risk management are extremely limited. In the Wye River and Separation Creek case little opportunity for community input has been provided due to privileging of technical approaches leading to minimisation of third party rights in the Bushfire Management Overlay (BMO). The Bushfire Management Overlay is an urban planning control that requires design and siting standards to be met during permit processes. In the current case, new standards relating to the BMO were introduced with no democratic or consultation processes. This is in stark contrast to the extensive participation rights afforded to citizens in other aspects of the Victorian Planning System (March, 2012). We argue that real benefit can be gained by developing approaches that directly facilitate action at the wider community level that develops and promotes risk justice and the public good, in contrast to current attention almost exclusively directed to individual homes.

7. This approach raises many difficult broader questions, such as who decides the level of risk that is acceptable for a particular community? Also, how should the standards be financed if the outcome is for public good and will it exclude the groups of people referred to earlier in this chapter who are already experiencing inequality and an absence of social justice, and may not be able to afford the higher standards?

5.10 Conclusions

The natural phenomenon of bushfires manifests as risks that result from human activities and values particularly in locations where extensive vegetation and urban areas interact in many parts of Australia, and is in effect, a human phenomenon in its implications. The risks posed by bushfires are considerable and can often overwhelm human capacities to deal with them. The challenge of bushfires exists at the intersection of multiple factors that drive risks, influenced by human and natural systems.

The application of a risk justice approach to the complex of tensions between individual rights and the public good relating to bushfire risks suggests a range of implications that intersect with the use of urban planning and building codes, as they are applied. Examination of the case of the Wye River–Jamieson Track bushfire that occurred on Christmas Day 2015 in Victoria, Australia, indicates that reconceived approaches to risk justice are needed to achieve just outcomes, particularly in circumstances where legacy risks exist due to past decisions that continue to have implications for risks over time. The existing emphases on individual property rights including the sale and purchase of land over time has a tendency to privilege rights and to diminish the opportunity to achieve collective rights and the public good as a shared level of risk exposure for all citizens. The implication is that there is a case to establish minimum standards across all settlements—acknowledging that the mechanisms to achieve such standards will be complex and potentially financially challenging. A key part of achieving risk justice will be via the integration of individual and overall actions of various actors including governments. However, such an approach will result in major changes in planning, accompanied by the need for wide acceptance and cooperation between all actors.

References

ABS Australian Bureau of Statistics. (2006). ABS Quickstats (Census). Retrieved 2016, from ABS.

Agyeman, J., Schlosberg, D., Craven, L., & Matthews, C. (2016). Trends and Directions in Environmental Justice: From Inequity to Everyday Life, Community, and Just Sustainabilities. *Annual Review of Environment and Resources, 41*(1), 321–340. https://doi.org/10.1146/annurev-environ-110615-090052

Albrechts, L. (2010). More of the Same Is Not Enough! How Could Strategic Spatial Planning Be Instrumental in Dealing with the Challenges Ahead? *Environment and Planning B, Planning & Design, 37*(6), 1115–1127.

Albrechts, L. (2011). Transformative Practices. In S. Oosterlynck, L. Albrechts, & F. Moulaert (Eds.), *Strategic Spatial Projects: Catalysts for Change*. London: Taylor & Francis.

Alexander, D. E. (1999). *Natural Disasters*. London: Kluwer.

Alexander, E. R. (2002). The Public Interest in Planning: From Legitimation to Substantive Plan Evaluation. *Planning Theory, 1*(3), 226–249.

Audi, R. (1998). *The Cambridge Dictionary of Philosophy* (1997 ed.). Cambridge: Cambridge University Press.

Australian Institute for Disaster Resilience. (2015). National Emergency Risk Assessment Guidelines: Handbook 10. Retrieved August 2, 2019, from https://knowledge.aidr.org.au/resources/handbook-10-national-emergency-risk-assessment-guidelines/

Badlan, R., Sharples, J., Evans, J., & McRae, R. (2017). *The Role of Deep Flaming in Violent Pyroconvection*. Paper Presented at the 22nd International Congress on Modelling and Simulation, Hobart, Tasmania, Australia.

Beck, U. (1986). *Risk Society: Towards a New Modernity*. London: Sage.

Beck, U. (2008). World at Risk: The New Task of Critical Theory. *Development and Society, 1*, 1–21.

Blanchi, R., Leonard, J., Haynes, K., Opie, K., James, M., & Dimer de Oliveira, F. (2014). Environmental Circumstances Surrounding Bushfire Fatalities in Australia 1901–2011. *Environmental Science & Policy, 37*(March), 192–203.

Blanchi, R., Leonard, J., & Leicester, R. (2006). Bushfire Risk at the Rural-Urban Interface. Retrieved August 2, 2019, from http://bushfirecrc.com/sites/default/files/managed/resource/bushfire_risk_at_the_rural_urban_interface_-_brisbane_2006_0.pdf

Bolin, B., & Kurtz, L. C. (2018). Race, Class, Ethnicity, and Disaster Vulnerability. In H. Rodríguez, W. Donner, & J. Trainor (Eds.), *Handbook of Disaster Research* (pp. 113–129). Cham: Springer.

Bulkeley, H., Edwards, G. A. S., & Fuller, S. (2014). Contesting Climate Justice in the City: Examining Politics and Practice in Urban Climate Change Experiments. *Global Environmental Change, 25*, 31–40.

Bureau of Meteorology. (2016). Monthly Summary for Australia, Product Code IDCKGC1A00, Tuesday 5 January 2016 (2015). Retrieved March 2016, from http://www.bom.gov.au/climate/current/month/aus/archive/201512.summary.shtml

Campbell, H., & Marshall, R. (2002). Utilitarianisms Bad Breath: A Re-evaluation of the Public Interest Justification for Planning. *Planning Theory, 1*(1), 163–187.

Caney, S. (2010). Climate Change and the Duties of the Advantaged. *Critical Review of International Social and Political Philosophy, 13*, 203–228.

CFA. (2013). Leaving Early: Bushfire Survival Planning Template. Retrieved August 2, 2019, from https://www.cfa.vic.gov.au/documents/20143/98951/4713_CFA_Pullout_LEAVING_web.pdf/9ba3051e-993b-794c-5ed7-ce9a286f6ddc

Cohen, J. D. (2000). Preventing Disaster: Home Ignitability in the Wildland-Urban Interface. *Journal of Forestry, 98*(3), 15–21.

Cohen, J. D. (2008). The Wildland-Interface Fire Problem: A Consequence of the Fire Exclusion Paradigm. *Forest History Today, 20,* 20–26.

Country Fire Authority. (2016). Community Update: Lorne—Jamieson Track. Retrieved from https://www.facebook.com/cfavic/posts/10153898618164416

Crichton, D. (1999). The Risk Triangle. In J. Ingleton (Ed.), *Natural Disaster Management* (pp. 102–103). London: Tudor Rose.

Dahl, R. A. (1982). *Dilemmas of Pluralist Democracy: Autonomy Vs. Control.* New Haven: Yale University Press.

Dash, N., Peacock, W. G., & Morrow, B. H. (1997). And the Poor Get Poorer: A Neglected Black Community. In B. H. Morrow et al. (Eds.), *Hurricane Andrew: Ethnicity, Gender, and the Sociology of Disasters* (pp. 206–225). New York: Routledge.

Dworkin, R. (1977). *Taking Rights Seriously.* London: Duckworth.

Elliott, J. R., & Pais, J. (2006). Race, Class, and Hurricane Katrina: Social Differences in Human Responses to Disaster. *Social Science Research, 35*(2), 295–321.

Emergency Management Victoria. (2015). Fundamentals of Emergency Management (Class 1 Emergencies). Melbourne. Retrieved from https://www.emv.vic.gov.au/responsibilitiesmanaging-emergencies/the-fundamentals-of-emergency-management

Folger, R. (1987). Distributive and Procedural Justice in the Workplace. *Social Justice Research, 1*(2), 143–159.

Gill, A. M., & Stephens, S. L. (2009). Scientific and Social Challenges for the Management of Fire-Prone Wildland–Urban Interfaces. *Environmental Research Letters, 4,* 1–10.

Gonzalez-Mathiesen, C., & March, A. (2018). Establishing Design Principles for Wildfire Resilient Urban Planning. *Planning Practice & Research,* Taylor & Francis Journals, *33*(2), 97–119.

Habermas, J. (1984). *The Theory of Communicative Action V1* (T. McCarthy, Trans., Vol. I). London: Beacon Press.

Hillard, L. (2016). *Wye River: Brigade (Country Fire Authority),* Autumn 2016, 4–6.

Hillier, J. (2010). Strategic Navigation in an Ocean of Theoretical and Practice Complexity. In J. Hillier & P. Healey (Eds.), *The Ashgate Research Companion to Planning Theory: Conceptual Challenges for Spatial Planning* (pp. 447–480). Farnham: Ashgate.

Hopkins, L. D. (2001). *Urban Development: The Logic of Making Plans.* Washington: Island Press.

Leonard, J., Opie, K., Blanchi, R., Newnham, G., & Holland, M. (2016). *Wye River/Separation Creek Post-Bushfire Building Survey Findings.* Report to the Victorian Country Fire Authority, April 2016. Retrieved August 2, 2019, from https://publications.csiro.au/rpr/pub?pid=csiro:EP16924

Leventhal, G. S., Karuza, J., & Fry, W. R. (1980). Beyond Fairness: A Theory of Allocation Preferences. In G. Mikuta (Ed.), *Justice and Social Interaction* (pp. 165–218). New York: Springer-Verlag.

Lukasiewicz, A., & Baldwin, C. (2017). Voice, Power, and History: Ensuring Social Justice for All Stakeholders in Water Decision-Making. *Local Environment, 22*(9), 1042–1060.

March, A. (2003). Outward Views: Rights and Utility in Victorian State Planning. *Urban Policy and Research, 21*(3), 263–279.

March, A. (2012). *The Democratic Plan: Analysis and Diagnosis.* Farnham: Ashgate Publishing.

March, A., & Kornakova, M. (2017). Urban Planning for Disaster Recovery. In A. March & M. Kornakova (Eds.), *Urban Planning for Disaster Recovery.* London: Elsevier.

March, A., & Rijal, Y. (2015). Reducing Bushfire Risk by Planning and Design: A Professional Focus. *Planning, Practice and Research, 30*(1), 33–53.

Mattei, U. (2000). *Basic Principles of Property Law.* London: Greenwood Press.

May, P. J. (1991). Addressing Public Risks: Federal Earthquake Policy Design. *Journal of Policy Analysis and Management, 10*(2), 263–285. https://doi.org/10.2307/3325175

Meerow, S., Newell, J. P., & Stults, M. (2016). Defining Urban Resilience: A Review. *Landscape and Urban Planning, 147,* 38–49.

Mikkola, E. (2008). Forest Fire Impacts on Buildings. In J. Heras, C. Brebbia, D. Viegas, & V. Leone (Eds.), *Modelling, Monitoring and Management of Forest Fires.* Southampton: WIT Press.

Mileti, D. (1999). *Disasters by Design: A Reassessment of Natural Hazards in the United States.* Washington: Joseph Henry Press.

Neave, M., Rossiter, C., & Stone, M. (1994). *Sackville and Neave Property Law: Cases and Materials.* Sydney: Butterworths.

Padilla, A. B. M., & Vega-García, C. (2011). On the Comparative Importance of Fire Danger Rating Indices and Their Integration with Spatial and Temporal Variables for Predicting Daily Human-Caused Fire Occurrences in Spain. *International Journal of Wildland Fire, 20*(1), 46–58.

Price, O., & Bradstock, R. (2013). Landscape Scale Influences of Forest Area and Housing Density on House Loss in the 2009 Victorian Bushfires. *Plos One, 8*(8), e73421-1–e73421-6.

Ramsay, C., & Rudolph, L. (2003). *Landscape and Building Design for Bushfire Areas*. Melbourne: CSIRO Publishing.

Rawls, J. (1971). *A Theory of Justice*. Oxford: Oxford University Press.

SAMSHSA. (2017). Greater Impact: How Disasters Affect People of Low Socioeconomic Status. Retrieved August 2, 2019, from https://www.samhsa.gov/sites/default/files/programs_campaigns/dtac/srb-low-ses.pdf

Schlosberg, D., & Collins, L. B. (2014). From Environmental to Climate Justice: Climate Change and the Discourse of Environmental Justice. *Wiley Interdisciplinary Reviews: Climate Change*, 5(3), 359–374. https://doi.org/10.1002/wcc.275

Sen, A. (2000). *Development as Freedom*. New York: Anchor Books.

Standards Australia. (2018). *AS 3959–2018: Construction of Buildings in Bushfire-Prone Areas*. Sydney: Standards Australia. SAI Global Publishers.

Stanley, J. R. (2014). Climate Change: A New Challenge for Social Policy. In A. McClelland & P. Smyth (Eds.), *Social Policy in Australia: Understanding for Action* (3rd ed.). Australia: Oxford University Press.

Stanley, J. R. (2017). Equity in Recovery. In A. March & M. Kornakova (Eds.), *Urban Planning for Disaster Recovery* (pp. 31–46). London: Elsevier.

Steffen, W., Dean, A., Rice, M., & Mullins, G. (2019). The Angriest Summer. Retrieved August 2, 2019, from https://www.climatecouncil.org.au/resources/angriest-summer/

Stephenson, C. (2010). The Impacts, Losses and Benefits Sustained from Five Severe Bushfires in South-Eastern Australia Fire and Adaptive Management. Retrieved from https://www.ffm.vic.gov.au/__data/assets/pdf_file/0010/21115/Report-88-The-Impacts-Losses-and-Benefits-Sustained-from-Five-Severe-Bushfires-in-SE-Aust.pdf

Teague, B., McLeod, R., & Pascoe, S. (2009). Interim Report: Victorian Bushfires Royal Commission. *2009 Victorian Bushfires Royal Commission*. Retrieved August 2, 2019, from http://royalcommission.vic.gov.au/Commission-Reports/Interim-Report.html

Thibaut, J., & Walker, L. (1975). *Procedural Justice: A Psychological Analysis*. Hillsdale, NJ: Lawrence Erlbaum Associates.

UNISDR. (2015). *Sendai Framework for Disaster Risk Reduction 2015–2030*. United Nations International Strategy for Disaster Reduction, 37 pp.

Victoria, E. M. (2018). Bushfire Safety Policy Framework. Retrieved August 2, 2019, from https://www.emv.vic.gov.au/publications/bushfire-safety-policy-framework

Woodruff, P. (2019). Growing Toward Justice. In M. LeBar (Ed.), *Justice*. Oxford Scholarship Online: Oxford University Press.

CHAPTER 6

Climate Change Adaptation Litigation: A Pathway to Justice, but for Whom?

Tayanah O'Donnell

6.1 Introduction

Litigation offers options for societal change via the application of statute, legal principle(s), and the sometimes subsequent creation of precedent, precedent being a judicial judgment that usually requires other courts to follow its findings. Litigation offers, in theory, one way to achieve 'justice'. But litigation is not without flaws: it is incremental and often piecemeal as it advances legal arguments based on specific facts, and precedent is often built slowly. However, the idea of litigation can result in more change than the litigation hoped to resolve. A common symbol of justice, as it relates to law, is that of a woman wrapped in biblical robes, blindfolded, wielding a sword in one hand while deftly holding a set of scales in the other. Blind Lady Justice, 'Justitia', is said to date back to the ancient Greek goddess Themis (Knox, 2014) and wears her blindfold to ensure impartiality, fairness, and equality for all before 'the law' (Knox, 2014). The argument goes, that because she does not see who is in front

---

T. O'Donnell (✉)
Fenner School of Environment and Society, Australian National University,
Canberra, ACT, Australia
e-mail: tayanah.odonnell@anu.edu.au

© The Author(s) 2020
A. Lukasiewicz, C. Baldwin (eds.), *Natural Hazards and Disaster Justice*, https://doi.org/10.1007/978-981-15-0466-2_6

of her, she is impartial in her quest for justice before the law. Justitia cannot be dissuaded or have her own bias corrupt her.

For the purposes of this chapter, thinking of Justitia is a useful way to open our minds to the pervasiveness of the ideas that frame how justice might be achieved via litigation (noting that there are other means of achieving justice, including distributive and procedural justice). Justitia appears in courtrooms, in books, and in popular media. She is a recognised symbol in the Western World of law and of the pathways to justice apparently to be yielded from law and legal processes including that of litigation. These ideas belie the 'law' as a system of objectivity which, despite heady inroads being made in advancing climate change policy through legal actions, numerous feminist and critical lenses have proven to be untrue and an over-simplification (McCormick et al., 2018; O'Donnell & Talbot-Jones, 2018).

Scholars who view law as co-constituted with the social, cultural, and political make us aware of alternatives to the mainstream ideology of an objective law (Blomley, 2014; Braverman, Blomley, Delaney, & Kedar, 2014; Delaney, 2010, 2016), and prompt us to rethink tendencies to conceive the social world in binaries (Graham, 2011). This is particularly so when we think about nature, place, or 'the environment' as having its own autonomy or more recently, legal rights (O'Donnell, 2016a; O'Donnell & Talbot-Jones, 2018). There remains, however, considerable space for delving into deliberate uses of law as a mode of social ordering (see Robinson & Graham, 2018; von Benda-Beckman, von Benda-Beckman, & Griffiths, 2009) including with deliberate engagement with ideas of litigation as a pathway to justice.

Thinking through the power of litigation as a social ordering mechanism is particularly important when exploring regulatory responses to climate change. This is because litigation is one of the more visible aspects of law. Where litigation is pursued as a deliberate process aimed to engineer particular social outcomes or to right real or perceived wrongs, litigation can be considered as a process that seeks to enforce rights or obligations as they arise under the law. But is this justice, and if so, for whom? The answer to this question is that yes, litigation is usually perceived as a process for obtaining justice, whether it be for humans as individuals, institutions, and governments. While this human-centric approach is being utilised more and more frequently to push climate change mitigation and adaptation responses; a human-centric approach also loses sight of the materiality of the Earth, for whom mitigation and adaptation responses are purporting to benefit. But this is changing (O'Donnell & Talbot-Jones, 2018).

Moreover, climate change litigation is rapidly becoming a field in its own right, comprising many different types of legal cases spanning numerous jurisdictions, and each concerned in some way with mitigation, adaptation, or sometimes both (Peel & Osofsky, 2019; Setzer & Vanhala, 2019). Indeed, climate change litigation has been building in recent years (Setzer & Vanhala, 2019). As the impacts of climate change are felt more and more (Simlinger & Mayer, 2019), such litigation is often designed to test the law's capacity to respond to loss and damage where it has failed to do so adequately in the past, or where regulatory reform and redress is slow or ineffective to curb emissions, or where the enabling of climate adaptation is thwarted by the co-constitutivity of law and our spatial world as seen through a legal geography paradigm (O'Donnell, 2016a, 2017, 2019a).

Climate change itself manifests as a series of climate extremes. It is these extremes which institutions such as the law will confront as humans seek to engineer particular social outcomes, and/or right wrongs. Increasingly there are inroads being made by litigious actions expanding the reach of civil law actions for the impacts of these climate extremes, where we are starting to see obligations on governments to foresee harm caused by climate change becoming more salient (Klein, 2015). Indeed, "*the courts are being asked to play an increasing role in apportioning responsibility for loss and damage resulting from climate change*" (Marjanac, Patton, & Thornton, 2017, p. 616), and attribution science is a key aspect of this. Attribution science is an emerging scientific discipline that links anthropogenic climate change with weather events by determining the extent to which the damage from such events can be attributed to human induced climate change (Marjanac & Patton, 2018). Marjanac and Patton's review of the expedient growth of event attribution science, and the role it will have in determining causation of damage in climate change litigation comes at a pertinent time, as climate litigation grows, and cases that by previous standards were not about climate change, may in future be centred on the impacts of climate change. Critically, this includes land use planning litigation, which is usually about the construction of land use planning law or policy decision-making processes (O'Donnell & Gates, 2013), though claims in negligence law have been argued to be more fraught, though more compelling (Burkett, 2013; Craig, 2018). Such processes, at law, demand a 'reasonable' decision, and/or reasonable foreseeability: this reasonableness will increasingly require a relevant decision maker to consider climate change in their decision-making process in order for that decision to be considered reasonable. Moreover, justice granted by and through law for the damage caused by climate change remains a strong incentive for litigants who are keen to recover mounting losses to commence and pursue actions. Most of these losses are to property, for which damages or other types of remedy are sought. Alternatively, sometimes it is the prevention of these losses that is sought—a position which seeks to retain the status quo, if this can even be possible when we are living through the greatest extinction event of our times.

This chapter takes as a departure point Rosemary Lyster's conceptualisation of climate justice, which at once is both a moral and ethical imperative for all living things on Earth (2015), including the Earth herself. This chapter argues that law does and will continue to play a significant role for how society addresses both the already-here and imminently forthcoming impacts and consequences of global climate change—it is one part of fulfilling this moral and ethical imperative. Exactly how law-driven change manifests itself, using a litigation lens, is a story which is beginning to unfold around the world (Peel, Osofsky, & Forester, 2017; Setzer & Vanhala, 2019). As such, a deeper, material focus is warranted (Graham et al., 2018), and not just because of the enormous problems climate change impacts pose to social, political, environmental, economic, and legal systems alike (consider further Barnett et al., 2014). It is at this juncture that this chapter explores ongoing litigation about coastal defences in Byron Bay, New South Wales (NSW), Australia. This case study illustrates the complexity of motivations behind land use planning (see also Robb, Payne, Stocker, Middle, & Trosic, 2017), and related litigation. The litigation discussed here usefully illustrates the types of future litigation that might be expected as sea levels rise and coastal climate change impacts are realised. The discussion that follows also illustrates the juxtaposition of public law mechanisms and private property interests, and the challenges of managing competing interests in dynamic material environments such as coastlines (Pollard, Spencer, & Brooks, 2018).

6.2 The Legislative Context

The national context for coastal management in Australia has long been fraught. For the past 25 years there have been ongoing iterations of law reform, progressive and reactive local government policy, political masquerading, state government leadership that appears as quickly as it disappears, and patchy national leadership (Harvey, 2016; Harvey & Clarke, 2019; O'Donnell, 2019a; O'Donnell, Smith, & Connor, 2019). There are numerous competing interests that challenge integrated coastal management (Frolich, Jacobson, Fidelman, & Smith, 2018) and climate change adaptation, including a policy context that has been preoccupied with private property rights and associated discourses. I discuss extensively a decade of New South Wales coastal law reform that is illustrative of the legal and political geographies for coastal management and governance, in O'Donnell et al. (2019). That paper traces in detail the legislative and policy landscape in the state jurisdiction of New South Wales, including the pertinent law reform measures (see also O'Donnell, 2016b) that arose in 2016 and came into force in 2018. It is, however, necessary to offer a brief synopsis of that law reform here, in order to contextualise the discussion of the Byron litigation.

Ten years ago, the principal statutory framework for the New South Wales coast was the *Coastal Protection Act 1979* (NSW), which operates to encourage the development and implementation of coastal management (Measham et al., 2011). This legislation has now been replaced by the Coastal Management Act 2016 (NSW), which is now the parent legislative framework specifically for coastal management. This Act's objective (section 3) is to 'manage the coastal environment of New South Wales in a manner consistent with the principles of ecologically sustainable development for the social, cultural and economic wellbeing of the people of the state' with 13 sub clauses. One of these contains a specific reference to climate change, as follows: section 3(f) 'to mitigate current and future risks from coastal hazards, taking into account the effects of climate change'. In addition to the *Coastal Management Act* are other legislative frameworks, including the *Environmental Planning and Assessment Act 1979* (NSW) (EPA Act). The EPA Act remains the principal statutory instrument that governs strategic planning and development assessment in New South Wales. It also contains various public interest requirements for development assessment and approvals. In January 2011 the *Coastal Protection and Other Legislation Amendment Act 2010* (NSW) came into force. This legislation amended the *Coastal Protection Act 1979* (NSW) by, among other things, allowing for the construction of Emergency Protection Works (EPWs) by a coastal landowner without development approval or immediate oversight by the Coastal Panel. Dissatisfaction arose over this arrangement, with EPWs unable to satisfy landowners who wanted more relaxed rules on their placement, and local authorities who wanted more stringent rules. Repeals of state-wide sea level rise planning benchmarks created still more uncertainty, bearing in mind that the implementation of unified coastal management plans across the state was ad hoc at best (Lyster, 2015; O'Donnell, 2019a; Productivity Commission, 2012; Thom, 2012).

On 13 November 2014, then New South Wales Minister for Planning Rob Stokes announced the coastal management law reforms that have since resulted in the Coastal Management Act (Stokes, 2014). The Act provides modern and comprehensive legislative framework, as argued by Harvey and Clarke (2019). The framework is also comprised of the Act, a new State Environmental Planning Policy (SEPP) and a new Coastal Management Manual. The Coastal Management Manual details the mandatory requirements and other elements of the preparation of Coastal Management Programmes (CMPS), which have replaced the previous coastal zone management plans, and work in concert with the Coastal Manual in regulating coastal development and shoreline management. Under this new framework, the Coastal Council continues to hold statutory independence, and advises the Minister on coastal management challenges and opportunities (Mitchell, 2016).

Despite this, the Coastal Panel was, and remains, an important mechanism for independent coastal management, in that the panel (now called the 'Coastal Council', as per the *Coastal Management Act 2016* (NSW)) has statutory authority to review coastal protection works across New South Wales. It also has specific powers to advise the relevant minister on coastal management matters (specified in the Act). When the new Coastal Management Act was awaiting proclamation (and therefore not in force), the Coastal Panel/Council was known as a 'Transitional Coastal Panel' (between 2016 and 2018). It was therefore the Transitional Coastal Panel that played an important role in the most recent Byron litigation judgments, Ralph Lauren Pty Ltd. v New South Wales Transitional Coastal Panel; Stewartville Pty Ltd. v New South Wales Transitional Coastal Panel; and Robert Watson v New South Wales Transitional Coastal Panel [2018] NSWLEC 207, handed down by his Honour Justice Brian Preston on 21 December 2018, discussed later in this chapter.

6.3 Byron Bay

Attempts to use the New South Wales land use planning system to embed climate change adaptation responses into local government policy have frequently been met with '*vociferous resistance*' from residents and property developers (Cronshaw, 2012, p. 1; Piper, 2012), and particularly from owners of beachfront private property. The consequence of this has been to show how local and state politics intervene into law making (Abel et al., 2011; Gibbs, 2016; Harvey & Clarke, 2019; Harvey & Smithers,

2018; O'Donnell, 2019a). What I will refer to hereinafter as the 'Byron litigation' is illustrative of these overlapping tensions and controversies. The Byron litigation, in its totality, has been broiling for two decades between the Byron Shire Council, which has attempted to implement its policy of coastal management, and beachfront property owners, who simultaneously want their properties protected from coastally related weather events by the local government, but do not want these protections to impede upon their private property rights. I narrow the scope of discussion to a particular focus on three substantive pieces of litigation: 2009–2010 litigation in the New South Wales Land and Environment Court; 2014–2016 litigation in the New South Wales Supreme Court; and some of the 2017–2019 litigation in the New South Wales Land and Environment Court.

6.4 The Byron Litigation

Belongil Beach is a precarious coastal settlement in the far north of New South Wales, Australia. On the beachfront are several houses; some are protected by coastal engineering options, such as sandbag walls or sand dune re-nourishment. These houses are backed by an estuary (Fig. 6.1).

It is in this material context that I discuss at length the 2009 litigation in O'Donnell (2016a). The 2009–2010 Vaughan litigation is comprised of several related litigious actions, following a particularly devastating coastal storm in May of that year. The storm caused the loss of a sandbag wall on Belongil Beach and behind that wall, the loss of several metres of private property along that strip of coast. Belongil Beach home owners, the Vaughans, attempted to respond to this by placing concrete boulders on the land outside of their property. The Vaughans argued that these concrete boulders were necessary for further protection of the now damaged sand dune and embankment. The Council, however, argued that the placement of concrete boulders contravened several policies, including their planned retreat policy. The Council successfully sought and obtained an injunction in the New South Wales Land and Environment Court which prevented the placement of the boulders. The Vaughans then brought an action in the same court and related to the same damage, seeking a hearing and judgment on the interpretation of development consent as to the quality and maintenance of the failed sandbag wall. This litigation settled in February 2010 with the parties agreeing that the Byron Shire Council owed an obligation under the development consent to repair and

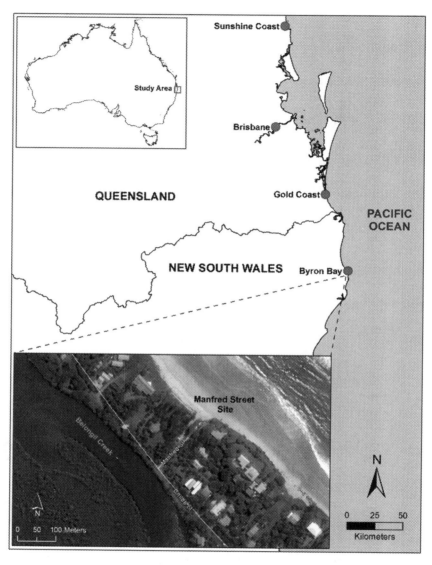

Fig. 6.1 The location of the Vaughan property, left of Manfred St Belongil Beach, Byron Bay

maintain the sandbag wall. Litigation was then commenced in the New South Wales Supreme Court, alleging negligence by the Council. All of this was occurring while the legislative and policy reform outlined above was underway.

The building of protective sea walls at Belongil has long been controversial, though the popularity of sea walls as an adaptation measure is increasing around the world. There are three primary reasons for their controversial status at Belongil. First, the building of walls as a coastal protection measure serves a very small subset of human communities, often to the great detriment of ecosystems, dynamic coastal marine systems, and to the public amenity of coastlines. Second, they prioritise the protection of private property that in turn can create a hierarchy of adaptation responses. Reliance on litigation to protect private property may enable a path dependency with its attendant maladaptive adaptation outcomes (Macintosh, 2013; Barnett et al., 2014; O'Donnell, 2019a). Third, building sea walls can result in other members of local communities feeling disenfranchised, as local governments are seen to be acting in the interests of this small subset of the local community (i.e. beachfront property owners), at great expense to the rest (Author Unknown, 2016; Lollback, 2016).

The legality of the now long standing original rock walls along Belongil Beach is questionable. By 2016, the Byron community, including local residents and rates payers that were not beachfront owners, were outraged by the ongoing litigation and costs to the Council, which were perceived to cater to only a select few properties (Lollback, 2016). The outcomes of the litigation in 2016 were categorised as a win for *'wealthy residents'* (Lovejoy, 2016), while other residents have retaliated with claims that ordinary rights also need protecting (Cornwall, 2018). The Transitional Coastal Panel (so named until the Act changing their name to Coastal Council was proclaimed) argued before the court that the original rock walls had originated without consent (see par [72], Ralph Lauren Pty Ltd. v New South Wales Transitional Coastal Panel; Stewartville Pty Ltd. v New South Wales Transitional Coastal Panel; Robert Watson v New South Wales Transitional Coastal Panel [2018] NSWLEC 207).

Because the wall has existed for so long, and its removal would now cause further damage to public and private lands, it is difficult and impractical for the wall to be removed. Nor can it be left to deteriorate. This signals a demise for a planned retreat in Belongil (Cornwall, 2019), notwithstanding the reported ongoing feelings of *'community betrayal'* documented and expressed in community groups (see, e.g., Feliu, 2016), or known detriment to the public beach (Lovejoy, 2018).

6.5 Justice for the Coast

The judgment in Ralph Lauren Pty Ltd. v New South Wales Transitional Coastal Panel; Stewartville Pty Ltd. v New South Wales Transitional Coastal Panel; Robert Watson v New South Wales Transitional Coastal Panel [2018] NSWLEC 207 ('Ralph Lauren') is the latest instalment in the long running Byron litigation in both the New South Wales Supreme Court and the New South Wales Land and Environment Court. This case involved the Transitional Coastal Panel (now the Coastal Council) and an application to the Transitional Panel from three Belongil residents for development consent to carry out works to repair deteriorating sea walls. The Panel ultimately refused the consent on several grounds. Notably, these included public interest considerations, including the restrictions of public access to the beach, and the detrimental impact to the coastal foreshore if the sea walls were expanded (see Ralph Lauren, par 4). The Panel's reasoning followed a two-fold legal argument. Firstly, approving the works would not only result in an uncoordinated approach to coastal erosion responses at Belongil Beach, but would confer a private benefit at the expense of the broader public. Secondly, if not contrary to the public interest, the consent should only operate for a limited period of time, because a longer permit would result in a continued occupation of public land.

In adjudicating on an appeal to a decision made by the Panel, Chief Justice Brian Preston of the Land and Environment Court found against granting consent to the landowners to construct the rock walls, on several grounds. Firstly, the 'public interest' preconditions in s55 M1(a) (i) of the Coastal Management Act 1979 (NSW) were not satisfied, in that the walls would be built on public land, and therefore impede public access and amenity or use of the public beach. The public's ability to escape coastal events, or to access the beach through public thoroughfares, or to otherwise enjoy the public space of Belongil Beach would be impaired by the proposed works. This is because the size of the planned walls was substantial, and in his Honour's words, '*significant*'. For each of the properties, the wall would be between 40 and 60 metres in total length, several metres wide, and several metres high. Some of the buried components of the walls would be exposed at low tide. The '*apparent height*' (Ralph Lauren, par 118–121) of the proposed walls would be more severe during low tide or during coastal erosion events. Further, Preston CJ found that:

> *The existing sea walls are not lawful. No development consent has been sought or obtained for the carrying out of the existing sea walls on the beach in front of each of the landowners' properties. By law, the sea walls should not exist on the beach.* (Ralph Lauren, par 127)

The court did not need to consider in detail the time limited consent, because it found in favour of the Panel on the public interest grounds, as against the interpretation of then-applicable law and land use planning policy (Coastal Protection Act 1979 (NSW), s55 M; Byron Shire Council Local Environment Plan 1988, cl. 88(3)).

This finding is significant for a tranche of litigation that has been ongoing for over two decades in this particular geographical location. Though his Honour did not centre the material environment in this judgment (rather, he focussed on the public interest of human accessibility, amenity, and safety, among other factors), his judgment recognises that the dynamism of the coast is relevant to the placement or otherwise of protective works, as foreshadowed in s55 M of the *Coastal Protection Act 1979* (NSW). Importantly for the coast, these public interest considerations are replicated in the new *Coastal Management Act 2016* (NSW). Preston's decision in Ralph Lauren hints at broader opportunities to reimagine law, not as separate from its material environment, but as central to it (see also Bartel, McFarland, & Hearfield, 2014; Graham, Davies, & Godden, 2018; O'Donnell, 2016a; O'Donnell & Talbot-Jones, 2018). However, justice in law, and for the environment, still has a way to go.

Peel et al. (2017) argue that climate change litigation will see a broadening as climate change impacts cut across various lawsuits in a myriad of ways (consider also Peel and Osofsky, 2019). Climate litigation in all its forms will continue to poke, prod, and provoke societal change, with rapid advancement in attribution science and general duties of care rapidly coming together. This will have specific consequences for how and when climate science is used as expert material in supporting legal arguments. Critically, there is now an emphasis on causative issues including with respect to proving cause-and-effect for climate change impacts, with specific consequences for the reasonable foreseeability test under negligence law (Marjanac & Patton, 2018). The Byron Bay litigation, though consumed as it is primarily on specific interpretations of land use planning law, offers a contextualised case study through which we can see how climate change litigation might unfold in future claims.

6.6 Conclusion

It is difficult to implement policies that balance competing interests over short-term and long-term time horizons. This is especially so in light of anticipated climate change impacts. Balancing these priorities—especially through a litigation pathway—necessitates that there be a winner and a loser, due to the adversarial nature of litigation. In human-centred outcomes, achieved through institutions such as law, it is the dynamism of the material environment and the changing scale of this dynamism caused by climate change that is too often neglected. This will have profound conceptual and practical implications for the achievement of climate change justice. In the Byron litigation, the most recent judicial decision has focussed on the importance of the material environment, access to, and amenity of the land use planning legal system. There is growing momentum in the recognition of legal personhood for nature (O'Donnell & Talbot-Jones, 2018). Moreover, public interest considerations ought to shift our focus to achieving justice for all things on Earth, not just for the wealthiest or the loudest litigants (O'Donnell & Gates, 2013).

Though more recent inroads have seen an active reorientation towards stewardship of the Earth, there remains a critical observation that is not addressed: legality is as much about power as is it is about the letter of the law (O'Donnell, 2019b). Using litigation as a tool to achieve 'justice' is likely to continue, but this approach is not without risk of adverse or ad hoc regulatory or social outcomes that are often to the detriment of the natural environment. Moreover, this is a risk most prevalent where individuals seek to use public law to protect individual rights.

References

Abel, N., Gorddard, R., Harman, B., Leitch, A., Langridge, J., Ryan, A., & Hevenga, S. (2011). Sea Level Rise, Coastal Development and Planned Retreat: Analytical Framework, Governance Principles and an Australian Case Study. *Environmental Science and Policy, 14*(3), 279–288.

Author Unknown. (2016, August 23). Shifting Sands for Councils on Responsibility for Past Projects. *The Australian*. Retrieved May 11, 2019, from https://www.theaustralian.com.au/subscribe/news/1/?sourceCode=TAWEB_WRE170_a_GGL&dest=https%3A%2F%2Fwww.theaustralian.com.au%2Fopinion%2Fshifting-sands-for-councils-on-responsibility-for-past-projects%2Fnews-story%2F74910f502ed45a20de2eb5cc4ec307fd&memtype=anonymous&mode=premium&v21suffix=49-b

Barnett, J., Graham, S., Mortreux, C., Fincher, R., Waters, E., & Hurlimann, A. (2014). A Local Coastal Adaptation Pathway. *Nature Climate Change, 4*(12), 1103–1108.

Bartel, R., McFarland, P., & Hearfield, C. (2014). Taking a De-binarised Envirosocial Approach to Reconciling the Environment vs Economy Debate: Lessons from Climate Change Litigation for Planning in NSW, Australia. *Town Planning Review, 85*(1), 67–95.

Blomley, N. (2014). Disentangling Law: The Practice of Bracketing. *Annual Review of Law and Social Science, 10*(1), 133.

Braverman, I., Blomley, N., Delaney, D., & Kedar, A. (Eds.). (2014). *The Expanding Spaces of Law: A Timely Legal Geography.* Stanford Law Books.

Burkett, M. (2013). Duty and Breach in an Era of Uncertainty: Local Government Liability for Failure to Adapt to Climate Change. *George Mason Law Review, 20*(3), 775–802.

Byron Shire Council Local Environment Plan 1988

Coastal Management Act 2016 (NSW).

Coastal Protection Act 1979 (NSW).

Coastal Protection and Other Legislation Amendment Act 2010 (NSW).

Cornwall, D. (2018, December 26). Frustrated Byron Bay Beach Residents Hit Another Legal Wall. *The Australian.* Retrieved June 6, 2019, from https://www.theaustralian.com.au/nation/politics/frustrated-byron-bay-beach-residents-hit-another-legal-wall/news-story/bf84167dec4be3f0ae8fa1abab540d06

Cornwall, D. (2019, March 14). Coastal Retreat Option Buried in State's Shifting Sands. *The Australian.* Retrieved June 6, 2019, from https://www.theaustralian.com.au/nation/politics/coastal-retreat-option-buried-in-states-shifting-sands/news-story/8a00b5a8778a82e22cc78aaa3bf978e3

Craig, R. K. (2018). *California Climate Change Lawsuits: Can the Courts Help with Sea-Level Rise, and Who Knew What When?* Utah Law Faculty Scholarship. Retrieved from https://dc.law.utah.edu/scholarship/120

Cronshaw, D. (2012, March 18). Mayor Draws Line Over Sea Level Rise Attacks. *Newcastle Herald.* Retrieved April 10, 2019, from https://www.theherald.com.au/story/114517/mayor-draws-line-over-sea-rise-attacks/

Delaney, D. (2010). *The Spatial, the Legal and the Pragmatics of World-Making: Nomospheric Investigations.* Routledge.

Delaney, D. (2016). Legal Geography III: New Worlds, New Convergences. *Progress in Human Geography, 41*(5), 667–675.

Environmental Planning and Assessment Act 1979 (NSW).

Feliu, L. (2016, April 8). Byron Councillors 'Betray Community' on Rock Walls. *Echo NetDaily.* Retrieved June 11, 2019, from https://www.echo.net.au/2016/04/byron-councillors-betray-community-on-rock-walls/

Frolich, M. F., Jacobson, C., Fidelman, P., & Smith, T. F. (2018). The Relationship Between Adaptive Management of Social-Ecological Systems and Law: A Systematic Review. *Ecology and Society, 23*(2). Retrieved March 11, 2019, from https://www.ecologyandsociety.org/vol23/iss2/art23/

Gibbs, M. T. (2016). Why Is Coastal Retreat So Hard to Implement? Understanding the Political Risk of Coastal Adaptation Pathways. *Ocean and Coastal Management, 130*, 107–114.

Graham, N. (2011). *Lawscape: Property, Environment, Law*. Routledge.

Graham, N., Davies, M., & Godden, L. (2018). Broadening Law's Context: Materiality in Socio-Legal Research. *Griffith Law Review, 26*(4), 480–510.

Harvey, N. (2016). The Combination Lock-in Effect Blocking Integrated Coastal Zone Management in Australia: The Role of Governance and Politics. In A. Chircop, S. Coffen-Smout, & M. McConnell (Eds.), *Ocean Yearbook 30* (pp. 1–31). Brill Neijoff.

Harvey, N., & Clarke, B. (2019). 21st Century Reform in Australian Coastal Policy and Legislation. *Marine Policy, 103*, 27–32.

Harvey, N., & Smithers, S. (2018). How Close to the Coast? Incorporating Coastal Expertise into Decision-Making on Residential Development in Australia. *Ocean and Coastal Management., 157*, 237–247.

Klein, J. (2015). Potential Liability of Governments for Failure to Prepare for Climate Change. SSRN Electronic Journal (Sabin Center for Climate Change Law, Columbia Law School, August 2015). Retrieved from http://web.law.columbia.edu/sites/default/files/microsites/climate-change/klein_-_liability_of_governments_for_failure_to_prepare_for_climate_change.pdf

Knox, B. A. (2014). The Visual Rhetoric of Lady Justice: Understanding Jurisprudence Through 'Metonymic Tokens'. *Inquiries Journal, 6*(5), 1.

Lollback, R. (2016, April 6). 'Urgent' Call to Stop Controversial Rock Wall. *The Northern Star*. Retrieved May 2, 2019, from https://www.northernstar.com.au/news/urgent-call-to-stop-controversial-rock-wall/2987885/

Lovejoy, H. (2016, August 24). How a Handful of Wealthy Belongil Residents Won the Right to Keep Their Seawalls. *Echo NetDaily*. Retrieved June 6, 2019, from https://www.echo.net.au/2016/08/handful-wealthy-belongil-residents-won-right-keep-seawalls/

Lovejoy, H. (2018, April 18). No Assurance Beach Will Remain if Belongil Rock Walls Approved. *Echo NetDaily*. Retrieved May 2, 2019, from https://www.echo.net.au/2018/04/no-assurance-beach-will-remain-belongil-rock-walls-approved/

Lyster, R. (2015). *Climate Justice and Disaster Law*. Cambridge: Cambridge University Press.

Macintosh, A. (2013). Coastal Climate Hazards and Urban Planning: How Planning Responses Can Lead to Maladaptation. *Mitigation and Adaptation Strategies for Global Change, 18*(7), 1035.

Marjanac, S., & Patton, L. (2018). Extreme Weather Event Attribution Science and Climate Change Litigation: An Essential Step in the Causal Chain? *Journal of Energy & Natural Resources Law, 36*(3), 265–298.

Marjanac, S., Patton, L., & Thornton, J. (2017). Acts of God, Human Influence and Litigation. *Nature Geoscience, 10*, 616–619.

McCormick, S., Glicksman, R. L., Simmens, S. J., Paddock, L., Kim, D., & Whited, B. (2018). Strategies in and Outcomes of Climate Change Litigation in the United States. *Nature Climate Change, 8*, 829–833.

Measham, T. G., Preston, B. L., Smith, T. F., Brooke, C., Gorddard, R., Withycombe, G., & Morrison, C. (2011). Adapting to Climate Change Through Local Municipal Planning: Barriers and Challenges. *Mitigation and Adaptation Strategies for Global Change, 16*(8), 889.

Mitchell, S. (2016, May 4). New South Wales, *Parliamentary Debates*, Legislative Council.

O'Donnell, E., & Talbot-Jones, J. (2018). Creating Legal Rights for Rivers: Lessons from Australia, New Zealand, and India. *Ecology and Society, 23*(1). https://doi.org/10.5751/ES-09854-230107

O'Donnell, T. (2016a). Legal Geography and Coastal Climate Change Adaptation: The Vaughan Litigation. *Geographical Research, 54*(3), 301–312.

O'Donnell, T. (2016b). New South Wales Coastal Law Reform: A Preliminary Assessment. *Australian Environment Review, 31*(3), 178–181.

O'Donnell, T. (2017). Climate Adaptation on the Australian East Coast. In M. H. Bruun, P. J. Cockburn, B. S. Risager, & M. Thorup (Eds.), *Contested Property Claims: What Disagreement Tells Us about Ownership* (1st ed., pp. 151–165). Oxon: Routledge.

O'Donnell, T. (2019a). Coastal Management and the Political-Legal Geographies of Climate Change Adaptation in Australia. *Ocean and Coastal Management, 175*, 127–135.

O'Donnell, T. (2019b). Contrasting Land Use Policies for Climate Change Adaptation: A Case Study of Political and Geo-legal Realities for Australian Coastal Locations. *Land Use Policy, 88*, 104145.

O'Donnell, T., & Gates, L. (2013). Getting the Balance Right: A Renewed Need for the Public Interest Test in Addressing Coastal Climate Change and Sea Level Rise. *Environment and Planning Law Journal, 30*(3), 220–235.

O'Donnell, T., Smith, T. F., & Connor, S. (2019). Property Rights and Land Use Planning on the Australian Coast. In E. C. H. Keskitalo & B. L. Preston (Eds.), *Research Handbook on Climate Change Adaptation Policy* (pp. 403–416). Cheltenham: Edward Elgar.

Peel, J., Osofsky, H., & Forester, A. (2017). Shaping the 'Next Generation' of Climate Change Litigation in Australia. *Melbourne University Law Review, 41*, 793–844.

Peel, J., & Osofsky, H. M. (2019). Litigation as a Climate Regulatory Tool. In C. Voigt (Ed.), *International Judicial Practice on the Environment: Questions of Legitimacy* (pp. 311–336). Cambridge: Cambridge University Press.

Piper, G. (2012, March 15). New South Wales, *Parliamentary Debates*, Legislative Assembly, 9792.

Pollard, J. A., Spencer, T., & Brooks, S. M. (2018). The Interactive Relationship Between Coastal Erosion and Flood Risk. *Progress in Physical Geography, 20*(2), 1–12.

Productivity Commission. (2012, July 10). *Barriers to Effective Climate Change Adaptation, Transcript of Proceedings* (Craik, W., Coppel, J., & Byron, N.), Sydney. Retrieved 2019, from https://www.pc.gov.au/inquiries/completed/climate-change-adaptation/public-hearings/20120710-climate-change-adaptation-sydney-transcript.pdf

Ralph Lauren Pty Ltd v New South Wales Transitional Coastal Panel; Stewartville Pty Ltd v New South Wales Transitional Coastal Panel; and Robert Watson v New South Wales Transitional Coastal Panel [2018] NSWLEC 207.

Robb, A., Payne, M., Stocker, L., Middle, G., & Trosic, A. (2017). Development Control and Vulnerable Coastal Lands: Examples of Australian Practice. *Urban Policy and Research.* https://doi.org/10.1080/08111146.2018.1489791

Robinson, D. F., & Graham, N. (2018). Legal Pluralisms, Justice and Spatial Conflicts: New Directions in Legal Geography. *Geographical Journal, 184,* 3–7. https://doi.org/10.1111/geoj.12247

Setzer, J., & Vanhala, L. (2019). Climate Change Litigation: A Review of Research on Courts and Litigants in Climate Government. *Wiley Interdisciplinary Reviews: Climate Change, 10*(3), 1–19. https://doi.org/10.1002/wcc.580

Siddle v NSW Transitional Coastal Panel [2018] NSWLEC 1383.

Simlinger, F., & Mayer, B. (2019). Legal Responses to Climate Change Induced Loss and Damage. In R. Mechler, L. M. Bouwer, T. Schinko, S. Surminski, & J. Linnerooth-Bayer (Eds.), *Loss and Damage from Climate Change: Concepts, Methods and Policy Options* (pp. 179–204). Cham, Switzerland: Springer.

Stokes, R. (2014, November 13). *Address to the NSW Coastal Conference 2014.* Retrieved from http://www.environment.nsw.gov.au/resources/coasts/coastreforms-minister-speech-13nov14.pdf

Thom, B. (2012). Climate Change, Coastal Hazards and the Public Trust Doctrine. *Macquarie Journal of International and Comparative Environmental Law, 8*(2), 21–41.

von Benda-Beckman, F., von Benda-Beckman, K., & Griffiths, A. (2009). *Spatializing Law: An Anthropological Geography of Law in Society.* Ashgate.

CHAPTER 7

Looking to Courts of Law for Disaster Justice

Michael Eburn

7.1 Introduction

The concept of disaster justice, identified at the start of this book, suggests that disasters are not natural phenomena but the intersection of a hazard with human vulnerability. That vulnerability, in turn, is the product of social decisions about where and how we live. Some of those choices people make for themselves but some are made by others. People may choose to live in an area prone to wildfire but their vulnerability to wildfire can be increased by the choices of others. Neighbours who fail to maintain their properties (Eburn & Cary, 2017) or who light fires in catastrophic weather conditions, electricity supply companies that fail to maintain poles and wires, local governments that limit a person's ability to adequately prepare their property in order to make it defensible (Timbs v Shoalhaven City Council [2004] NSWCA 81) all contribute to a person's vulnerability to risk. Where people suffer a loss, they may look to the courts of law to deliver a form of disaster justice.

M. Eburn (✉)
ANU College of Law, Australian National University,
Canberra, ACT, Australia
e-mail: michael.eburn@anu.edu.au

© The Author(s) 2020
A. Lukasiewicz, C. Baldwin (eds.), *Natural Hazards and Disaster Justice*, https://doi.org/10.1007/978-981-15-0466-2_7

This chapter will consider the legal consequences after some notable disasters and the implications of a desire to look to law for 'justice'. It is argued that the state should consider alternatives to adversarial legal process in order to deliver justice to all those involved in disasters.

7.2 Poor Outcomes Are Not Necessarily Disasters

Not every tragedy is a disaster. A house fire that causes the loss of a family home or a motor vehicle accident that leads to death is a tragedy and a disaster for those involved but for the community it is not a disaster. For the emergency services—fire brigades, ambulance services, police and rescue services—these events are part of their normal day to day operations and well within their training, resources and capacity.

At the other extreme, a catastrophic disaster is one that overwhelms the resources of the state and threatens the ongoing effective operation of government (EMA, 2010; Gissing, Eburn, & McAneney, 2018). Emergency Management Australia says (EMA, 2018, p. 5):

> A catastrophic disaster is what is beyond our current arrangements, thinking, experience and imagination (i.e. that has overwhelmed our technical, non-technical and social systems and resources, and has degraded or disabled governance structures and strategic and operational decision making functions).

It is arguable that Australia has never suffered a catastrophic disaster, the most likely candidate being Cyclone Tracey that destroyed Darwin, the capital of the Northern Territory on Christmas Eve 1974. Effective local government was destroyed and an official from the Commonwealth government flew in to take command of the response to the disaster.

Between the incidents that constitute the 'business as usual' of the emergency services and the catastrophic disasters are other large-scale events described variously as emergencies or disasters. For the purposes of this chapter these are referred to as disasters. Australia has had its share of these events. Examples include:

- The 1983 Ash Wednesday fires that burned across Victoria and South Australia claiming 75 lives and 3700 buildings;
- The 2003 bushfires that burned into Canberra, the national capital, claiming four lives and 500 homes;

- The 2009 Black Saturday fires in Victoria that claimed 173 lives and over 2000 homes;
- The 2011 flooding in Victoria and Queensland coupled with the impact of Cyclone Yasi, also in Queensland.

Without attempting to develop a comprehensive definition of a disaster what distinguishes these events from day to day emergencies or incidents is the long time period of events and the inability of the emergency services to bring the hazard under control. No amount of firefighting effort was going to extinguish the fires of 1983, 2003 or 2009 at least not at the time of their largest impact. During these events those in the path of the hazard—the fire or flood—must fend for themselves as the event overwhelms the capacity of the emergency services to contain the hazard or even respond to calls for assistance. The notion of disaster, at least as discussed in this chapter, implies a large event that overwhelms the resources normally available for emergency response, even if it does not reach the scale of a catastrophic disaster and threaten the very operation of government.

7.3 Looking to the Courts of Law to Deliver Disaster Justice

Following any event, whether it is a disaster or a simple accident, the most common 'cause of action' will be a claim alleging that the defendant negligently caused the event and that this, in turn, caused the person making the claim to suffer losses. Those losses may be for damage to property or damages for personal injuries or death.

In a study on why people sue their medical practitioners, Vincent, Phillps, and Young (1994) said:

> *Four main themes emerged from the analysis of reasons for litigation: concern with standards of care—both patients and relatives wanted to prevent similar incidents in the future; the need for an explanation—to know how the injury happened and why; compensation—for actual losses, pain and suffering or to provide care in the future for an injured person; and accountability—a belief that the staff or organisation should have to account for their actions.*

For the sake of the discussion we assume that people suing after a disaster have similar concerns and objectives (see also Reilis, 2009).

The reality is, as the discussion below demonstrates, that many of those objectives will not be met. The only remedy that a court can give is an order that the defendant pay to the claimant the value of their losses doing the best to put a dollar value on many things that simply cannot be measured in money terms—loss of sentimental personal possessions, loss of enjoyment of life, loss of security, and so on, cannot be compensated by money but a judge must do his or her best to put a dollar value on those things. Even with those limitations, of the four factors identified by Vincent, Phillps and Young, 'compensation' is effectively the only remedy offered by the civil courts.

Where people want to 'prevent similar incidents in the future' civil litigation may not be an effective tool. In nearly all cases it will be an insurer, not the defendant—in legal jargon the tortfeasor—who pays. A finding of liability may mean there are higher premiums in the future but that is not certain. The amounts awarded may be large from the perspective of the claimant but may not represent significant amounts to a large business or their insurer. There may be little actual consequences for the tortfeasor. Other factors such as better regulation, community pressure or the criminal law may have a greater deterrent effect than the threat of a civil suit for damages.

With respect to obtaining an explanation and accountability that may or may not happen. Whilst preparing a court case parties must put their version of events in their court pleadings and make relevant documents and witnesses available to the other side. A person who has suffered injury or losses may get some sense of the other side's explanation from those court ordered documents and processes, but they may not. The process itself is mitigated by legal counsel and is limited to dealing with the matters raised by the legal proceedings which may exclude much that a person who wanted to an 'explanation' would consider relevant.

A plaintiff—that is the person making the claim—may want their 'day in court' in order to put the chief executive of the electrical company in the witness box, to hear him or her explain why they did not spend more money on maintenance and to have the chance to explain how their actions affected the community, but that may not happen. Most cases settle out of court. The defendant's insurer makes an offer that counsel for the plaintiffs advises them to take. There are penalties in terms of cost orders that may be made if a person rejects an offer but then fails to achieve a significantly better result in court. Every case has risks, so even the most determined plaintiff has to be made to understand that there is a risk that they will lose with massive personal costs. The courts have systems in place to encourage parties to reach their own compromise. The entire system is set up to settle the claim rather than to have the matter resolved in court.

The reality of this situation can be seen in the litigation that followed the Black Saturday bushfires of February 2009. On that day bushfires burned in Victoria claiming 173 lives and over 2000 homes. The 2009 Victorian Bushfires Royal Commission reported on 'the 15 most damaging, or potentially damaging, fires that burned on 7 February' (Victoria, 2010, Summary, p. 2). The Kilmore-East fire caused the greatest loss of life and property, claiming 119 lives and 1242 homes (Victoria, 2010, Volume 1, p. 70). The Royal Commission found that:

> *The fire started after the conductor between poles 38 and 39 failed and the live conductor came into contact with a cable stay supporting pole 38. This contact caused arcing that ignited vegetation near the base of pole 38. An electrical fault was recorded at 11:45.*
>
> *The conductor failed as a result of fatigue on the conductor strands very close to where a helical termination was fitted to the conductor at pole 39 ... The fatigue of the conductor strands was partly caused by the helical termination being incorrectly seated ... causing stress to the conductor ... The conductor was probably 43 years old.*

A line inspection carried out in February 2008 had failed to identify the incorrectly seated helical fitting.

A finding that the fire was caused by an incorrectly placed conductor that was not detected during a line inspection would give potential plaintiffs confidence that this fire was due to the negligence of the authority that owned or inspected the pole and wire. There was a class action over this fire. (A class action allows a single representative plaintiff to sue on behalf of the class of people affected by an event.) The class action resulted in a settlement of $500 million—at the time the largest single verdict in an Australian court (ABC, 2014). This is not however an example of the defendant's being 'held to account' or required to give a satisfactory explanation of their conduct.

> *SP AusNet said the settlement was without admission of liability by the company and other parties.*
>
> *It said it believed it was likely to win the lawsuit, but the uncertainty, complexity and scale of the case lead them and other parties to settle.*

> "SP AusNet extends its deepest sympathy to those who suffered losses in the Black Saturday bushfires," the statement said.
>
> "SP AusNet's position has been, and continues to be, that the conductor which broke and which initiated the fire was damaged by lightning, compromising its fail-safety design in a manner which was undetectable at the time."
>
> "It is a tragedy that the conductor eventually failed on one of the worst days imaginable."
>
> "SP AusNet's management of its network did not involve any negligence."

More can be learned from the published judgements of the Victorian Supreme Court. In Matthews v SPI Electricity & Ors (Ruling No 3) [2011] VSC 399, the judge Mr Justice Forrest had to consider whether or not an associate judge could read various documents from, or provided to, the 2009 Victorian Bushfires Royal Commission to assist her to determine what documents were and were not relevant to the court proceedings. The 'state parties' (the Country Fire Authority, the Department of Sustainability and Environment and the State of Victoria (representing Victoria Police)) agreed to the Associate Judge seeing some documents but only if their agreement was not considered "*evidence in the proceeding [and] does not constitute an admission that the information is accurate, relevant or admissible*" (Matthews v SPI Electricity & Ors (Ruling No 3) [2011] VSC 399, [11]). The Royal Commission had cost the government many millions of dollars, but the government and its agencies were not prepared to admit that the conclusions reached by the Commission were 'accurate'.

Before the case could proceed in court it settled. A class action is brought by a representative plaintiff on behalf of people who suffered losses in the same event, as such it can affect the legal rights of people who are not actively engaged in the litigation. To ensure that those people's rights are properly considered, settlement of a class action must be approved by the Court. In Matthews v AusNet Electricity Services Pty Ltd. & Ors [2014] VSC 663 Justice Osborn of the Supreme Court of Victoria approved the settlement.

The terms of the settlement were that $60 million was for the claimant's legal fees. Three-eighths of the remaining balance was to meet claims for personal injury or death and five-eighths for property damage and economic losses. It was estimated that personal injury claimants would receive about 70% of their total claims, and those claiming for property and economic losses would obtain about 33% of their total claims.

In approving the claim, the Court has to consider whether or not the settlement is a fair compromise for everyone in the class not just fair as between the representative plaintiff and the defendants. In this case the court accepted that this was a fair compromise. In coming to that conclusion the court noted that the case, despite taking some 208 days of hearing, 100 witnesses, over 10,000 documents as evidence, had not yet reached a stage where the trial judge had determined whether or not any of the defendants had been negligent. The issue of negligence had been contested so the plaintiff faced a risk that if they did not settle, and the case proceeded, the plaintiff could lose and get nothing. Justice Osborn said (Matthews v AusNet Electricity Services Pty Ltd. & Ors [2014] VSC 663, [292]):

> *Once it is understood that each of the claims made by the plaintiff faced some real risk of complete failure, it is difficult to conclude otherwise than that the proposed settlement is within the range of reasonable compromise.*

With this settlement the plaintiff (and the members of the class action) would receive significant damages whereas, if the matter went to trial they may get nothing, or much less. The settlement would bring the action to an end whilst continuing with the litigation may take a number of years (Matthews v AusNet Electricity Services Pty Ltd. & Ors [2014] VSC 663, [309]) and would be complicated by likely appeals, possibly to the High Court of Australia ([310]). It was, on the other hand, expected that with the settlement, individual claims will be resolved within 18 months ([330]).

If we recall the reasons for suing identified by Vincent, Phillps and Young (Vincent et al., 1994), above, they were a desire to prevent similar incidents in the future, to obtain an explanation of how the injury, or in this case the fire, happened, to obtain compensation and to hold the organisations to account for their actions. At the end of this litigation there may have been a $500 million payout but there is still no definitive conclusion as to what caused the electrical assets to start the fire so no explanation as to cause and no ruling that any of the defendants were negligent and even the compensation was doubtful. People were expected to be compensated at about two-thirds of their total losses and payments were subject to them then proving their losses to the lawyers that had previously represented them. The settlement was compounded by delay (ABC, 2016a, 2016b).

Settlement is not unusual. Other claims over Black Saturday settled (Rowe v AusNet Electricity Services Pty Ltd. & Ors [2015] VSC 232; Mercieca v SPI Electricity Pty Ltd. & Ors [2012] VSC 204) as did litigation over Ash Wednesday fires (Woodend Water v. Hyan 1990 VIC LEXIS 1106). (There were many other claims from Ash Wednesday where there were interlocutory judgements but no final judgment has been found, suggesting that these cases, too, settled—May v Electricity Trust of South Australia (1993) SASC 4149; Seas Sapfor Forests v Electricity Trust of South Australia (1993) SASC 4004 and (1996) SASC 5718; and SA Electricity v Union Insurance [1997] SASC 6241. In Ballantyne v Electricity Trust of South Australia (1993) SASC 4275 the substantive claim was resolved by commercial arbitration but that does appear to have constituted a finding that the Electricity Trust was liable for two Ash Wednesday fires). Litigation against the Australian Capital Territory for negligence in response to the 2003 fires also settled (Andrews & Doherty, 2012; Byrne, 2012). Litigation against New South Wales over the same fires resulted in a verdict in favour of the defendants and no compensation for the plaintiffs. The first case that has seen a court say to an electricity supply authority—you are responsible for this bushfire, you were negligent, it was your fault, was determined in 2019 (Daniel Herridge & Ors v Electricity Networks Corporation T/As Western Power [No 4] [2019] WASC 94).

Apart from a near universal failure to reach a definitive conclusion on fault, these cases come at an incredible expense. In two Black Saturday class actions, the plaintiff's lawyers received $20 million (Rowe v AusNet Electricity Services Pty Ltd. & Ors [2015] VSC 232) and $60 million (Matthews v AusNet Electricity Services Pty Ltd. & Ors [2014] VSC 663). Those costs orders were the amount from the settlement to be used to pay the plaintiff's or claimant's costs. Those costs do not include the legal costs incurred by the defendant companies and the State of Victoria. It follows that these reported costs of $80 million represent a very small percentage of the actual costs involved in both the litigation.

There is also a time factor to consider. The fires that burned into Canberra began on 8 February 2003 and reached Canberra on 18 February 2003. The decision of the trial judge, finding no liability on the part of the State of New South Wales was handed down on 17 December 2012, two months short of the fire's tenth anniversary (Electro Optic Systems Pty Ltd. and West v NSW [2012] ACTSC 184). The Black Saturday bushfires occurred on 7 February 2009; as noted above, claimants were still waiting for compensation in 2016 (ABC 2016a, 2016b). In

2018 the litigation from the 2011 Brisbane floods is ongoing. 'Justice delayed is justice denied' (Burstyner & Sourdin, 2014).

Looking to the courts for justice is slow, expensive, unlikely to reach any firm conclusion as to fault, will leave those affected under-compensated and is unlikely to meet any of the non-monetary objectives that plaintiffs identify when asked why they are pursuing legal action (Reilis, 2009; Vincent et al., 1994).

7.3.1 Post-Event Inquiries

Apart from litigation there is usually some, or many, post-event inquiries into disasters. Following the 2003 Canberra fires there was the McLeod Review into the Operational Response (McLeod, 2003) and a coroner's inquest and inquiry (Doogan, 2006) as well as the lengthy litigation. Following the 2009 Black Saturday fires an extensive Royal Commission investigated the cause of and response to those fires (Victoria, 2010). A 2010 senate inquiry identified that, since 1939, "*there have been at least 18 major bushfire inquiries in Australia*" (Senate, 2010). Other research identified 55 post-natural disaster (not just bushfire) reviews and inquiries since 2009, producing 1336 recommendations (Cole, Dovers, Gough, & Eburn, 2018).

Notwithstanding their commitment to a no-blame process, these inquiries, particularly when led by lawyers, often fall back on adversarial or court-like process where witnesses are examined and cross-examined by lawyers representing interested parties. Whilst not all inquiries adopt adversarial practices (see, for example, Keelty, 2011 and Ferguson, 2016), where they do, witnesses are called to answer questions put to them by counsel. It is counsel that makes submissions to the Commissioners or coroner as to what inferences and findings are open on the evidence that has been led. Where there is ambiguity it is up to the tribunal to determine where responsibility lies and more importantly, who is responsible for ensuring that the same circumstances do not arise in the future. At the end of the process the Commission or the coroner makes findings and may make recommendations that may or may not be adopted by stakeholders. Looking to these types of inquiries for 'justice' may also be fraught and lead to disappointment (Eburn & Dovers, 2015, 2017).

Post-event inquiries are bound by their terms of reference to inquire into the causes of and response to a disaster. They are not authorised or able to make findings as to fault, determine the value of non-economic losses nor can they order anyone to 'make good' (either by compensation or remediation) the losses caused by a disaster.

7.4 A New Approach: Adopting Restorative Practices

Restorative justice is an increasing feature of the criminal justice system in Australia and around the world. The aim of restorative justice is to deal with the harms caused by crime by allowing victims to face offenders and explain the impact of their behaviour and to give offenders the chance to account for their behaviour and to reach agreement on how they may make good (to the extent that is possible) the damage caused by their behaviour (Johnstone, 2003).

Whilst responding to fires and floods is not an issue of criminal law (even if the fire is caused by arson), there are similar issues. The event causes massive harms in loss of property, life and a sense of security. People are traumatised by the losses and the impact on their lives. They may feel that the state agencies failed them in the preparation and planning and the response. Responders are also members of the affected communities so volunteer emergency service personnel who are responding on behalf of their community may feel let down if their actions are not valued or honoured by the community or if they feel that their agency did not properly support them or allow them to take actions that they thought were required. Equally staff from agencies such as land management agencies and local government authorities also live in the affected communities and can be both victims as well as receiving blame and criticism for their actions. Just as crimes cause harm that needs to be repaired, so do significant natural hazard events.

Restorative practices take many forms. What is common, and what makes the practices restorative, is the conscious decision to put those affected, rather than the event itself, at the centre of the process. Restorative practices ensure that those affected are involved in designing the process and taking responsibilities for implementing the learning (BNHCRC, 2018).

7.4.1 Sharing Responsibility

The 2009 Victorian Bushfires Royal Commission took the view that effective disaster management involved a shared responsibility (Victoria, 2010, 4, Final Report Vol II, [9.1]). The notion of shared responsibility has also been adopted in the National Strategy for Disaster Resilience (COAG, 2011). However merely stating that there is a shared responsibility does not define who is responsible for what. The Royal Commission may make recommendations on the balance of shared responsibility and governments may adopt policies to encourage individuals, communities, business and the non-government sector to accept responsibility but neither the inquiry or legislative process allows for communities or individuals to expressly 'accept' or own the responsibility that others think does, or should, belong to them. Adopting restorative practices would be (McCold, 2000):

> *a cooperation soliciting approach that encourages a process of acceptance of responsibility, facing the consequences by making amends to individuals and relationships, and encouraging re-acceptance into the community.*

Braithwaite and Strang (2001, p. 10) say that restorative justice

> *does not subordinate emotion to dispassionate justice, as in the blindfolded icon of justice balancing the scales. Nor does restorative justice subordinate emotion to rational bureaucratic routines. Space is created in civil society for the free expression of emotions, however irrational they may seem ... where there is moral ambiguity over right and wrong in a conflict, [the authors] ... prefer allowing the ambiguity to stand rather than coerced allocation of responsibility. Speaking to participants in advance of a conference inviting them to own as much responsibility as they feel able to volunteer can be enough to trigger a virtuous circle of owning responsibility instead of a vicious circle of denial and blaming the other.*

McLaughlin et al. (2003, p. 7) say that restorative justice restores the role of community:

> *A corollary of the critique of the overwhelming power held by state bureaucracies and agencies is restorative justice's view of the role of 'community' in formal legal processes. Marginalization or exclusion of community from the processes of determining outcomes ... explains the failure of statutory legal practices .... Restoring the historical place of community is central to the founding propositions of most restorative justice proponents.*

The move to the development of resilient communities with shared responsibility for planning and preparing for inevitable hazard events should be extended to shared responsibility for reviewing those events and considering what lessons can be drawn for future action. Restorative practices would increase community involvement in reviewing a fire or other hazard and owning with government, their responsibility for decisions that contributed to the impact of the event.

7.4.2 The Use of Restorative Justice Beyond Criminal Law Is Not Unique and Is Growing

The use of restorative principles is expanding beyond traditional criminal law. Restorative principles lie behind attempts at peacebuilding (Llewellyn & Philpott, 2014) and post-conflict inquiries in South Africa, Rwanda, Northern Ireland (Daly, 2004) and East Timor (Braithwaite, Charlesworth, & Soares, 2012). Restorative justice practices have also been suggested as an appropriate response for industrial disasters (Cooper, 2008). Nova Scotia, Canada, is currently holding its first restorative public inquiry. According to the Nova Scotia Home for Colored Children Restorative Inquiry website (n.d.; emphasis added):

> *A traditional public inquiry is focused on uncovering facts and laying blame. We need to understand not only what happened, but why it happened and why it matters for all Nova Scotians. We need a process shaped by restorative principles that does no further harm, includes all voices and seeks to build healthy and just relationships so we can learn and act together.*
>
> *The Restorative Inquiry will look at the past with a focus on future solutions: not only preventing any more harm, but making meaningful changes that will help us treat each other more justly and equitably in the future.*

Two of the goals of the inquiry are to 'Support collective ownership, shared responsibility and collaborative decision-making' and to learn 'what happened, what matters about what happened for the future, who was affected and how, and the contexts, causes and effects of what happened' (Nova Scotia Home for Colored Children Restorative Inquiry, 2015, p. 6). The Inquiry process (ibid., p. 9) involves three elements of work related to its overall objectives:

- Relationship building
- Learning and understanding
- Planning and action.

These goals and work elements would be fitting in an inquiry into a complex event such as the 2003 Canberra fires or the Black Saturday fires of February 2009.

7.5 Compensation

As noted, traditional inquires cannot allocate compensation, only a court can do that. However, an administrative compensation scheme, without the need to prove fault would not require litigation, would be 'partly restorative' (McCold, 2000, p. 401) and may go some way to locating 'justice'. Failure to address issues of compensation will fail to redress losses caused by natural hazard events and will leave aggrieved persons to seek compensation before the courts with the return to adversarial proceedings and huge costs.

Following the 9/11 terrorist attacks on the United States, the US government introduced a no-fault compensation scheme that was open to everybody who was killed or injured. By agreeing to take part in the no-fault scheme people waived their right to sue. An attorney was appointed to manage the scheme and to assess the value of each person's claim based on the normal legal principles for the quantification of damages (Feinberg, 2005). Without going into the details of the scheme or how damages were calculated, it is apparent that the scheme was focused on the needs of those who had suffered loss and removed the need to spend time and millions of dollars to prove legal culpability or blame.

Given that the insurance companies that pay out to cover insured losses following Australia's bushfires have, through the collection of premiums, wise investment and reinsurance, the means to meet their obligations then it must be time to consider some sort of no-fault catastrophic compensation scheme and divert the money that is currently being spent on legal costs to improving community resilience.

One solution may be to use the settlement scheme adopted in Victoria as a model for future disasters. All insurers who are at risk, for example, those that offer household insurance, insurance for critical infrastructure and government insurers or self-insurers could contribute to a fund that can be used to pay out compensation following a significant disaster that meets a prescribed threshold in terms of losses or is a declared disaster for the purposes of the scheme. In a scheme such as this, insurers may pay out in circumstances where liability could not be established but the cost and time savings would be significant and could justify that exposure.

Government involvement in such schemes is not unprecedented. Already the Australian government provides some support following a disaster. The Government pays a disaster recovery allowance to provide some income support for people whose employment is affected by a disaster (Social Security Act 1991 (Cth) ss 1061KA to 1061KE). There is also a one-off Government Disaster Recovery Payment for people adversely affected by a disaster (ss 1061K-1061PAAE). These social security payments are emergency and short-term support rather than compensation for all the losses, including non-economic losses that a person may suffer in a disaster.

In the event of a declared terrorism incident, the Commonwealth Government, through the Australian Reinsurance Pool Corporation will compensate an insurance company for amounts paid to an insured for damages caused by the terrorism incident (Terrorism Insurance Act 2003 (Cth) ss 6–8, 35 and 37). A similar scheme could, if necessary, operate to provide cover for declared disaster events. Such a scheme would recognise that, by definition, a disaster occurs when the resources of the state that are normally sufficient to respond to a hazard are overwhelmed and the losses reflect not just the hazard but the choices that have been made on how resources have been allocated to disaster mitigation and response. It might also help communities to rebuild by creating a sense of shared responsibility for the losses. In 2011, following the Queensland floods, Prime Minister Julia Gillard said (Commonwealth, 2011):

> *As Australians, we stick together. United in mateship. United in our shared desire to help those in need. This [Tax Laws Amendment (Temporary Flood Reconstruction Levy)] bill formalises that desire to help. Beyond the legal and budgetary language, it simply says this:*
> *You won't be alone. We will get through this together. We won't let go.*

Providing compensation, not just emergency relief, for those affected by disasters, without having to spend five to ten years in complex litigation at costs that run into the hundreds of millions of dollars, would carry the same message.

It is beyond the scope of this chapter to consider the constitutional, legal and financial issues that would be involved in establishing such a scheme. The issue is raised here because failure to address issues of funding recovery will also fail to address necessary issues for those seeking justice. Although the need to consider these issues is raised, they will require further research.

7.6 Conclusion

After a disaster many people look to courts and court-like institutions to deliver 'justice'. Cost, delay and contested facts all conspire to frustrate the desire for justice. Where a tribunal, whether a court or inquiry, is left to make sense of the evidence and deliver a verdict or recommendation it is inevitable that some stakeholders will feel that the tribunal failed to deliver the outcome that they wanted or saw as essential to justice. Looking to the courts (in all their forms) for 'justice' is a fraught and risky exercise.

This chapter suggested that a better way to look for justice would be to adopt 'restorative practice' inquiries—that is inquiries that focus on the consequences of an event, its impact upon people and relationships and which seeks to understand why an event was important for those involved and how those involved can make sense, learn from and take responsibility for identifying and implementing learning from the event. Essential to the restorative process is consideration of making good financial losses without the need to blame. A disaster is in part the product of society's choices. An event overwhelms a community and its resources because of choices that have been made about how resources are allocated between competing demands. If a person suffers a loss because of those community decisions, it is fitting that the community makes good those losses as part of the restorative process.

Having identified the drawbacks in looking to the courts, and having suggested an alternative approach, the chapter ends. It is not possible to set out what the ideal 'restorative inquiry' would look like as it depends on the event and the needs of those affected. Part of the process should be to allow those affected to determine how they want an inquiry or review to work and what the important factors that need to be considered are. Trying to define 'the way' to do it would be the antithesis of a restorative practice. Equally it is beyond the scope of this chapter to identify an ideal catastrophic event compensation fund. This chapter suggests the need for universal disaster insurance and so opens the door for further research to consider how such a scheme could be structured and funded.

References

Andrews, L., & Doherty, M. (2012, September 20). Fire Litigation Ends for ACT. *Canberra Times* (Online). Retrieved January 11, 2019, from https://www.canberratimes.com.au/national/act/fire-litigation-ends-for-act-20120920-268d2.html

Australian Broadcasting Corporation (ABC). (2014, July 15). Black Saturday Bushfire Survivors Secure $500 Million in Australia's Largest Class Action Payout. *ABC News* (Online). Retrieved January 11, 2019, from https://www.abc.net.au/news/2014-07-15/black-saturday-bushfire-survivors-secure-record-payout/5597062

Australian Broadcasting Corporation (ABC). (2016a, February 7). Black Saturday Bushfire Compensation Delay from Maurice Blackburn Lawyers Worries Victims. *ABC News* (Online). Retrieved January 11, 2019, from https://www.abc.net.au/news/2016-02-06/black-saturday-compensation-delayed-by-maurice-blackburn/7145792

Australian Broadcasting Corporation (ABC). (2016b, August 5). Heartbreak and Frustration as Black Saturday Victims Wait for Payout. *Background Briefing* (Online). Retrieved January 11, 2019, from https://www.abc.net.au/radionational/programs/backgroundbriefing/black-saturday-survivors-critical-as-class-action-drags-on/7692988

Braithwaite, J., Charlesworth, H., & Soares, A. (2012). *Networked Governance of Freedom and Tyranny: Peace in Timor-Leste*. ANU E Press.

Braithwaite, J., & Strang, H. (2001). Introduction: Restorative Justice and Civil Society. In H. Strang & J. Braithwaite (Eds.), *Restorative Justice and Civil Society* (pp. 1–13). Cambridge University Press.

Burstyner, N., & Sourdin, T. (2014). Justice Delayed Is Justice Denied. *Victoria University Law and Justice Journal*, 4(1), 46–60.

Bushfire and Natural Hazards Cooperative Research Centre (BNHCRC). (2018, June 13). *Restorative Inquiries and Natural Disasters*. A Report on the Symposium Held at the University of Newcastle (2019).

Byrne, E. (2012, September 20). Bushfire Litigation Ends for ACT Govt. *ABC News* (Online). Retrieved January 11, 2019, from https://www.abc.net.au/news/2012-09-20/fire-litigation-over-for-act-government/4272298

Cole, L., Dovers, S., Gough, M., & Eburn, M. (2018). Can Major Post-Event Inquiries and Reviews Contribute to Lessons Management? *Australian Journal of Emergency Management*, 33, 234–239.

Commonwealth. (2011, February 10). Parliamentary Debates, House of Representatives, 381 (Julia Gillard).

Cooper, D. (2008). Thinking About Justice 'Outside of the Box': Could Restorative Justice Practices Create Justice for Victims of International Disasters? *New England Law Review*, 42(4), 693–700.

Council of Australian Governments (COAG). (2011). National Strategy for Disaster Resilience, Commonwealth of Australia. Retrieved January 14, 2019, from https://knowledge.aidr.org.au/media/2153/nationalstrategyfordisasterresilience.pdf

Daly, K. (2004). Restorative Justice: The Real Story. In D. Roche (Ed.), *Restorative Justice* (pp. 85–109). Ashgate.

Doogan, M. (2006). The Canberra Firestorm: Inquests and Inquiry into Four Deaths and Four Fires Between 8 and 18 January 2003 (ACT Coroners Court).

Eburn, M., & Cary, C. (2017). You Own the Fuel, but Who Owns the Fire? *International Journal of Wildland Fire, 26*(12), 999–1008.

Eburn, M., & Dovers, S. (2015). Learning Lessons from Disasters: Alternatives to Royal Commissions and Other Quasi-Judicial Inquiries. *Australian Journal of Public Administration, 74*(4), 495–508.

Eburn, M., & Dovers, S. (2017). Reviewing High-Risk and High-Consequence Decisions: Finding a Safer Way. *Australian Journal of Emergency Management, 32*(4), 26–29.

Emergency Management Australia (EMA). (2010). National Catastrophic Disaster Plan (NATCATDISPLAN), Commonwealth of Australia.

Emergency Management Australia (EMA). (2018). Australian Disaster Preparedness Framework, Commonwealth of Australia.

Feinberg, K. R. (2005). *What Is Life Worth? The Unprecedented Effort to Compensate the Victims of 9/11*. Public Affairs, New York.

Ferguson, E. (2016). *Reframing Rural Fire Management*. Report of the Special Inquiry into the January 2016 Waroona Fire, Government of Western Australia.

Gissing, A., Eburn, M., & McAneney, J. (2018). *Planning and Capability Requirements for Catastrophic and Cascading Events*, Bushfire and Natural Hazards CRC. Retrieved January 15, 2019, from https://www.bnhcrc.com.au/publications/biblio/bnh-4795

Johnstone, G. (2003). *Restorative Justice: Ideas, Values, Debates* (pp. 1–2). Willan Publishing.

Keelty, M. J. (2011). *A Shared Responsibility*. The Report of the Perth Hills Bushfire February 2011 Review, Government of Western Australia.

Llewellyn, J., & Philpott, D. (2014). Restorative Justice and Reconciliation: Twin Frameworks for Peacebuilding. In J. J. Llewellyn & D. Philpott (Eds.), *Restorative Justice, Reconciliation and Peacebuilding* (pp. 14–36). Oxford University Press.

McCold, P. (2000). Toward a Holistic Vision of Restorative Juvenile Justice: A Reply to the Maximalist Model. *Contemporary Justice Review, 3*(4), 357–414.

McLaughlin, E., Fergusson, R., Hughes, G., & Westmarland, L. (2003). Introduction: Justice in the Round—Contextualising Restorative Justice. In E. McLaughlin, R. Fergusson, G. Hughes, & L. Westmarland (Eds.), *Restorative Justice: Critical Issues* (pp. 1–17). SAGE.

McLeod, R. (2003). Inquiry into the Operational Response to the January 2003 Bushfires in the ACT (ACT Government).

Reilis, T. (2009). *Perceptions in Litigation and Mediation*. Cambridge University Press, Chapter 2.

Senate. (2010). Select Committee on Agricultural and Related Industries: The Incidence and Severity of Bushfires across Australia, Commonwealth of Australia.

The Nova Scotia Home for Colored Children Restorative Inquiry. (2015). *Mandate & Terms of Reference (Nova Scotia)*. Retrieved January 14, 2019, from https://restorativeinquiry.ca/sites/default/files/u4/nshcc-restorative-inquiry-report.pdf

The Nova Scotia Home for Colored Children Restorative Inquiry. (n.d.). Retrieved January 14, 2019, from https://restorativeinquiry.ca/

Victoria 2009 Victorian Bushfires Royal Commission. (2010). Final Report, Parliament of Victoria. Retrieved from http://royalcommission.vic.gov.au/Commission-Reports/Final-Report.html

Vincent, C., Phillips, A., & Young, M. (1994). Why Do People Sue Doctors? A Study of Patients and Relatives Taking Legal Action. *The Lancet, 343*(8913), 1609–1613.

CASES

Ballantyne v Electricity Trust of South Australia (1993) SASC 4275.
Daniel Herridge & Ors v Electricity Networks Corporation T/As Western Power [No 4] [2019] WASC 94.
Electro Optic Systems Pty Ltd and West v NSW [2012] ACTSC 184.
Matthews v AusNet Electricity Services Pty Ltd & Ors [2014] VSC 663.
Matthews v SPI Electricity & Ors (Ruling No 3) [2011] VSC 399.
May v Electricity Trust of South Australia (1993) SASC 4149.
Mercieca v SPI Electricity Pty Ltd & Ors [2012] VSC 204.
Rowe v AusNet Electricity Services Pty Ltd & Ors [2015] VSC 232.
SA Electricity v Union Insurance [1997] SASC 6241.
Seas Sapfor Forests v Electricity Trust of South Australia (1993) SASC 4004.
Seas Sapfor Forests v Electricity Trust of South Australia (1996) SASC 5718.
Timbs v Shoalhaven City Council [2004] NSWCA 81.
Woodend Water v. Hyan 1990 VIC LEXIS 1106.

LEGISLATION

Terrorism Insurance Act 2003 (Cth).

CHAPTER 8

How to Be Fair in Prioritizing Support in the Aftermath of Disasters: Pakistan's Housing Reconstruction Challenges Following the 2010 Flood Disaster

Steven Schilizzi and Muhammad Masood Azeem

8.1 Introduction

In 2010, Pakistan was hit by the worst flood in the century. A total of 1.6 million homes were damaged and 0.9 million utterly destroyed or washed away (World Bank, 2010), meaning that more than 5 million people were left homeless. This is equivalent to a quarter of the Australian population. The task of rebuilding was, and always is, a formidable one for cash-constrained developing countries like Pakistan. This holds both for individual households who have lost their home and for the government

S. Schilizzi (✉)
UWA School of Agriculture and Environment, The University of Western Australia, Perth, WA, Australia
e-mail: steven.schilizzi@uwa.edu.au

M. M. Azeem
Centre for Agribusiness, UNE Business School, University of New England, Armidale, NSW, Australia

© The Author(s) 2020
A. Lukasiewicz, C. Baldwin (eds.), *Natural Hazards and Disaster Justice*, https://doi.org/10.1007/978-981-15-0466-2_8

that might wish to help them. Because resources and funding are limited, government must set priorities and often make hard choices. If help is to be offered to the victims, what is to be rebuilt first and what later? Who is to be helped first and whom later? Should the amount of help differ between victims and if so, how?

These questions raise challenges given that households' ability to cope with post-disaster consequences differ widely, this being particularly true of their capacity to rebuild. In developing countries, a large proportion of them are too poor to be able to afford reconstruction in a reasonable amount of time, if at all. During that time, they will typically have nowhere to live other than in temporary tents, where hygiene conditions are known to be less than ideal and disease outbursts are a constant risk. Of course, poorer households usually live in cheaper houses than those of richer households. Yet, the share of income needed for everyday survival is higher for the poor, so that they are less able to afford rebuilding the same home than the rich are. But should they be rebuilding the same house? If government support is forthcoming, should its policy aim to rebuild what existed before the flood? Or should it build better and safer?

As the 2010 Pakistan flood tragically demonstrated, the cheaper houses used by the poor are much more likely to be swept away than the better-built houses used by the richer people. Cheap basic houses may for example not withstand any flood greater than a one-in-three-year event, whereas good-quality houses may withstand one-in-thirty-year events. As a result, rebuilding the same house means reproducing the greater vulnerability suffered by the poor; worse, reproducing the inequalities in vulnerability to future disasters. From an ethical standpoint, such a policy goal can appear highly questionable. The government, or any other helping agency, thus faces an ethical dilemma. If it should not just replace what existed before, then what should it do? Given a limited budget, providing more help here means providing less there. At the same time, ideally any organization spending large amounts of scarce funds will want to make sure that money buys best value. Value for money set against prioritizing who gets how much defines a typical equity-efficiency trade-off.

As if to make matters worse, in the case of Pakistan, without government support no household category can afford the best quality house, least vulnerable to future floods. And if equity or justice is a concern, equity criteria or norms come in many colours, and decision makers, invoking different ones, will typically disagree on which solution is most equitable or just. This results in a thorny problem for any organization looking for the best use of its funds in a post-disaster social reconstruction program.

The upshot here is this: given a limited budget, a disaster-relief reconstruction program must choose between supporting greater overall protection and resilience while favouring the richer households and supporting a greater number of poorer households while achieving less overall protection. This is because, per dollar spent, better houses achieve better resilience in terms of flood protection. This also means that government can either support more cheap houses for the poor and achieve less total protection, or support more safe houses for the poor but reach a smaller number of them. In some sense, this is also a quality vs. quantity trade-off: quality of houses vs. number of houses for the poor. Underlying this problem is the interplay of three factors: households' individual budgets, the government's budget, and the costs of house reconstruction. In more general terms: the victims' ability to cope, the helping entity's limited resources, and the cost of reduced vulnerability to future disasters.

The above considerations lead to the following question: how can a disaster relief program use limited resources in a way that will most reduce vulnerability to future disasters while at the same time differentiating its actions to cater for those people most in need? Given that the poorest people are both the most vulnerable and the least able to cope with future disasters, any relief program will, one way or another, face a question of justice: a one-size-fits-all will not work. As further discussed in the concluding chapter of this book, the justice at stake here comes in the form of distributional equity, one of the three forms examined in Lukasiewicz and Baldwin (2017): how can one share or distribute limited resources between different people whose heterogeneous means do not match their heterogeneous needs? Not only is value-for-money pitted against distributional equity, but different equity norms, upheld by competing stakeholders, are also pitted against each other. This chapter investigates how one may methodically address this problem and what challenges await solutions that can satisfy stakeholders with different and potentially conflicting priorities. Background to drivers of post-disaster reconstruction policies.

The frequency of occurrence and magnitude of disasters have increased significantly over the last decade in both developed and developing countries. Housing is usually the element that is the most severely affected by the disaster in developing countries (Ahmed, 2011). As per Barakat (2003), the impact of disaster on built environment is 20 times higher in developing countries as compared to that in developed countries. Disasters such as the 2010 earthquake and tsunami in Chile left about 800,000 homeless (Romero & Albornoz, 2016) and the 2009 earthquake in Italy left about 66,000 people displaced (Mazza, Chiara Pino, Peretti, Scolta, & Mazzarelli, 2014). Likewise, more than 1.6 million housing units were affected by the 2010 flood in Pakistan. The number of completely destroyed houses was estimated at 0.9 million and the total cost of long-term housing reconstruction was estimated at approximately US$ 2 billion (World Bank, 2010). As per Patel and Hastak (2013), the delay in provision of housing leads to mental stress for survivors. Given the tragic impact of disaster on housing and communities, it is not surprising that the affected communities tend to prioritize housing reconstruction as the most urgent need (Delaney & Shrader, 2000). Housing reconstruction is therefore the key initiative of governments and international agencies (Ahmed, 2011).

In theory, the objective of the post-disaster reconstruction is to 'build back better', meaning that post-disaster buildings should be better buildings than the pre-disaster times in their ability to withstand future disasters and safeguard lives (Vahanvati & Mulligan, 2017). For instance, in the context of Pakistan, The World Bank recommended 'build back better' for areas prone to flood disaster and 'build back better/safer' for areas prone to flood and seismic risk, on the grounds that these are economically efficient solutions (World Bank, 2010). In practice, implementing post-disaster housing reconstruction is highly complex, despite the well-meaning intentions of local governments and international agencies. The reasons for the complexity involved in the post-disaster reconstruction are many, including weak institutional arrangements in developing countries and the lack of transparency, accountability, and coordination among donors, governmental agencies, NGOs, communities, and other stakeholders.

Each of the various stakeholders involved in housing reconstructions may have different values, norms, and interests (Opdyke, Lepropre, Javernick-Will, & Koschmann, 2017). For instance, donors and government agencies may have preferences for 'building back better' keeping in view the technical and financial factors (Kim & Olshansky, 2014; Mannakkara & Wilkinson, 2014; Oliver-Smith, 1990). However, building post-disaster houses is not merely a technical affair, as it also relates to the social and cultural preferences of affected communities and their perception of fairness and transparency in housing allocation (Ophiyandri, Amaratunga, Pathirage, & Keraminiyage, 2013). According to Mazza et al. (2014), disaster-affected communities feel discontented if they are not adequately informed about the choices to be implemented in housing reconstruction. Rahmayati (2016) report that the post-disaster housing reconstruction projects are insensitive and ignorant of the concerns of disaster-stricken communities. The same author finds that the idea of 'building back better' was not successful in mobilizing the Acehnese communities in Indonesia and instead proposed 'building the same as before is better than building back better' (p. 357).

The governments and donor-agencies acknowledge that post-disaster housing reconstruction should include both efficiency and fairness (equity) aspects because equity in the distribution of houses is at least as important as the efficiency in the design of houses (World Bank, 2010). However, in practice most of the housing reconstruction approaches are primarily governed by the efficiency concerns (Duyne Barenstein, 2015; Hayles, 2010; Oliver-Smith, 1996; Tran, 2015). The majority of the earlier studies on housing reconstruction also deal mainly with the issue of maximizing the economic efficiency of investment through engineering approaches and building materials used in the construction (Oliver-Smith, 1990; Powell, 2013). Equity concerns in housing reconstruction usually receive little attention from both policy makers and researchers. In this study, we incorporate the missing piece in the policy design by examining the post-disaster housing reconstruction by how distributive equity and economic efficiency, as well as the trade-offs between them, affect the chosen policy.

8.2 Analysing the Problem

To address the dilemmas that the reconstruction program raises, we shall use, to fix ideas, actual data for Pakistan's 2010 flood disaster, based on a large database describing about 90,000 households in the Punjab region of Pakistan (Azeem, 2016). To keep things simple and clear, we have rounded off many numbers, but kept realistic orders of magnitude. We start by considering a government budget of 100 million Rs. (rupees) targeting a specific region in rural Punjab, with 1000 households. An equal distribution of this budget to all households, irrespective of wealth or type of house, would yield 100,000 Rs. each. This figure can serve as a benchmark for evaluating what follows.

Households can be categorized into five groups by wealth or income, each being associated with a specific type of house pre-existing to the disaster. This is a simplification but approximates reality quite well. Table 8.1 summarizes the corresponding information.

Table 8.1 Household types

Household category	Number of households	Type and cost of house	House resilience	Net income	Net income as	Ratio to poorest
	(Total 1000)	Before flood	(in years)	(yearly)	% of cost	Category
Very poor	330	U = 100 k Rs.	3	25,550	26%	1
Poor	240	U = 100 k Rs.	3	49,275	49%	2
Middle	180	M = 180 k Rs.	10	86,870	48%	3.5
Mid-rich	150	M = 180 k Rs.	10	153,300	85%	6
Rich	100	S = 415 k Rs.	30	364,635	88%	14

U = Unsafe; M = Mid-safe; S = Safe house
k = Thousand; Rs. = Rupees
An unsafe house, called a Kacha or Kutcha unit (option 1) has its foundation, plinth, and superstructure made of mud. A mid-safe house, called a Pakka or Pucca unit (option 2) has its foundation and plinth (up to 6 feet high) made of burnt brink. In this case, the superstructure is made of adobe or mud wall. A safe house or multiple hazards resistant unit (option 3) has foundation, plinth, and superstructure made of burnt bricks (World Bank, 2010)
Net income is the amount of money available to households for spending and saving after deducting annual expenditures on variable expenses such as food. House resilience denotes its ability to withstand flood shocks of various magnitudes, for example a 1 in 3, 10, or 30 years' flood

The data reveals the following. Poorer households number more than richer ones. There are three types of houses and the safest type costs more than four times the most vulnerable type, but at the same time is ten times safer when measured in terms of expected resilience to future floods (the rarer a flood over time, the more severe it is). The richest category has a net income 14 times greater than the poorest. And although the poorest have cheaper houses, their net income is less able to cover the costs of rebuilding one than that of the richer: the poorest income can cover only a quarter of the costs while the rich can on average cover about 90% of the cost.

The total cost of 'build as before' will be around Rs. 158 million for 1000 houses. However, the available budget of Rs. 100 million means that government can afford about 60% of the total cost of reconstruction. The budget constraint and the differences between households' affordability will affect how the government (or any other helping agency) allocates its available budget. In Pakistan, the government tends to directly fund the reconstruction of houses rather than giving money to households that they can then spend, in principle, on new housing. The decision facing the government is therefore how much to subsidize a given type of house for a given social category. To keep things simple, let us consider just four levels of subsidies: 25%, 50%, 75%, and 100% of the cost of rebuilding a house. Which level the government will choose will depend on its goals. These can be at least three, depending on whether it wants to maximize the number of poor households with a new house, the total number of houses or the total amount of flood protection, measured by the number of houses built multiplied by their expected resilience (3, 10, or 30 years).

Solving the problem therefore requires the following minimum knowledge:

1. The available overall budget
2. The reconstruction costs of each house type and their safety value (flood resilience)
3. The relevant social categories, their average disposable budgets, and their relative numbers
4. The different subsidy policies envisaged

Solving this problem involves trade-offs even before worrying about equity. To see this, consider how many houses can be built in the two extreme cases: when all the costs are borne by the government (equivalent to a 100% subsidy) and when all the costs are borne by households (equivalent to no subsidy at all, or 0%). In the first case, with its budget of 100 million Rs., the government can build 1000 'unsafe' houses, or 556 'mid-safe' houses, or 239 'safe' houses. Table 8.2 shows the result if

Table 8.2 Number of houses afforded by households

Household categories	House		
	Unsafe	Mid-safe	Safe
Very poor	0	0	0
Poor	0	0	0
Low-middle	0	0	0
High-middle	230	0	0
Rich	365	203	0
Affordable by hhds	595	203	0
Affordable by govt	1000	556	239
As % of government's	59%	36%	0%

hhds = households; govt = government

instead households must shoulder all the costs, each type of house taken separately. Because of the undesirability of temporary tents, we assume that reconstruction takes place within one year.

In both cases, either a small percentage of the safer houses or none at all is affordable. Note that if the government wants to fund all the 1000 houses, it can only do so with poor-quality (unsafe) houses: this reflects the conditions of a rural developing country. Typically, left to their own means, all households except the richest category are stuck with the most basic house types, leaving them as vulnerable as before to future flood disasters. Even the richest category cannot afford to rebuild the safest type of house. These outcomes have assumed an 'either-or' approach; clearly, some mix of funding must be better. But which mix?

Table 8.3 explores the results of various funding mixes between the government and individual households. Results are simplified to binary yes/no outcomes (i.e. affordable or not).

These results present stakeholders with a choice rather than with a solution. Which one would they prefer? Would they choose the maximum number of houses that are

1. affordable by households? Then they build 1502 but only unsafe houses;
2. affordable with the government's budget? Then they build 4000 but only unsafe houses;
3. affordable by the poorest households? Then they build 1333 and again only unsafe houses!

If stakeholders want to include the safest houses, they need a subsidy of at least 75%, and then only the two richest household categories will be able to afford them.

However, these results have assumed the same subsidy rate for all social categories; they are therefore likely to be artificial. Instead, any government or aid organization would want to vary the rate depending on how much a household can afford relative to the cost of rebuilding: for instance, 100% for the poorest, 0% for the richest, and 50% for those in between. But this cannot be done without ushering in, explicitly or implicitly, an equity or justice norm. The next section examines how we can do this.

Table 8.3 Subsidy rates and reconstruction outcomes

Hhld categories	0% subsidy benchmark			With 25% subsidy			With 50% subsidy			With 75% subsidy		
	Unsafe	Mid-safe	Safe	Unsafe	Mid-safe	Safe	Unsafe	Mid-safe	Safe	Unsafe	Mid-safe	Safe
Very poor	No	No	No	No	No	No	No	No	No	Yes	No	No
Poor	No	No	No	No	No	No	No	No	No	Yes	Yes	No
Low-middle	No	No	No	Yes	No	No	Yes	No	No	Yes	Yes	No
High-middle	Yes	No	No	Yes	Yes	No	Yes	Yes	No	Yes	Yes	Yes
Rich	Yes	Yes	No	Yes	Yes	Yes	Yes	Yes	Yes	Yes	Yes	Yes
Hhld-affordable	595	203	0	1001	440	116	1502	661	174	1333	741	319
Govt affordable	1000	556	239	4000	2222	957	2000	1111	478	1333	741	319
% govt affordable	59%	36%	0%	25%	20%	12%	75%	59%	36%	100%	100%	100%

Hhld = household; govt = government

8.3 THE IMPACT OF EQUITY: EFFICIENCY TRADE-OFFS

Although the literature suggests one can define anywhere between one and two dozen equity norms, depending on how they are categorized, let us limit ourselves here to three to keep things simple. Adding more would not change the nature of the problem (Schilizzi & Black, 2009). Let us consider equality, ability-to-pay (AtP), and vertical equity (VE) (Cazorla & Toman, 2000; Ringius, Torvanger, & Underdal, 2002). Equality defines an equal share of the government's budget given to all households, irrespective of their income; ability-to-pay defines a share that is inversely proportional to net income; and vertical equity is AtP adjusted so that the richest get no subsidy. Given the above data, they translate into the following allocation shares to each of the five social categories, starting with the poorest on the left:

Equality   {20%; 20%; 20%; 20%; 20%}
Ability-to-pay (AtP) {49%; 25%; 14%; 8%; 3%}
Vertical equity (VE) {55%; 26%; 13%; 6%; 0%}

In the equality principle, we distribute houses in a way that each category of household receives 20% of the available houses. In the case of proportionality or ability to pay, the poor are favoured more than the rich. Vertical equity is the more extreme form of the proportionality principle in which the 'very poor' category receives 55% and the 'rich' category receives nothing, or 0%.

However, when we say 'equality', we must remember Nobel laureate Amartya Sen's famous question: "*Equality of what?*" Should each social category receive an equal share of the government's available budget, or should every household receive an equal amount of funding? If the number of households differ in each category, the outcomes will differ. Table 8.4 shows by how much they will differ in our case. In the top panel, government chooses the first option; in the lower panel, it chooses the second. As always in this type of problem, when percentages are equal, amounts are not, and vice-versa, so that unequal weights yield equal outcomes and equal weights yield unequal outcomes!

Clearly, the final allocation to each household differs markedly from that to each category: because the number of poor households is greater than rich ones, the household-based distribution, as the last column shows, is 'flatter' (more equal) than the category-based one. In the top panel, for example, an equal percentage allocation leads to greater abso-

8 HOW TO BE FAIR IN PRIORITIZING SUPPORT IN THE AFTERMATH...

Table 8.4 Amartya Sen's "equality of what?"

Category based	Before	Category weights	Allocation	After	Ratio to VP 'after'
Very poor (VP)	25,550	20%	60,606[a]	86,156[b]	1.00
Poor	49,275	20%	83,333	132,608	1.54
Low-middle	86,870	20%	111,111	197,981	2.30
Upper-middle	153,300	20%	133,333	286,633	3.33
Rich	364,635	20%	200,000	564,635	6.55

Household based	Before	Individual weights	Allocation	After	Ratio to VP 'after'
Very poor (VP)	25,550	33%	100,000	125,550	1.00
Poor	49,275	24%	100,000	149,275	1.19
Low-middle	86,870	18%	100,000	186,870	1.49
Upper-middle	153,300	15%	100,000	253,300	2.02
Rich	364,635	10%	100,000	464,635	3.70

[a] We multiplied 20% (= 1 of 5 categories) with the total government budget of 100 million and divided this by the number of households in the respective category. For example, (0.2 * 100,000,000)/330 = 60,606. The 330 represents the number of households in the 'Very Poor' category
[b] The number 86,156 is obtained simply by adding 25,550 (net income before allocation) and 60,606 (the allocation of funds based on category weight)

Table 8.5 The role of different equity norms on final household budgets

	Equality	AtP	VE
Very poor	125,550	*210,944*	*223,693*
Poor	**149,275**	145,405	144,827
Lower-middle	**186,870**	141,398	134,609
Upper-middle	**253,300**	184,199	173,882
Rich	**464,635**	*377,626*	*364,635*

Note: The highest final budget for each social category is highlighted **in bold**, while the two highest for each equity norm are highlighted in *italics*

lute allocations to the richer (compare 200,000 to only 60,606); in the lower panel, allocations of 100,000 are equal for all categories, but not for all households. If household equality is the goal, then the second option will be preferred. However, household-based allocations require more data than category-based, in particular on categories' relative numbers of households. This is why, in practice, the first option is often preferred under the (perhaps misleading) name of equality. Yet, the two are not necessarily in conflict, as Table 8.5 implies.

Table 8.5 shows the role of the three equity norms (equality, AtP, and VE) on households' final capacity to rebuild a house; that is, the sum of their original net income and the subsidy. The details underlying these results can be found in the Appendix (Table 8.7), where we can see that category weights direct allocations while household weights target households' final budget for reconstruction. If government targets only the poorest category, then the VE norm will be preferred. However, by doing this, the second poorest category, together with the lower middle, will be among the least favoured (equality does better for them). In fact, an equal allocation of funds to each household ends up favouring the richest categories most. Interestingly, AtP and VE both favour the poorest as well as richest categories (shown in italics). This is because the poorest receive a large allocation while the richest, though receiving little or none, are rich enough to start with. With these two norms, the three middle categories end up worst off. This outcome is not completely solvable until each social category is itself weighted relative to each other, reflecting their relative importance to the policy makers.

It turns out that if the government subsidizes at less than 100% house reconstruction costs, the two poorest categories fall by the wayside and will not be able to afford a new house. This would represent 57% of the (normalized) population of the Punjab! Thus, if the maximum number of poorer households are to benefit from the reconstruction program, the government should subsidize 100% of rebuilding costs. Of course, in that case, neither the number of safe houses nor the amount of flood resilience will be maximized. Table 8.6 shows what such a policy would lead to.

Just as in Table 8.5, VE favours the poor most, but even with a 100% subsidy, not all of them acquire a house: in the best of cases, only 84% do, and only 37% could have a 'safe' house. If equality is the allocation norm,

Table 8.6 Percent of poorer households with a safer new house, given a 100% subsidy

	Vertical equity	*Ability-to-pay*	*Equality*
Very poor			
Mid-safe house	84%	51%	35%
Safe house	37%	22%	15%
Poor			
Mid-safe house	69%	58%	48%
Safe house	30%	25%	20%

then, even with a 100% subsidy, only 15% of the poorest can acquire a safe house. The reason, of course, is that, given the very limited contribution from these households, and the fact that other households may receive some support, government funds are used to the maximum and are insufficient to fund more houses. However, though not shown in Table 8.6, the data in this case show that, subject to the three equity norms, the higher the subsidy rate, the more housing is achievable. This is unlikely to be the general case, but it is the case here where the rural population is overall very poor.

8.4　Conclusion

In any real situation, disaster relief is subject to resource and funding constraints. This study suggests that concerns for equity or justice in how disaster relief is distributed to people with different needs and different abilities to cope will strongly depend on how constraining the resources and funding are. This is true for any country, but particularly true for poorer developing countries. At least three factors affect the outcome: the severity of the constraints, the importance of achieving value for money with every dollar spent, and the (possibly implicit) choice of equity norm for a fair and just outcome. This sets the stage for a difficult problem involving difficult trade-offs, and it is no wonder that most if not all governments struggle and grope their way towards some kind of ad-hoc solution.

The choice between equity norms is inherently a political decision and will typically reflect the existing system of personal rights and entitlements as perceived by stakeholders, whose role in disaster relief is examined in Chap. 5. This choice can not only determine the policy, but also lead to unexpected outcomes, like equal allocations favouring the richest or ability-to-pay favouring the extreme at the expense of the middle categories. In addition, the choice will either dampen or amplify disaster impacts on different sections of the community who are already subject to differential impacts, thereby reducing or increasing their vulnerability and resilience to future disasters.

However, the importance of this choice is itself subject to the severity of the constraints weighing on the relief program. It appears that the more constrained the situation, the less sensitive is the decision to equity or justice; in other words, the outcome may remain the same no matter which equity norm is chosen. This also affects the relevance and the effectiveness of applying the precautionary principle. The reason is that, as a rule, the more constrained the situation, the stronger is the criterion of value for money (if money is the limiting resource) relative to the preferred equity criterion. Relief programs that are less reliant on, or need fewer external resources (e.g. a rich community with sufficient self-funding) are less likely to over-emphasize value for money; but then, equity concerns are also less likely to matter much.

That equity and justice are affected by resource constraints has profound implications for who controls the resources and who is making the decisions. It may happen that equity or justice will be, or appear to be, totally sacrificed when the policy maker is severely resource constrained and must seek greatest value for money, whatever this value might be. In such cases, it can be beneficial for the sake of justice to transfer the relief program to another, less resource constrained decision-making entity that is able, if willing, to implement a more equitable relief program. The economic point of view adopted in this chapter can thus bring another perspective to the other topics covered in this book, each of which can suffer from limited resources.

Appendix

Table 8.7 Details behind in-text Table 8.5 (The role of different equity norms on allocations and final household budgets)

	Equality	*AtP*	*VE*	*Equality*	*AtP*	*VE*
	A/ individual weights = goal			*B/ category weights = means*		
Very poor	20%	49%	55%	**33%**	61%	80%
Poor	20%	25%	26%	24%	23%	16%
Low-middle	20%	14%	13%	18%	10%	4%
Upper-middle	20%	8%	6%	15%	5%	1%
Rich	20%	3%	0%	10%	1%	0%
	C/ resulting hhld allocations			*D/ cumulative hhld budget*		
Very poor	100,000	185,394	198,143	**125,550**	210,944	223,693
Poor	100,000	96,130	95,552	149,275	145,405	144,827
Low-middle	100,000	54,528	47,739	186,870	*141,398*	*134,609*
Upper-middle	100,000	30,899	20,582	253,300	184,199	173,882
Rich	100,000	12,991	0	464,635	**377,626**	**364,635**

Hhld = household; AtP = Ability-to-pay; VE = Vertical equity

REFERENCES

Ahmed, I. (2011). An Overview of Post-Disaster Permanent Housing Reconstruction in Developing Countries. *International Journal of Disaster Resilience in the Built Environment*, 2(2), 148–164.

Azeem, M. M. (2016). *From Static One-Dimensional Poverty to Vulnerability to Multi-Dimensional Poverty: Towards Greater Effectiveness of Social Protection in Pakistan*. PhD dissertation, The University of Western Australia, 202 p.

Barakat, S. (2003). Housing Reconstruction After Conflict and Disaster. *Humanitarian Policy Group, Network Papers*, 43, 1–40.

Cazorla, M., & Toman, M. (2000). *International Equity and Climate Change Policy*. Climate Policy Brief No. 27, Resources for the Future, Washington, DC (22 p.).

Delaney, P., & Shrader, E. (2000). *Gender and Post-Disaster Reconstruction: The Case of Hurricane Mitch in Honduras and Nicaragua*. Decision Review Draft. Washington, DC: LCSPG/LAC Gender Team, The World Bank.

Duyne Barenstein, J. E. (2015). Continuity and Change in Housing and Settlement Patterns in Post-Earthquake Gujarat, India. *International Journal of Disaster Resilience in the Built Environment*, 6(2), 140–155.

Hayles, C. S. (2010). An Examination of Decision Making in Post Disaster Housing Reconstruction. *International Journal of Disaster Resilience in the Built Environment*, 1(1), 103–122.

Kim, K., & Olshansky, R. B. (2014). *The Theory and Practice of Building Back Better*. Taylor & Francis.

Lukasiewicz, A., & Baldwin, C. (2017). Voice, Power, and History: Ensuring Social Justice for All Stakeholders in Water Decision-Making. *Local Environment*, 22(9), 1042–1060.

Mannakkara, S., & Wilkinson, S. (2014). Re-Conceptualising 'Building Back Better' to Improve Post-Disaster Recovery. *International Journal of Managing Projects in Business*, 7(3), 327–341.

Mazza, M., Chiara Pino, M., Peretti, S., Scolta, K., & Mazzarelli, E. (2014). Satisfaction Level on Quality of Life Post-Earthquake Rebuilding. *International Journal of Disaster Resilience in the Built Environment*, 5(1), 6–22.

Oliver-Smith, A. (1996). Anthropological Research on Hazards and Disasters. *Annual Review of Anthropology*, 25(1), 303–328.

Oliver-Smith, A. (1990). Post-Disaster Housing Reconstruction and Social Inequality: A Challenge to Policy and Practice. *Disasters*, 14(1), 7–19.

Opdyke, A., Lepropre, F., Javernick-Will, A., & Koschmann, M. (2017). Inter-Organizational Resource Coordination in Post-Disaster Infrastructure Recovery. *Construction Management and Economics*, 35(8–9), 514–530.

Ophiyandri, T., Amaratunga, D., Pathirage, C., & Keraminiyage, K. (2013). Critical Success Factors for Community-Based Post-Disaster Housing Reconstruction Projects in the Pre-Construction Stage in Indonesia. *International Journal of Disaster Resilience in the Built Environment*, 4(2), 236–249.

Patel, S., & Hastak, M. (2013). A Framework to Construct Post-Disaster Housing. *International Journal of Disaster Resilience in the Built Environment*, 4(1), 95–114.

Powell, P. J. (2013). Post-Disaster Reconstruction: A Current Analysis of Gujarat's Response After the 2001 Earthquake. In *Beyond Shelter After Disaster: Practice, Process and Possibilities* (pp. 40–53). Routledge.

Rahmayati, Y. (2016). Reframing 'Building Back Better' for Post-Disaster Housing Design: A Community Perspective. *International Journal of Disaster Resilience in the Built Environment*, 7(4), 344–360.

Ringius, L., Torvanger, A., & Underdal, A. (2002). Burden Sharing and Fairness Principles in International Climate Policy. *International Environmental Agreements: Politics, Law and Economics*, 2, 1–22.

Romero, H., & Albornoz, C. (2016). Socio-Political Goals and Responses to the Reconstruction of the Chilean City of Constitución. *Disaster Prevention and Management*, 25(2), 227–243.

Schilizzi, S., & Black, J. (2009). *Breaking Through the Equity Barrier in Environmental Policy*. Land and Water Australia, Canberra, 172 p. Retrieved from http://lwa.gov.au/products/pn21249

Tran, T. A. (2015). Post-Disaster Housing Reconstruction as a Significant Opportunity to Building Disaster Resilience: A Case in Vietnam. *Natural Hazards*, 79(1), 61–79.

Vahanvati, M., & Mulligan, M. (2017). A New Model for Effective Post-Disaster Housing Reconstruction: Lessons from Gujarat and Bihar in India. *International Journal of Project Management*, 35(5), 802–817.

World Bank. (2010). *Pakistan Floods 2010: Preliminary Damage and Needs Assessment Project*. Washington, DC: World Bank.

PART III

Vulnerability and Resilience

CHAPTER 9

Equitable Access to Formal Disaster Management Programmes: Experience of Residents of Urban Informal Settlements in Bangladesh

Mohammad Shahidul Hasan Swapan, Md. Ashikuzzaman, and Md. Sayed Iftekhar

9.1 Introduction

Bangladesh is one of the most disaster-prone countries in the world (Ahmed & Kelman, 2018). It was ranked in the first and second position in the Climate Change Vulnerability Index (CCVI), produced by the risk analysis company Verisk Maplecroft, in 2015 and 2016 respectively. The Germanwatch Global Climate Risk Index (CRI) 2018 placed Bangladesh in the sixth position in the list of the countries most affected by climate

---

M. S. H. Swapan
School of Design and the Built Environment, Curtin University, Bentley, WA, Australia
e-mail: m.swapan@curtin.edu.au

Md. Ashikuzzaman (✉)
Development Studies Discipline, Khulna University, Khulna, Bangladesh
e-mail: ashikuzzaman.md@ku.ac.bd

© The Author(s) 2020
A. Lukasiewicz, C. Baldwin (eds.), *Natural Hazards and Disaster Justice*, https://doi.org/10.1007/978-981-15-0466-2_9

change in the period 1997–2016 (Eckstein, Künzel, & Schäfer, 2018). Most of its large, densely settled population of 165 million is at significant risk to different forms of natural hazards. In terms of direct effect, impact on economic activity, damage or destruction of assets, the disasters which have been most important since independence in 1971 are exceptionally widespread riverine flooding; severe tropical cyclones and associated coastal storm surges; river bank erosion; and drought. It was estimated that 139,000 people were killed during the 1991 cyclone and 31 million were directly affected by the 1998 floods. During the 1997–2016 period, Bangladesh suffered damages worth over US$ 2.31 billion, equal to 0.67% of GDP, wrought by 187 natural catastrophes (Eckstein et al., 2018). Their cumulative impact affected the socio-economic condition and livelihoods of people forcing the national economic growth rate to decline. The effects are more severe due to widespread poverty, with around 23% of the population living in extreme poverty (Garschagen et al., 2016). These people are typically living and working in disaster-prone coastal zones of the country. The coastal zones of Bangladesh, comprising 32% of the land area, are particularly vulnerable to natural hazards (Iftekhar, 2006), which makes them an important study area for disaster justice.

Bangladesh has developed a widespread, complex and collaborative disaster management framework since the devastating cyclone of 1991 (Shaw, Mallick, & Islam, 2013). Traditionally, the government engages in delivering early warnings to the coastal areas for cyclones and enhancing relief networks as an emergency response to various disasters. A range of response mechanisms has been developed to tackle the mortality rate caused by catastrophic cyclones around coastal areas. This includes training of volunteers, establishing cyclone shelters and developing early warning systems. Recent data shows that only 17% of the total vulnerable population in the coastal areas has access to cyclone shelters (Hassan et al., 2013). Later, in response to global disaster policies (e.g., Hyogo Framework 2005–2015 and Sendai Framework 2015–2030), the Bangladeshi government gradually shifted focus towards more long-term resilience and recovery programmes to reduce disaster risks by incorporating increasing interventions of NGOs (Haque, Pervin, Sultana, & Huq, 2019).

Md. S. Iftekhar
UWA School of Agriculture and Environment, The University of Western Australia, Crawley, WA, Australia
e-mail: mdsayed.iftekhar@uwa.edu.au

Due to resource constraints, the country often relies on external sources to address disaster management and relief works. Bangladesh received significant foreign aid to deal with major catastrophes, for example over USD 600 million was obtained for emergency response including reconstruction and rehabilitation after the cyclone SIDR (2007) (Ozaki, 2016). Ozaki (2016) shows that only 13% of the total costs related to disaster were internally sourced during the 2000–2013 period. A significant portion of the total cost for recovery and rehabilitation projects in that period was secured from international donors and aid agencies as humanitarian aid and funding for disaster resilience and recovery. However, there is still a significant funding gap to recover from the loss. It leads to an increasing need for involving local NGOs in co-delivering post-disaster recovery and resilience outcomes (Mees, Crabbé, & Driessen, 2017).

A review of past studies shows that support from both government and NGOs has pros and cons. The extent of accessibility to this support varies with the nature of disaster and stages of response (e.g., pre-, during and post-disaster contexts) (Begum, 2012; Cash et al., 2013; Moroto, Sakamoto, & Ahmed, 2018; Paul, 2003). The government is often criticised for its top-down decision-making and bureaucratic approach which may hinder efficient service delivery. A number of studies also indicated corruption and mismanagement in disbursing relief works which hinder equitable access (Islam & Walkerden, 2015; Islam, Walkerden, & Amati, 2017; Paul, 2003; Paul & Hossain, 2013). On the other hand, a recent study by Moroto et al. (2018) on spatial distribution of NGOs working on disaster management reveals that "*NGOs did not generally locate in disaster-affected or disaster-vulnerable areas; rather, they located in areas that are more accessible or allowed other factors to affect their project location decisions*" (p. 261). This can again affect response mechanisms to serve actual vulnerable groups to various disasters.

Bangladesh now has a fairly well-developed institutional mechanism at the national and field levels for managing the consequences of natural disasters. Both government agencies and NGOs play a critical role in providing pre- and post-disaster supports. However, there is a gap in understanding how marginal communities living in informal settlements (slum areas) of urban areas experience disaster-related service provided by GOs and NGOs. According to the Bangladesh Bureau of Statistics (BBS), "*[a] [s]lum is a cluster of compact settlements of 5 or more households which generally grow very unsystematically and haphazardly in an unhealthy condition and atmosphere on government and private vacant land. Slums also exist in the owner based household premises*" (BBS, 2015). In 2015, around 35% of the total urban population of the six major cities lived in informal settlements (Sikder, Asadzadeh, Kuusaana, Mallick, & Koetter, 2015). The urban poor living in those areas suffer from a wide range of deprivations including income, employment, housing, and basic urban services. Islam, Angeles, Mahbub, Lance, and Nazem (2006) reported that about 90% of urban poor earn less than USD 2 and 95% are engaged in informal jobs. More than 80% of the houses in slum areas are either temporary shacks or in very poor condition (Sowgat, Wang, & McWilliams, 2017). The vulnerabilities of the urban poor to natural hazards are higher due to their limited income and assets, temporary and informal land tenure, lack of entitlements in accessing municipal services and lack of institutional and political protection (Alam, McGregor, & Houston, 2018). It is further aggravated by poor adaptive capacities (Haque, Dodman, & Hossain, 2014).

In the disaster management literature, it has been suggested that disaster risk reduction programmes in Bangladesh should prioritise the vulnerable section of the community (women, poor, person with disability and socially excluded people) (Rashid & Shafie, 2013) as they are likely to be the subjects of social injustice and have inequitable access to disaster management programmes. However, there are not many studies that have examined if there are any systematic differences in the types of disaster-related supports from government and non-government organisations received by urban informal dwellers. In this regard, this chapter investigates the extent of support received by households in informal settlements in a major coastal city in Bangladesh, Khulna. We refer to the tangible support for this study including shelter, food to training and monetary support to recover from the disaster in the longer term. Immediate family and informal networks are a major source of support during and post disaster periods (Islam & Walkerden, 2014); however, we do not include them here. Islam and Walkerden (2015) suggest that households' links with organisations (e.g., government and NGOs) form an important part of the disaster recovery system. This 'linking network' "*provide(s) opportunities for people to express their needs and receive recovery support*" (p. 1708). Realising the need to understand the role of linking networks in local disaster management, we focus on comparing the extent of support from government and non-government organisations with the objective of discerning any differences in access to disaster-related support by the urban poor living in informal settlements. In this chapter, we aim to answer two research questions: (1) are there any systematic differences in the types of disaster-related supports from GOs and NGOs received by urban informal dwellers? and (2) is there any difference in access to these services by different socio-economic groups? The findings will aid in understanding the dynamics of the population in accessing support and in recognising the systematic differences in how NGO-led support systems fill out the gap left by GOs.

9.2 Study Area

The study was conducted in Khulna, the third largest metropolitan city located in the southwest coastal belt of Bangladesh. The city has an area of 45.65 km^2 accommodating around 1,781,000 people in 2011, which is projected to grow over to 2,805,000 by 2025 (Haque et al., 2014). Khulna has a rapidly growing population with a higher urbanisation rate (5%) than the national average. The urbanisation is characterised by spontaneous spatial expansion and sporadic growth of informal settlements. A large number of the city population is poor who live in around 520 low-income informal settlements (Parvin, Alam, & Asad, 2016) (Fig. 9.1). The informal settlements are characterised by high-density living, inadequate or no drainage and waste management systems resulting in poor environmental and health conditions. Less than 5% of the residents in slums areas are connected to municipal water supply while around 80% of them have access to shared pit latrines (Islam et al., 2006). Haque et al. (2014) define the urban poor in Khulna city as having *"household income below BDT 3500 per month (approximately US$ 45), owning productive assets worth less than BDT 5000 (approximately US$ 65), lacking access to microfinance and consuming fewer than three meals per day"* (p. 115).

Extreme weather events and climatic hazards are frequent in the greater Khulna area including cyclones, tidal surges, and heavy rainfall. The surrounding districts of Khulna city are exposed to the Bay of Bengal, which attracted catastrophic cyclones in the past including Sidr (2007), Aila (2009) and Mahasen (2013). It accommodates a high volume of climate refugees from the nearest districts after major environmental catastrophes (Salauddin & Ashikuzzaman, 2012). Recent data shows that around 45% of the total households in the city are affected due to waterlogging caused

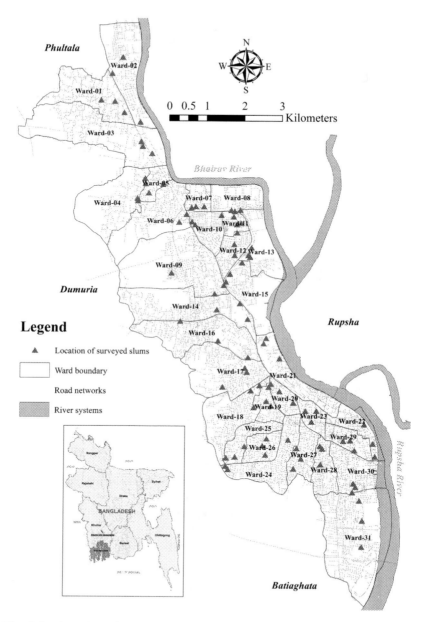

Fig. 9.1 Location of Khulna city and major informal settlements. (KCC, 2011)

by river floods and heavy rainfall (Rahman, 2005). Almost everyone in the city has been exposed to frequent extreme weather events during the last decade (Rooney et al., 2011).

9.3 METHODS AND MATERIALS

9.3.1 Data Collection

We surveyed a total of 93 slums out of 520 across Khulna city in Bangladesh using a structured questionnaire during October–November 2016. An initial list of urban settlements for individual jurisdiction was derived from the study report "Atlas of poor settlements in Khulna City Corporation" (KCC, 2011). A systematic random sampling method was used for the survey. With a 95% confidence level (and 2.5 confidence interval), a targeted sample was calculated as 1477 households. To get a wider picture of the informal settlements of Khulna city, we selected top three slums from each administrative unit (locally known as a 'ward') across the city. From each ward, three slums with the highest number of households were selected. The target number of households specified for individual wards was distributed across the selected slums in proportion to the number of households in each slum. In each selected slum, every fifth household was targeted to carry out the survey. In case of no response, the next household was approached. The survey stopped when the required number of responses was completed. We have received a total of 1122 valid responses.

An initial draft of the questionnaire was developed based on the learnings from relevant literature to design the categories of the local hazards (Shaw et al., 2013), typology of support needed during and post disaster periods (Islam & Walkerden, 2014, 2015) and finally the role of GOs and NGOs for short- and long-term recovery (Haque et al., 2014; Islam et al., 2017; Roy, Hulme, & Jahan, 2013). The draft questionnaire was piloted prior to the main survey with 30 households in Rupsha slum which is the largest slum in Khulna city. Based on the results from pre-testing, the questionnaire was revised and finalised. The respondents were asked about the types of support they have received before and after a disaster from government and non-government organisations. Summary statistics comparing different types of services have been presented. Regression models have been used to examine the relationship between the extent of support received and the socio-economic conditions.

9.4 Results and Discussion

9.4.1 Respondents' Characteristics

Among the respondents, almost half are below 40 years (46%) (Table 9.1). Men were overly represented in the sample, which is not surprising given that our target was to interview household heads. As mentioned by the respondents more than 90% of the households were male dominated. However, this also means that the results may be biased as women are likely to have different risk perceptions than men. Almost half of the respondents do not have any formal education. They also have low ownership of land parcels and houses.

Table 9.1 Descriptive statistics of demographic and socio-economic condition of the sample

Characteristics	Mean	Standard deviation
Total household income/month (BDT[a])	9475	4898
No. of children	2.03	1.23
No. of adult dependents	1.51	1.12
No. of income earners	1.42	0.63
No. of years living in the city	25.78	16.08
Proportion of male respondents (%)	83	
Proportion of female-headed household (%)	8	
Proportion of respondents with no or non-formal education (%)	46	
Proportion of respondents with land ownership (%)	14	
Proportion of respondents participating actively in any group activities (%)	38	
Proportion of respondents with access to microcredit (%)	42	
Age group (%)		
18–29	12	
30–39	34	
40–49	31	
50–59	14	
60–69	6	
70+	2	
Rating of own general health and fitness (%)		
Excellent	3	
Very good	8	
Good	41	
Fair	26	
Poor	22	

[a]1 BDT = 0.012 USD

9.4.2 Support to Recover from Disaster

The respondents identified the level of support they received from government agencies and NGOs at different stages of recovery (Table 9.2). It is apparent that government agencies play a key role in providing intermediate support such as emergency response and providing immediate shelter. However, there is a diminishing level of government support experienced by the respondents in terms of long-term recovery. Instead, NGOs have a prominent role in the long-term recovery of the informal settlers by providing microcredit and skill training. It is important to note that there is a strong positive relationship between the different types of supports received by the respondents (Table 9.3). For example, those who received immediate government support are also likely to receive immediate non-government support. This could indicate a potential overlap between GO's and NGO's disaster management programmes. The findings call for more co-produced and integrated initiatives to maximise support to the affected groups by engaging the key stakeholders. While the government agencies require support from NGOs to deliver more efficient services during the disaster period, state patronisation of local and NGOs-led long-term programmes could be more sustainable.

Table 9.2 Distribution of respondents by types of support received

Support	Recovery stage	GO		NGO	
		Mean	SD	Mean	SD
Warning	Immediate	22.55	41.81	11.41	31.81
Search and rescue	Immediate	8.73	28.25	6.68	24.99
Shelter	Immediate	13.81	34.52	8.20	27.45
Food	Immediate	5.61	23.03	9.54	29.39
Water and sanitation	Immediate	12.57	33.16	19.88	39.92
Cash	Immediate	2.67	16.14	7.40	26.18
Household items	Medium-term	2.23	14.77	5.44	22.68
Essentials for children	Medium-term	4.46	20.64	4.28	20.25
Building materials	Medium-term	4.37	20.45	4.10	19.84
New house	Medium-term	1.69	12.91	2.23	14.77
Livelihood support	Long-term	4.81	21.41	8.20	27.45
Microcredit or loan	Long-term	3.12	17.39	36.90	48.27
Training	Long-term	3.03	17.15	16.67	37.28

Table 9.3 Correlation between different types of support received by the respondents

		GO			NGO		
		Immediate	Medium-term	Long-term	Immediate	Medium-term	Long-term
GO	Immediate						
	Medium-term	0.51					
	Long-term	0.51	0.56				
NGO	Immediate	0.69	0.43	0.46			
	Medium-term	0.41	0.58	0.33	0.52		
	Long-term	0.42	0.30	0.33	0.51	0.37	

Note: Correlation coefficients are significant at 1% level of significance

9.4.3 Relationship Between Level of Supports and Socio-Economic Conditions

To understand the relationship between the level of support received by the respondents and their socio-economic conditions, we have conducted a set of regressions. We have grouped the individual support types into three recovery stages (immediate, medium-term and long-term). For each respondent, we have added the types of support they have received for each stage. The total number of support types was then used as the dependent variable. To explain the variation in support received, a set of socio-economic factors was used as explanatory variables in an ordered logit model set up.

In Table 9.4, the relationship between the extent of support received from the government organisations and the socio-economic conditions of the respondents has been presented. It can be seen that people who are not members of any group (including social clubs, charity or political parties) and from a female-headed household have lower odds of receiving higher levels of immediate support compared to the respective group. On the other hand, females from households with a greater number of children and without access to credit have higher odds of receiving a higher level of immediate support. Respondents who do not own any land, do not have access to credit, with poor health conditions and from households with more income earners have higher odds of receiving a higher level of medium-term support. Respondents with poorer health conditions and from households with a higher number of income earners have higher odds of getting a higher level of long-term support.

Table 9.4 The relationship between the extent of support received from government organisations and the socio-economic conditions of the respondents

Variables	Immediate support			Medium-term support			Long-term support		
	OR	P	SE	OR	P	SE	OR	P	SE
No formal education	0.84		(0.12)	0.71		(0.18)	1.08		(0.26)
Does not own land	1.01		(0.20)	3.57	***	(1.74)	1.10		(0.39)
Not member of any group	0.51	***	(0.08)	0.37	***	(0.10)	0.70		(0.19)
Poor health condition	1.20		(0.19)	1.75	**	(0.49)	2.37	***	(0.60)
Does not have access to credit	1.82	***	(0.28)	1.80	**	(0.50)	1.50		(0.39)
Female	2.69	***	(0.57)	3.98	***	(1.23)	1.13		(0.44)
Member of a female-headed household	0.55	*	(0.19)	0.68		(0.33)	1.80		(0.84)
Age >= 70 years	0.50		(0.23)	0.86		(0.67)	0.39		(0.41)
No. of years living in the area	1.00		(0.00)	0.98	**	(0.01)	0.98	***	(0.01)
Household income	1.00		(0.00)	1.00		(0.00)	1.00		(0.00)
No. income earners	1.00		(0.11)	1.38	*	(0.24)	1.39	*	(0.26)
No. of children	1.16	***	(0.06)	1.02		(0.10)	0.95		(0.08)
No. of adult dependents	0.91		(0.06)	0.90		(0.12)	0.95		(0.10)
Number	1211			1211			1211		
Wald	67	***		60	***		32	***	
LL	−1171			−376			−364		

Note: OR = Odds Ratio, Robust Standard Error. '***', '**' and '*' indicate significance at 1%, 5%, and 10% levels respectively. An odds ratio above 1 indicate positive association while a value below 1 indicates a negative association. For example, the odds of getting immediate support is 2.69 times higher for female respondents compared to male respondents

The relationship between the extent of support received from non-government organisations and the socio-economic conditions of the respondents also shows some consistent patterns (Table 9.5). It can be seen that females are likely to receive higher NGO support compared to males. Conversely, people with no formal education and/or those who are not members of any group have lower odds of receiving support from non-government organisations. Respondents from households with more children have higher odds of getting higher levels of support from NGOs.

Our results indicate that the role of disaster management programmes in Bangladesh in alleviating existing social injustice is ambiguous. It seems that having no formal education and not being a member of any group are

Table 9.5 The relationship between the extent of support received from non-government organisations and the socio-economic conditions of the respondents

Variables	Immediate support			Medium-term support			Long-term support		
	OR	P	SE	OR	P	SE	OR	P	SE
No formal education	0.75	**	(0.11)	0.76		(0.17)	0.61	***	(0.08)
Does not own land	1.16		(0.21)	2.50	**	(0.96)	1.10		(0.17)
Not member of any group	0.60	***	(0.09)	0.43	***	(0.10)	0.80		(0.11)
Poor health condition	1.19		(0.21)	1.47		(0.35)	1.62	***	(0.24)
Does not have access to credit	1.14		(0.17)	2.13	***	(0.51)	0.59	***	(0.08)
Female	2.28	***	(0.46)	2.69	***	(0.75)	1.91	***	(0.37)
Member of a female-headed household	0.58	*	(0.18)	0.60		(0.28)	0.86		(0.25)
Age >= 70 years	0.55		(0.24)	0.00	***	(0.00)	0.81		(0.32)
No. of years living in the area	1.00		(0.00)	0.99		(0.01)	0.99		(0.00)
Household income	1.00	**	(0.00)	1.00		(0.00)	1.00		(0.00)
No. income earners	1.06		(0.11)	1.15		(0.18)	0.96		(0.10)
No. of children	1.11	**	(0.06)	1.09		(0.09)	1.16	***	(0.05)
No. of adult dependents	0.91		(0.06)	0.89		(0.10)	1.05		(0.06)
Number	1211			1211			1211		
Wald	53	***		2633	***		81	***	
LL	−1149			−464			−1130		

Note: OR = Odds Ratio, Robust Standard Error. '***', '**' and '*' indicate significance at 1%, 5%, and 10% levels respectively

detriments to accessing both government and non-government support. People with lower human capital (education) and social capital (group membership) are likely to have lower access to disaster support perpetuating the status quo. Females have comparatively higher access than their male counter-parts which is on par with the government policy of female empowerment and could potentially reduce gender inequality. However, the results show that female-headed households are likely to receive lower support indicating the inability of the disaster programmes to reach the truly vulnerable section of the community. Therefore, it seems that the current disaster management programmes in Bangladesh are missing the opportunity to become thoroughly inclusive in providing support to vulnerable communities and reducing existing inequalities in the socio-economic condition of the vulnerable urban dwellers.

9.5 Conclusion

Both government and non-government organisations are playing a critical role in providing disaster-related support. As expected, governments' strong role in providing warning and shelter and NGOs' role in providing microcredit and skill development training have been identified by the respondents. It has been revealed that NGOs are partaking in disaster management as an effective partner which further signifies their involvement in maximising the support to the affected communities. In this regard, NGOs can enhance their immediate support network by achieving partnerships with government departments. On the other hand, government organisations could consider mainstreaming successful project ideas driven by NGOs. This will, in turn, strengthen long-term disaster management initiatives using existing resources. The results from this analysis further reveal that there are some socio-economic factors or characteristics which potentially influence the access to disaster-related supports from government and non-government organisations by urban informal settlers.

We acknowledge that there are other socio-political factors that could influence their perception and actual access to support. It may include political connection, prevailing patron-client networks and informal transaction among urban actors, however, this has not been investigated. Further, given that our sample comprises predominantly male respondents separate studies targeting only female urban dwellers should be carried out to understand the gendered perspective of accessing pre- and post-disaster support.

References

Ahmed, B., & Kelman, I. (2018). Measuring Community Vulnerability to Environmental Hazards: A Method for Combining Quantitative and Qualitative Data. *Natural Hazards Review, 19*, 04018008.

Alam, A., McGregor, A., & Houston, D. (2018). Photo Response: Approaching Participatory Photography as a More Than Human Research Method. *Area, 50*, 256–265.

BBS. (2015). *Census of Slum Areas and Floating Population 2014*. Bangladesh Bureau of Statistics Dhaka.

Begum, K. (2012). *Challenges of Mainstreaming Climate Change Adaptation in Bangladesh*, Master of Public Policy, Politics and Policy, University of Tasmania, Hobart.

Cash, R. A., Halder, S. R., Husain, M., Islam, M. S., Mallick, F. H., May, M. A., … Rahman, M. A. (2013). Reducing the Health Effect of Natural Hazards in Bangladesh. *Lancet, 382*, 2094–2103.

Eckstein, D., Künzel, V., & Schäfer, L. (2018). *Global Climate Risk Index 2018: Who Suffers Most from Extreme Weather Events? (Weather Related Loss Events in 2016 and 1996 to 2016)*, Briefing Paper, Germanwatch, 2018. Germanwatch, Bonn.

Garschagen, M., Hagenlocher, M., Comes, M., Dubbert, M., Sabelfeld, R., Lee, Y. J., … Neuschäfer, O. (2016). *World Risk Report 2016*. Bündnis Entwicklung Hilft (Alliance Development Works), and United Nations University–Institute for Environment and Human Security (UNU-EHS), Berlin and Bonn, Germany.

Hassan, A., Islam, A., Chakravorty, N., & Al Hossain, B. M. T. (2013). Disaster Risk Reduction Investment and Reduction of Response Cost in Bangladesh. In R. Shaw, F. Mallick, & A. Islam (Eds.), *Disaster Risk Reduction Approaches in Bangladesh* (pp. 331–341). Tokyo: Springer.

Haque, A. N., Dodman, D., & Hossain, M. M. (2014). Individual, Communal and Institutional Responses to Climate Change by Low-Income Households in Khulna, Bangladesh. *Environment and Urbanization, 26*, 112–129.

Haque, M., Pervin, M., Sultana, S., & Huq, S. (2019). Towards Establishing a National Mechanism to Address Losses and Damages: A Case Study from Bangladesh. In *Loss and Damage from Climate Change* (pp. 451–473). Springer.

Iftekhar, M. (2006). Conservation and Management of the Bangladesh Coastal Ecosystem: Overview of an Integrated Approach. *Natural Resources Forum, 30*, 230–237.

Islam, N., Angeles, G., Mahbub, A., Lance, P., & Nazem, N. (2006). *Slums of Urban Bangladesh: Mapping and Census 2005*. Center for Urban Studies (CUS), Dhaka.

Islam, R., & Walkerden, G. (2014). How Bonding and Bridging Networks Contribute to Disaster Resilience and Recovery on the Bangladeshi Coast. *International Journal of Disaster Risk Reduction, 10*, 281–291.

Islam, R., & Walkerden, G. (2015). How Do Links Between Households and NGOs Promote Disaster Resilience and Recovery?: A Case Study of Linking Social Networks on the Bangladeshi Coast. *Natural Hazards, 78*, 1707–1727.

Islam, R., Walkerden, G., & Amati, M. (2017). Households' Experience of Local Government during Recovery from Cyclones in Coastal Bangladesh: Resilience, Equity, and Corruption. *Natural Hazards, 85*, 361–378.

KCC. (2011). *Atlas of Poor Settlements in Khulna City Corporation (Urban Partnerships for Poverty Reduction—Mapping Urban Poor Settlements in UPPR Project Towns)*. Khulna City Corporation.

Mees, H., Crabbé, A., & Driessen, P. P. (2017). Conditions for Citizen Co-production in a Resilient, Efficient and Legitimate Flood Risk Governance Arrangement. A Tentative Framework. *Journal of Environmental Policy & Planning, 19*, 827–842.

Moroto, H., Sakamoto, M., & Ahmed, T. (2018). Possible Factors Influencing NGOs' Project Locations for Disaster Management in Bangladesh. *International Journal of Disaster Risk Reduction, 27*, 248–264.

Ozaki, M. (2016). *Disaster Risk Finance in Bangladesh* (ADB South Asia Working Paper Series, 46). Asian Development Bank (ADB), Manila.

Parvin, A., Alam, A., & Asad, R. (2016). A Built Environment Perspective on Adaptation in Urban Informal Settlements, Khulna, Bangladesh. In M. Roy, S. Cawood, M. Hordijk, & D. Hulme (Eds.), *Urban Poverty and Climate Change: Life in the Slums of Asia, Africa and Latin America* (pp. 73–91). London and New York: Routledge.

Paul, B. K. (2003). Relief Assistance to 1998 Flood Victims: A Comparison of the Performance of the Government and NGOs. *Geographical Journal, 169*, 75–89.

Paul, S. K., & Hossain, M. N. (2013). People's Perception about Flood Disaster Management in Bangladesh: A Case Study on the Chalan Beel Area. *Stamford Journal of Environment and Human Habitat, 2*, 72–86.

Rahman, M. (2005). *Cities and Climate Change: Preparation of Risk and Vulnerability Maps of Khulna City, Bangladesh*. Retrieved from http://www.clacc.net/knowledge/Documents/Cities&ClimateChange

Rashid, A. M., & Shafie, H. (2013). Gender and Social Exclusion Analysis in Disaster Risk Management. *Disaster Risk Reduction Approaches in Bangladesh*, 343–363.

Rooney, P., Schonhofen, C., Jachnow, A., Hossain, I., Vogt, C., & Linden, A. (2011). Climate Change and the Urban Poor: Support of the German Development Cooperation to a City in Bangladesh. In K. Otto-Zimmermann (Ed.), *Resilient Cities: Cities and Adaptation to Climate Change-Proceedings of the Global Forum 2010*. Bonn: Springer Science & Business Media.

Roy, M., Hulme, D., & Jahan, F. (2013). Contrasting Adaptation Responses by Squatters and Low-Income Tenants in Khulna. *Bangladesh, Environment and Urbanization, 25*, 157–176.

Salauddin, M., & Ashikuzzaman, M. (2012). Nature and Extent of Population Displacement Due to Climate Change Triggered Disasters in South-Western Coastal Region of Bangladesh. *International Journal of Climate Change Strategies and Management, 4*, 54–65.

Shaw, R., Mallick, F., & Islam, A. (2013). *Disaster Risk Reduction Approaches in Bangladesh*. Springer.

Sikder, S. K., Asadzadeh, A., Kuusaana, E. D., Mallick, B., & Koetter, T. (2015). Stakeholders Participation for Urban Climate Resilience: A Case of Informal Settlements Regularization in Khulna City, Bangladesh. *Journal of Urban and Regional Analysis, 7*(1), 5–20.

Sowgat, T., Wang, Y. P., & McWilliams, C. (2017). Pro-poorness of Planning Policies in Bangladesh: The Case of Khulna City. *International Planning Studies, 22*, 145–160.

CHAPTER 10

Children's Experiences of Disaster: A Case Study from Lombok, Indonesia

Harriot Beazley

10.1 Introduction

Indonesia sits on the "Pacific Ring of Fire" and is one of the most disaster-prone regions globally, where active tectonic plates collide and many volcanic eruptions, earthquakes and tsunamis occur (UNESCAP, 2015). Over a period of three weeks in July–August 2018, three massive earthquakes hit the island of Lombok in Eastern Indonesia. A 6.4-magnitude earthquake struck North Lombok on 29 July 2018, and just one week later a 7.0-magnitude quake struck the same area again. Exactly two weeks later another major 6.9-magnitude earthquake hit Eastern Lombok, on 19 August 2018. The rupturing fault lines near the epicentre were recorded by NASA to have lifted the ground by 25 centimetres, while in other places it dropped by 15 centimetres (ABC, 2018). The earthquakes triggered landslides on the volcano Mount Rinjani, although no increased volcanic activity was recorded. Tsunami warnings were raised after the second earthquake as the epicentre was out to sea, but a

H. Beazley (✉)
School of Social Science, University of the Sunshine Coast, Sippy Downs, QLD, Australia
e-mail: hbeazley@usc.edu.au

© The Author(s) 2020
A. Lukasiewicz, C. Baldwin (eds.), *Natural Hazards and Disaster Justice*, https://doi.org/10.1007/978-981-15-0466-2_10

small wave of 13 centimetres did not cause any damage or loss of life. The earthquakes were the strongest to hit Lombok in recorded history, or living memory.

The districts of North and East Lombok were the most effected by the earthquakes. By the end of August the death toll was officially confirmed as 563 people, more than 7770 were injured and 417,000 were left homeless, forced to live in makeshift shelters (Carter, 2019). Most of those killed and injured were struck by collapsed buildings. Officials stated that up to 90% of structures across North Lombok and North East Lombok were destroyed or damaged (World Vision, 2018). According to official statistics the three earthquakes destroyed an estimated 229,000 houses, 45 schools, 78 mosques, 1 hospital, 4 health facilities and 3818 public facilities, including roads, bridges, traffic lights, cemeteries and sports grounds (Jakarta Post, 2018; Massola, Rosa, & Rompies, 2018). Thousands of other houses were damaged, and 430 schools were closed due to structural damage, and the inability of teachers to return to work (Carter, 2019; World Vision, 2018).

10.2 Child-Centred Disaster Risk Reduction

Children are among those most at risk from natural disasters, and constitute a third to half of fatalities arising from natural events (Beazley & Ball, 2018; Save the Children, 2010; WHO, 2011). Children, especially those living in poverty or in marginal and underdeveloped environments, shoulder a disproportionate share of the burden produced by disasters, both in the short and long term, especially when child protection is not integrated as a priority in Disaster Risk Reduction (DRR) policy (Pfefferbaum, Pfefferbaum, & Van Horn, 2018; Save the Children, 2011; UNISDR, 2015). Dangers and negative impacts include immediate injury and death, illness and malnutrition (Alexander & Magni, 2013). Disasters also impact the mental health of children, who are more likely to experience post-traumatic stress disorder (PTSD), behavioural problems and depression (Jia et al., 2010; Peek, 2008). Psychological trauma and emotional distress are often caused through family separation and death, loss of home and abrupt changes to daily life, including the ability to attend school and access an education (Save the Children, 2010). Children are also vulnerable to exploitation after disasters, and their vulnerability is increased when they are orphaned, displaced or disabled (Delaney, 2006; Peek, 2008, pp. 1–19; Save the Children, 2011). Girls are particularly vulnerable to exploitation following natural disasters, and in Indonesia a direct correlation between the intensity of natural disasters and child marriage has recently been documented, as a response to adverse effects on welfare resulting from natural disasters (Kumala Dewi & Dartanto, 2019). The marriage of daughters is often a household strategy to cope with the financial shocks of natural disasters, by reducing the size and expenditure of a household (Kumala Dewi & Dartanto, 2019). In Lombok a girl is also more likely to be married underage if both her parents are working overseas, as it reduces the financial burden for her carers (Beazley, Butt, & Ball, 2018).

In spite of children's vulnerability, however, disaster management practices are usually dominated by programs and strategies directed at adults, with children regarded as passive victims having a limited role to play (Amri, Haynes, Bird, & Ronan, 2018, p. 241). In Indonesia recent research has shown that the voices and needs of children are often not heard in post-disaster settings (Beazley, 2016). Consequently, children's rights are frequently overlooked during or after a disaster, including their right to safety and protection, and health and education, with adverse long-term consequences for the children, their families and communities. As a result of these neglected issues it has become increasingly evident that children's rights must play a central role in earthquake recovery. When considering the impact of disasters on children it is therefore important not to disregard broader issues of justice and children's rights, as outlined in the United Nation's *Convention on the Rights of the Child*, to which Indonesia is a signatory (UN, 1989). Children's rights must be considered in all disaster management processes. To inform Child-Centred Disaster Risk Reduction (CCDRR) policy, and to protect children's rights, it is therefore necessary to appreciate the *"underlying factors that influence children's vulnerabilities, risk and protective factors"*, before, during and after major disasters (Amri et al., 2018, p. 249).

This chapter aims to respond to these concerns by focussing on the underlying factors that impacted children's rights during the 2018 earthquakes in Lombok, Indonesia. Adopting a child-focussed research approach, the chapter explores children's experiences of the earthquakes, and how their rights could have been better protected. The aim of the research was to understand the impact of the earthquakes on children and young people in the worst affected areas of the island, including their psychological recovery, five months after the event. The research also aimed to assess the integration of CCDRR prior to the earthquakes, and disaster management and child protection strategies that were implemented post-disaster. These areas of research are important to build the resiliency of children and their communities for future natural disasters, and to reduce their vulnerabilities before, during and after disaster events. The chapter ends by outlining how participatory, rights-based research with children, young people and their communities is vital to ensure that disaster planning and post-disaster responses are appropriate, and that children's rights are not violated in the name of protection.

10.3 Methodology

The research for this chapter was conducted in January 2019, five months after the major earthquakes in August 2018. The first author has a close ongoing connection to the island, having lived and worked in Lombok on an AusAID Women's Health and Family Welfare program in 2000–2001, and returning regularly for visits and various research projects. From 2014–2017 she worked on an SSHRC (Canada) funded research project focussed on the impact of transnational migration on children of migrants left behind in remote communities of East Lombok (Ball, Butt, & Beazley, 2017; Beazley et al., 2018; Butt, Ball, & Beazley, 2017).

For this research, interviews were conducted in the capital city, Mataram, with government employees working with Bappeda (National Planning Agency), established contacts with local people affected by the earthquake, and with non-government organisations (NGOs) and individual volunteers involved in the relief effort in North and East Lombok. Research was also conducted in two villages in East Lombok which were affected by the earthquakes, and where the author has been conducting research for almost 20 years. In these villages interviews were conducted with the village heads, hamlet heads, school teachers, parents and young adults. Local and international newspaper articles, grey literature and NGO reports about the earthquakes were also reviewed for this study.

10.4 Research Findings

Lombok is an arid and remote island situated in Eastern Indonesian. It is characterised by high population growth, low wages, low education and employment, low status of women and girls, food insecurity, falling agricultural productivity, poor communications and poor health, including high infant mortality (Ball et al., 2017). The district of East Lombok is the third highest sending areas of transnational migrants in Indonesia, with impoverished conditions motivating an entrenched culture of unskilled and undocumented male migration to Malaysia, to the plantation sector, and unskilled female migration to the Middle East, to work as domestic workers (Khoo, Platt, Yeoh, & Lam, 2014; Lindquist, 2010). Transnational migration from the island is estimated at 200,000 people per year, although high rates of undocumented or 'irregular' migration mean exact figures are unknown (ILO, 2013). Overseas migrants strive to return to the island as a 'success', which includes bringing home gifts, buying household goods and building a concrete and brick house in a 'modern' style (instead of the traditional style made from rattan and bamboo, with thatched roofs).

The physical evidence of the contribution of transnational migrant workers can be seen across the island—the permanent concrete houses built from the money earned working abroad. These houses are known locally as *Rumah Malaysia* or *Rumah Arab* ('Malaysian Houses' or 'Arab Houses'), depending on where the money has come from to build them. In the villages where the research was conducted, almost every household had a family member who had worked, or was currently working, overseas to build a concrete house on their return (Ball et al., 2017). The new homes were usually built from cheap, poorer materials by the returned migrant workers and their family members—even the grandparents are involved. The houses had brick or concrete block walls, held together with weak mortar, and supported by little or no frames. They were usually built on a concrete floor base, with plastered walls, wooden beams, and clay tiles loosely secured on roofs, and flimsy or no ceilings. These structures do not have solid foundations and are heavy when they fall; "*If one had to design a system of construction for easy collapse and maximum injuries, this would be the perfect model*" (MacRae, 2018).

Most of those killed during the earthquakes were struck by falling debris in their own concrete homes, while an estimated 40 people died while praying in a mosque during Friday prayers (Paulus & Smith, 2018). Survivors struggled to access clean water and food, as water pipes were broken and shops were closed. After the earthquakes, survivors and other patients sheltered under makeshift tents, and relief calls were made for blankets and basic staples, such as rice, noodles, vegetables and nappies. Volunteers arrived from all over Indonesia to assist in the relief effort. The injured were treated outdoors in tents because hospitals were also damaged in the earthquakes.

10.4.1 Disaster Preparedness and Recovery

Despite Indonesia's propensity for natural hazards, the post-earthquake rescue and relief efforts were reportedly extremely slow in Lombok. Local people, volunteers and NGO relief personnel all complained about the lack of disaster preparedness, and the lack of equipment, or a clear recovery plan. For weeks after the earthquake the road from the capital, Mataram, to the North of the island was blocked with traffic, and ambulances and emergency services could not get through. Damaged roads and a lack of heavy machinery to move debris hindered rescue efforts, a clear indication that the Indonesian authorities were unprepared for the earthquake. A number of agencies were involved in rescue and relief operations, including the police, the military, government agencies, NGOs and domestic volunteers. However, because of a lack of equipment to clear damaged roads, local rescue teams were delayed in reaching many of the worst affected areas in North Lombok (Paul, 2018). There were only two helicopters sent from the National Disaster Management Agency (BNPB), which were not enough to cover the huge area, much of which is densely covered forest.

During discussions about the recovery process informants were upset with the Indonesian government for refusing to 'upgrade' the disaster from a 'local' to 'national' emergency, and to appeal for international assistance. According to the Vice President, the earthquakes were not a national emergency and the post-disaster handling process conducted by the provincial government was adequate (Nugraha, 2018). BNPB also claimed that Indonesia had sufficient resources and experience in handling natural disasters, and did not need outside help. These claims were hotly contested by local volunteers and NGOs.

Further, many families complained that in spite of President Joko Widodo visiting Lombok three times after the earthquakes, each time promising to send government aid for every home owner to rebuild their homes, they were yet to see any of the financial assistance five months later. The President promised 50 million (A$5000) per household for homes that were destroyed, 20 million (A$2000) for homes that were partially destroyed and 10 million (A$1000) for those who had suffered minor damage. As one local resident complained: "*[W]e have not received even 1000 Rupiah (10 cents), nothing*".

As a result of the lack of promised funds, people speculated that it was due to the local provincial government holding onto the money, or some other form of disaster-related corruption. These rumours were not helped when two government officials from the Ministry of Religious Affairs were found guilty, during the research period, for allegedly demanding and accepting illegal levies from earthquake-affected mosques when distributing Ministry money to rebuild the mosques on the island (Nugraha, 2019). In another incident a Mataram councillor was arrested for demanding money from an official who oversaw post-disaster plans to rebuild several elementary and junior high school buildings in the city (Nugraha, 2019).

In the context of disaster justice, disaster funds are susceptible to corruption due to poor supervision and a lack of accountability. As an official from Indonesia Corruption Watch (ICW) stated: *"During disasters, people are mostly focusing on the victims or how to rebuild infrastructure as fast as possible, while forgetting about monitoring the disbursement of disaster funds. Such conditions are ripe for self-enrichment"* (Nugraha, 2019).

The other issue that hindered the recovery process, and which was frequently mentioned as hampering the disbursement of promised government funds, was the Sulawesi earthquake and tsunami in September 2018, killing 4340 people, and the West Java volcanic eruption and tsunami in December 2018, which killed close to 500 people. Many informants described how all the relief organisations and NGOs immediately left Lombok when these more severe disasters happened, and never returned. As one middle aged man bemoaned: *"Since Sulawesi, the people of Lombok have been forgotten"* (pers. comm., 2019).

As a result of the slow recovery and rebuilding of houses, Lombok was still struggling to recover from the earthquake at the time of the research. Temporary schools were still running classes in tents, waiting for school structures to be assessed and classes to resume inside the buildings. Some villages in the North of the island still had no power or running water, and communities living 'deep in the forest' had not yet been reached by relief efforts. Thousands of families were still living in leaky tents, which were susceptible to high winds, rain and disease. One woman told of how a month after the three earthquakes she had to receive chemotherapy treatment in a tent outside the hospital, while a mother gave birth in a bed next to her (pers. comm., 2019). This was because hospital staff were too scared to treat people inside the building, for fear of another quake.

10.4.2 Five Months On

The two villages in this research were only 38 kilometres apart, but due to their geographic locations they were impacted by the earthquakes quite differently. The first village, Village A, is high up in the Southern foothills of Mount Rinjani, further away from the earthquake epicentres, and ten kilometres from the local town. In this village many of the houses were damaged, with large cracks running through the walls and ceilings. Although only two houses had totally collapsed during the earthquakes, the entire village of 500 people had left their houses and camped out in the open air, for fear of another quake coming at night, burying them while they slept. The adults in this village said that they had been sleeping outside under tarpaulins for months, and that the school had only recently re-opened after being closed for many months, as the teachers (from other towns and villages) had been unable to return.

Village B is located 38 kilometres north of Village A in North East Lombok, only 10 kilometres from the epicentre of the second earthquake. This village is on the coast—across the strait from the island of Sumbawa. The first thing that was instantly apparent entering Village B was the devastation and destruction along either side of the tarmac road. It was like a scene from a war movie—as though a bomb had been dropped, wiping out everything in all directions. Almost every concrete house was destroyed. Only the concrete floors remained. The walls of the houses had simply fallen outwards, like flat walls on toy houses, and the roof tiles had crashed down on the concrete floors below. Everyone was living in tents or makeshift wooden structures provided by various international relief organisations, pitched in front of the rubble and ruins of what used to be their homes. Families had their beds and possessions on the street, as they said the tents and the temporary shelters were too hot to sleep inside. People complained about the tin roofs of the hundreds of houses provided by one international NGO—as it was not possible to sleep inside them in the 35-degree Lombok heat. A woman with a baby in Village B described how she could not take her baby home as planned after the birth, as her house had been destroyed. All her baby had known in the past five months was the inside of their stifling tent.

Everywhere there were piles and piles of small broken bits of concrete and plaster rubble, where it had been dumped and not collected—along the beaches, in vacant lots, along the roads and in the fields. Pieces of furniture, plastic packets of relief food, clothing and children's shoes were scattered through the rubble. As we passed a collapsed minimarket (where apparently several people had died), dozens of children could be seen picking through the flattened debris, hoping to find a treasure from the former store.

In Village A, farmers reported their fields were cracked and covered in dust. Others described how sulphuric acid was seeping through the soil. A landslide had destroyed the steep path down to the river where women collected water and washed pots and clothes, and where children played. It was no longer safe for the children to go there, and they were instructed to stay in the village, close to their parents at all times. This loss of place and happy memories was a significant loss for the children, and their sense of well-being (Morrice, 2012).

In the coastal Village B, the second quake prompted a tsunami warning, forcing the entire village to flee their homes in the middle of the night. It was midnight, raining and pitch black as the electricity had cut out, and there was no moon. Everyone immediately joined forces, to first transport women and children in cars and on motorbikes to the top of the hill, where they had to wait all night in the rain without shelter. The tsunami alert was later called off, although families took their time to descend and return to their makeshift tents which they had constructed after the first quake. The next morning they woke up to the devastating scene of all their homes totally destroyed. Fortunately, no one was killed during the second quake in Village B, as everyone had been sleeping in tents outside their houses since the first quake two weeks before.

At the time of the research a minority of people who had access to funds had begun to very slowly rebuild their houses. Of interest was the fact that most of these people were rebuilding using traditional materials, with wood and rattan, as they were considered to be safer than the brick and concrete style previously favoured. It was also a government directive that people had to build *anti-gempa* (earthquake proof) houses themselves. The brick or concrete walls of these new houses were only permitted to be a meter high, and the rest of the building needed to be constructed using traditional methods, with wood, aluminium roofs and rattan walls. Many people commented on the scarcity of wood, rattan and aluminium available in Lombok, to meet this directive.

In Village A, one family still slept inside their house. This house was once considered poor and backward as it was a traditional Sasak style house with rattan walls and a thatch roof. The father and mother of the house had decided not to migrate to Malaysia or Saudi Arabia for work, and instead the father had local employment as a salesman. The children in the family appeared happy and the father proudly claimed that his house was an *anti-gempa* house which seemed to give him immense pleasure, as for so long people had looked down on his family for being poor. Unlike all the houses around him when the earthquakes happened the walls did not crack or fall down, they just swayed. He did observe, however, that the house was not wind-proof, and he held fears that it may blow away in a big storm. In other parts of the same village, and also in Village B, men and women reported that they were preparing to go overseas again to work, as they needed the money to rebuild. The impact of transnational migration on children left behind by their migrant parents is well documented (Ball et al., 2017; Beazley et al., 2018; Butt et al., 2017). These children are normally cared for by their grandparents, extended families or neighbours, which can lead to a multitude of significant concerns regarding their protection, especially during a crisis.

10.4.3 Psychological Responses

In spite of Lombok's location between two tectonic plates, no one in the research locations remembered an earthquake as large as the ones they had endured. They were the first major earthquakes in Lombok in living memory, and had a significant impact on the psyche of the local people, especially the children. At the time of the research the island had experienced thousands of aftershocks, and the ground had continued to shudder on a weekly basis, with some tremors recorded as large as 5.4 magnitude. Many adults had downloaded an app on their mobile phones that recorded the size and location of every earthquake and aftershock, and whether there was a tsunami warning. The ongoing tremors made people on edge and more anxious by the relentless experience, with heightened anxiety about the stability of houses still standing.

Trauma of post-disaster is a form of PTSD, which is a maladaptive reaction to a traumatic experience or event (Tentama, Mulasari, Sukesi, & Haryono, 2014). Children are more susceptible to PTSD than adults, and it is one of the most common psychological outcomes of exposure to disaster in children (Salcioğlu & Başoğlu, 2008). A number of recent studies have shown how symptoms of distress following a traumatic event may be expressed in multiple ways and can take their toll on children's social and emotional development (Cahill, Beadle, Mitch, Coffey, & Crofts, 2010; Mutch & Gawith, 2014; Widyatmoko, Tan, Seyle, Mayawatu & Silver, 2011). Studies that have examined the impact of earthquake trauma on children reveal that exposure to unpredictable and uncontrollable earthquakes and aftershocks may lead to pervasive conditioned fears in the majority of survivors (Salcioğlu & Başoğlu, 2008; Widyatmoko et al., 2011). These pervasive fears can generalise to a wide range of situations, including entering buildings, being alone, sleeping alone and sleeping in the dark. Other post-disaster signs of distress in children have been recorded as decreased appetite, bed wetting, headaches, stomach-aches, excessive crying, separation anxiety from parents and listlessness (Margolin, Ramos, & Guran, 2010; Widyatmoko et al., 2011). Children of distressed parents also exhibit more distress themselves, as children look to their parents for cues how to respond to the disaster (Margolin et al., 2010) Distressed children may also exhibit below average performance at school, including poor concentration, and regressive behaviours.

The research conducted a brief assessment of children's behaviour and psychological problems, relying on teachers and parents as primary informants. In all the research locations there were parents who described how the whole family was still sleeping outside as at least one of their children was too scared to sleep inside, due to being *sakit trauma* (traumatised) by the earthquakes and ongoing tremors. One mother described how her child slept in the doorway as he knew that he would be more protected by the frame, and it was closer to run outside, if he had to. Teachers also told of some children being too scared to enter any building or to stay alone, or that they refused to leave their parents and go to school for this reason. Teachers also reported that parents were too scared to send their children to school. The teacher in Village B explained how the children's trauma was demonstrated by their erratic and uncontrollable behaviour and hyperactivity. Other children were constantly day-dreaming or withdrawn. The same female teacher also described how children in her class would just start singing loudly, banging on furniture, or screaming, as a result of what she described as 'trauma'.

During discussions parents described how their children suffered from nightmares, and how some young people reported feeling 'phantom quakes', when they thought the ground was shaking but it was not, which made them feel panicked and dizzy. In Village B one family of eight only cooked and sat inside their temporary emergency wooden home during the day, but were still sleeping outside at night, in one double bed. The mother of the house said she and her children were not *berani* (brave enough) to sleep in the house at night. Parents said that they all wanted to recover from the earthquake, but still had high levels of anxiety and the constant aftershocks made it very hard to move on.

10.4.4 Trauma Healing

After the quakes in Lombok, the National Disaster Management Agency (BNPB) reported that NGOs and government agencies were working with communities to restore their mental health (Kompas, 2018). In the months following the disaster dozens of international NGOs worked with government agencies to run a range of activities for hundreds of children, to cheer them up, raise their spirits, increase resilience and optimism, and to teach them to be motivated and to 'chase their dreams' after the earthquakes (Manafe, 2018). In Indonesia such activities are referred to as *trauma healing*, and have become popular with NGOs and post-disaster relief efforts since the 2004 tsunami. In different activities children played competitive sports, and participated in circus activities and arts therapy (art, music, dance and listening to fairy tales).

> *Listening to fairy tales can improve cognitive abilities, train the imagination, and encourage children's emotional development. The many benefits obtained from hearing stories, are used as 'trauma healing' for children affected by the earthquake in Lombok.* (Sewaka, 2018)

Research which is focussed on helping children adjust after trauma suggests that emotional processing can be assisted through arts-based activities, leading to absorption of emotional disturbances (Mutch & Gawith, 2014). Without such processing children can continue to have nightmares and show signs of distress (Cahill et al., 2010; Mutch & Gawith, 2014). In what can only be understood as an opportunistic and unprincipled move, a number of private companies (Walls & Garudafood) reported how they went to North Lombok to run activities for the 2641 children living in tents and to offer them 'solace food'—Paddlepops, noodles and jelly, which they then advertised in press releases (Meirina, 2018). Other relief organisations trained teachers in schools to build psychosocial support activities into their curriculum, to recognise signs of distress and to provide comfort and support (Save the Children, 2018).

The village head in village A explained how soon after the earthquakes, relief workers had come to the village to implement 'trauma healing' for the children outside the primary school, which involved playing games, dancing and singing songs, to give them an opportunity to process their experiences in a safe place, and to take their minds off the disaster and their feelings of fear. One NGO activity was the handing out of Indonesian flags to the children in Village A and telling them to wave the flags and chant together: "*kami tidak takut*" ("we are not afraid"), again and again. One teacher confided that she was running out of ideas for activities that were fun and that helped to lessen stress and anxiety. She also felt that after five months everything had become 'trauma healing' and that there were far too many activities labelled as 'healing', which she felt had become superficial and meaningless. However, research does suggest that if done correctly children are able to positively adjust after traumatic events, by being given the opportunity to process their experiences through certain therapeutic activities. In one newspaper article about *trauma healing* in Lombok children reported feeling calmer, and a child reported that she no longer trembled when she entered a building (Savitiri, 2018). However, the same article also warns of 'trauma building' where children's trauma is enhanced as a result of television companies and individuals continuing to share video clips and audio clips of harrowing incidents during the disaster, via TV, radio and social media (Savitiri, 2018). The ongoing aftershocks also re-traumatised children and adults, with people panicking whenever a tremor occurred.

10.5 Post-Disaster Education and Disaster Risk Reduction

As well as *trauma healing* activities children were taught the scientific reasons for the earthquakes and how to protect themselves when one begins. They learnt this by memorising a song with the title: "Kalau Ada Gempa" ('If there is an earthquake').

> *If you feel an earthquake protect the head*
> *If there is an earthquake go under a table*
> *If there is an earthquake move faraway from glass*
> *If there is an earthquake run into the open.*

This song was part of the disaster preparedness and disaster mitigation education programs that were rolled out in schools across Lombok in the months following the earthquakes in 2018 and 2019. It was clear from these activities that there had been no such CCDRR programmes prior to the earthquakes. Teachers and government officials confirmed that they had never educated children about how to respond in the event of an earthquake.

During interviews with teachers and community members, it was clear that no one, including children, had been prepared for the earthquakes, and they did not know how to respond to the disaster. Research shows that a lack of disaster preparedness or pre-disaster education increases the impact of any event, especially on children. If a child is trained in pre-disaster education then they are more resilient when an earthquake hits, and can recover more quickly. They can also teach their siblings and other family members how to contribute to disaster preparedness and response planning.

10.6 Conclusion

In disasters nearly all the rights of children are implicated—ranging from basic survival to freedom from abuse and exploitation, and access to health care and education. All too often, at the critical juncture following a major disaster, children are relegated to the margins. (Irish Red Cross, n.d.)

The aim of this paper has been to draw attention to the violations of children's rights during the 2018 earthquakes in North and East Lombok. From the data that was collected the immediate issues that children confronted after the disaster included their basic right to live and the right to enjoy good health. The earthquakes also presented a direct threat to children's right to safety, adequate shelter, and access to clean water, food and clothing. The right to adequate shelter includes privacy, security, lighting, ventilation and sanitation, none of which the children in the research enjoyed, five months after the disaster—a clear example of distributional injustice. Girls in post-disaster Lombok were particularly vulnerable to exploitation and abuse as they had to live in tents with other members of the community, and use public sanitation facilities. As a result of having to live in tents for extended periods of time children were exposed to cold, damp, heat, rain and wind, and the constant threat of disease. They also lost their right to education when their schools were destroyed or damaged, the teachers stopped coming, and they remained closed for months on end.

The fact that government and local communities were totally unprepared for the scale of the earthquakes and that disaster preparedness and disaster mitigation education programs were only implemented after the events was a serious violation of children's rights, and an example of procedural injustice. The Indonesian and provincial governments have a legal obligation to protect children and to ensure their safety and security in all disaster contexts. Preparing for and responding to the specific needs of children during and post-disaster is especially important, due to the long-term repercussions of delays in relief efforts. For example, a child who is unable to attend school for a significant period of time might never fully recover from that loss of an education: "*The consequences of a disaster that occurs during the formative years of a child can last decades*" (Irish Red Cross, n.d.).

This research has demonstrated the vulnerability of children during disasters, and how children's marginality can be exacerbated by the impact of natural disasters, especially for children living in impoverished communities with minimal resilience to external shocks. Children's social marginalisation means they are often not considered during disaster prevention or mitigation procedures, and they are certainly not able to participate or influence those processes. It is therefore vital that in the future, specific attention is paid to children's rights and child protection issues during the design and implementation of disaster preparedness, and response policy or programs.

Children and youth have been formally recognised as agents of change in the global commitment for Child-Centred forms of Disaster Risk Reduction (UNISDR, 2015). The CCDRR acknowledges children as participants in policy making and programmes, and actively engaged in decision making and planning processes for disaster prevention preparedness and response. Although outside the scope of this research, it was clear from the findings that the only way to address procedural justice in relation to children's needs in any disaster preparedness or mitigation planning is to include children in these processes. In this regard both Amri et al. (2018, p. 9) and Beazley (2016) state that to better protect children in disasters, there is a need for child-centred participatory research to inform future policy. This is because children have the right to express their opinions in all matters that affect them. This is related to children's right to participate and for decisions to be made in the best interests of the child. As Save the Children (2010, p. 4) has stated:

> *Children should not be seen as victims, but actors in addressing the impacts of natural disasters ... on their lives and the life of their community. Policy makers and local authorities need to listen to children and see them as part of the future solutions. Children need to be involved in initiatives to build up their knowledge and resiliency thereby reducing the impacts of disasters, especially on the most vulnerable.*

References

ABC. (2018). *Indonesia Earthquake Lifted Lombok by 25 Centimeters, Scientists Say.* Retrieved from https://www.abc.net.au/news/2018-08-11/lombok-lifted-25-centimetres-by-indonesia-earthquake/10109604

Alexander, D., & Magni, M. (2013). Mortality in the L'Aquila (Central Italy) Earthquake of 6 April 2009. *PLoS Currents, 5*.

Amri, A., Haynes, K., Bird, D. K., & Ronan, K. (2018). Bridging the Divide Between Studies on Disaster Risk Reduction Education and Child-Centred Disaster Risk Reduction: A Critical Review. *Children's Geographies, 16*(3), 239–251.

Ball, J., Butt, L., & Beazley, H. (2017). Birth Registration and Protection for Children of Transnational Labor Migrants in Indonesia. *Journal of Immigrant & Refugee Studies, 15*(3), 305–325.

Beazley, H. (2016). Inappropriate Aid: The Experiences and Emotions of Tsunami 'Orphans' Living in Children's Homes in Aceh, Indonesia. In M. Blazek & P. Kraftl (Eds.), *Children's Emotions in Policy and Practice: Mapping and Making Spaces for Childhood* (pp. 34–51). Basingstoke: Palgrave Macmillan.

Beazley, H., & Ball, J. (2018). Children Youth and Development. In A. McGregor, L. Law, & F. Miller (Eds.), *Routledge Handbook of Southeast Asian Development* (pp. 211–223). London: Routledge.

Beazley, H., Butt, L., & Ball, J. (2018). 'Like It, Don't Like It, You Have To Like It': Children's Emotional Responses to the Absence of Transnational Migrant Parents in Lombok, Indonesia. *Children's Geographies, 16*(6), 591–603.

Butt, L., Ball, J., & Beazley, H. (2017). Migrant Mothers and the Sedentary Child Bias: Constraints on Child Circulation in Indonesia. *Asia Pacific Journal of Anthropology, 18*(4), 372–388.

Cahill, H., Beadle, S., Mitch, J., Coffey, J., & Crofts, J. (2010). *Adolescents in Emergencies.* Parkville: Youth Research Centre, The University of Melbourne.

Carter, S. (2019, March 14). Education in the Face of Disaster. *Palladium.* Retrieved from https://thepalladiumgroup.com/news/Education-in-the-Face-of-Disaster

Delaney, S. (2006). *Protecting Children from Sexual Exploitation and Sexual Violence in Disaster and Emergency Situations.* Bangkok: EPCAT International.

International Labour Organisation (ILO). (2013). *Better Protecting Indonesian Migrant Workers Through Bilateral and Multilateral Agreements.* Retrieved from http://www.ilo.org/jakarta/info/public/pr/WCMS_212738/lang%2D%2Den/index.htm

Irish Red Cross. (n.d.). *Children's Rights in Disasters.* Retrieved from https://reliefweb.int/sites/reliefweb.int/files/resources/IDL-Information-Sheet-No.6-Children-in-Disasters-March-2018.pdf

Jakarta Post. (2018). *Another 7-Magnitude Quake Hits Lombok, August 19th.* Retrieved from https://www.thejakartapost.com/news/2018/08/19/breaking-another-7-magnitude-quake-hits-lombok.html

Jia, Z., Tian, W., He, X., Liu, W., Jin, C., & Ding, H. (2010). Mental Health and Quality of Life Survey Among Child Survivors of the 2008 Sichuan Earthquake. *Quality of Life Research, 19*(9), 1381–1391.

Khoo, C. Y., Platt, M., Yeoh, B., & Lam, T. (2014). *Structural Conditions and Agency in Migrant Decision-Making: A Case of Domestic and Construction Workers from Java, Indonesia.* Sussex: Migrating Out of Poverty Research Programme Consortium.

Kompas. (2018). *'Trauma Healing' Children of Lombok Earthquake Victims, from Competition to Singing Together.* Retrieved from https://newsbeezer.com/indonesiaeng/trauma-healing-children-of-lombok-earthquake-victims-from-competition-to-singing-together/

Kumala Dewi, L. P. R., & Dartanto, T. (2019). Natural Disasters and Girls Vulnerability: Is Child Marriage a Coping Strategy of Economic Shocks in Indonesia? *Vulnerable Children and Youth Studies, 14*(1), 24–35.

Lindquist, J. (2010). Labour Recruitment, Circuits of Capital, and Gendered Mobility: Reconceptualizing the Indonesian Migration Industry. *Pacific Affairs, 83*(1), 115–132.

MacRae, G. (2018). Lombok Earthquakes: Different Building Designs Could Lessen Future Damage. *The Conversation.* Retrieved June 2019, from http://theconversation.com/lombok-earthquakes-different-building-designs-could-lessen-future-damage-101440

Manafe, D. (2018). *Dongeng jadi Trauma Healing bagi Anak Korban Gempa Lombok* (Fairy Tales Become Trauma Healing for the Children of the Lombok Earthquake Victims), Berita Satu. Retrieved June 2019, from https://www.beritasatu.com/nasional/505004/dongeng-jadi-trauma-healing-bagi-anak-korban-gempa-lombok

Margolin, G., Ramos, M. C., & Guran, E. L. (2010). Earthquakes and Children: The Role of Psychologists with Families and Communities. *Professional Psychology: Research and Practice, 41*(1), 1–9.

Massola, J., Rosa, A., & Rompies, K. (2018, August 6). Multiple Deaths After Powerful Earthquake Rocks Lombok and Bali. *Sydney Morning Herald.* Retrieved from https://www.smh.com.au/world/asia/powerful-enough-to-put-us-on-the-floor-earthquake-hits-lombok-bali-20180805-p4zvod.html

Meirina, Z. (2018). GarudaFood berikan penyembuhan trauma bagi anak Lombok. *Antara NTB.* Retrieved from https://mataram.antaranews.com/nasional/berita/735773/garudafood-berikan-penyembuhan-trauma-bagi-anak-lombok?utm_source=antaranews&utm_medium=nasional&utm_campaign=antaranews

Morrice, S. (2012). Heartache and Hurricane Katrina: Recognising the Influence of Emotion in Post-Disaster Return Decisions. *Area, 45*(1), 33–39.

Mutch, C., & Gawith, E. (2014). The New Zealand Earthquakes and the Role of Schools in Engaging Children in Emotional Processing of Disaster Experiences. *Pastoral Care in Education, 32*(1), 54–67.

Nugraha, P. (2018, August 21). Govt Won't Declare Lombok Quakes 'National Disaster'. *Jakarta Post*.

Nugraha, P. (2019, January 17). NTB Religious Affairs Officers Nabbed for Allegedly Demanding Share of Post-Disaster Reconstruction Fund. *Jakarta Post*. Retrieved from https://www.thejakartapost.com/news/2019/01/17/ntb-religious-affairs-officers-nabbed-for-allegedly-demanding-share-of-post-disaster-reconstruction-fund.html

Paul, B. (2018, September 22). Lombok Earthquakes Reveal That Indonesia's Disaster Management Is Shaky. *East Asia Forum: Politics and Public Policy in East Asia and the Pacific*. Retrieved from https://www.eastasiaforum.org/2018/09/22/lombok-earthquakes-reveal-that-indonesias-disaster-management-is-shaky/

Paulus, I. & Smith, N. (2018). Indonesia Earthquake: At Least 98 Dead as Tourist Island of Lombok Shaken by 6.4-Magnitude Tremor. *Daily Telegraph*. Retrieved from https://www.telegraph.co.uk/news/2018/07/29/indonesia-earthquake-least-three-dead-tourist-island-lombok-shaken/

Peek, L. (2008). Children and Disasters: Understanding Vulnerability, Developing Capacities, and Promoting Resilience—An Introduction. *Children, Youth and Environments, 18*(1), 1–29. Retrieved from http://www.jstor.org/stable/10.7721/chilyoutenvi.18.issue-1 [Google Scholar].

Pfefferbaum, B., Pfefferbaum, R. L., & Van Horn, R. L. (2018). Involving Children in Disaster Risk Reduction: The Importance of Participation. *European Journal of Psychotraumatology, 9*(2). Retrieved from https://www.ncbi.nlm.nih.gov/pmc/articles/PMC5804784/

Salcioğlu, E., & Başoğlu, M. (2008). Psychological Effects of Earthquakes in Children: Prospects for Brief Behavioral Treatment. *World Journal of Paediatrics, 4*(3), 165–167.

Save the Children. (2010). *Living with Disasters and Changing Climate: Children in Southeast Asia Telling Their Stories about Disasters and Climate Change*. Retrieved from https://www.preventionweb.net/files/submissions/15087_Livingwithdisastersweb.pdf

Save the Children. (2011). *Legacy of Disasters: The Impact of Climate Change on Children*. London: Save the Children, UK. Retrieved from https://resourcecentre.savethechildren.net/node/3986/pdf/3986.pdf

Save the Children. (2018). *Save the Children Raises Concerns for Indonesians on Earthquake-Devastated Lombok*. Retrieved from https://www.savethechildren.org.au/media/media-releases/Save-the-Children-raises-concerns-for-Indonesians

Savitiri, N. (2018). *Mengembalikan Senyuman Dan Keberanian Korban Pasca Bencana di Indonesia*. Australian Broadcasting Corporation (Returning the Smiles and Courage of Post-Disaster Victims in Indonesia). Retrieved from https://www.abc.net.au/indonesian/2018-10-19/mengembalikan-senyuman-dan-keberanian-korban-pasca-bencana/10366688

Sewaka, A. (2018). *Dongeng Bantu Sembuhkan Trauma Anak-anak Korban Gempa Lombok* (Fairy Tales Help Heal the Trauma of Children of the Lombok Earthquake Victims). Retrieved from https://www.haibunda.com/parenting/20181118190508-62-28615/dongeng-bantu-sembuhkan-trauma-anak-anak-korban-gempa-lombok

Tentama, F., Mulasari, S. A., Sukesi, T. W., & Haryono, W. (2014). The Effectiveness of Trauma Healing Methods to Reduce Post-Traumatic Stress Disorder (PTSD) on Teenage Victims of Mount Merapi Eruption. *International Journal of Research Studies in Psychology, 3*(3), 101–101.

UNESCAP. (2015). *Overview of Natural Disasters and Their Impacts in Asia and the Pacific*, 1970–2014. United Nations Economic and Social Commission for Asia and the Pacific. Retrieved from https://www.unescap.org/resources/overview-natural-disasters-and-their-impacts-asia-and-pacific-1970-2014

UNISDR (United Nations International Strategy for Disaster Reduction). (2015). *Sendai Framework for Disaster Risk Reduction 2015–2030*. Retrieved from http://www.wcdrr.org/uploads/Sendai_Framework_for_Disaster_Risk_Reduction_2015-2030.pdf

United Nations. (1989). *Convention on the Rights of the Child*. Retrieved from https://www.unicef.org/sites/default/files/2019-04/UN-Convention-Rights-Child-text.pdf

United Nations International Strategy for Disaster Risk Reduction (UNISDR). (2015). *UNISDR Says the Young Are the Largest Group Affected by Disasters*. Retrieved from https://www.unisdr.org/archive/22742/

WHO. (2011). *Disaster Risk Management for Health Fact Sheets: Child Health*. Retrieved from https://www.who.int/hac/events/drm_fact_sheet_child_health.pdf?ua=1

Widyatmoko, S., Tan, E. T., Seyle, D. C., Mayawatu, E. H., & Silver, R. C. (2011). Coping with Natural Disasters in Yogyakarta, Indonesia: The Psychological State of Elementary School Children as Assessed by Their Teachers. *School Psychology International, 32*(5), 484–497.

World Vision. (2018). *Indonesia Quakes and Tsunamis: Facts, FAQs*. Retrieved from https://www.worldvision.org/disaster-relief-news-stories/2018-indonesia-earthquake-facts

CHAPTER 11

How a Failure in Social Justice Is Leading to Higher Risks of Bushfire Events

Janet Stanley

11.1 Introduction

Bushfires are one of the more frightening and potentially damaging extreme events that threaten lives and livelihoods. While Australia has experienced many severe bushfires since European settlement, aggravating factors are leading to an increase in the occurrence of bushfires, a longer fire season, now extending from September to April, and more extreme fires that are very difficult to extinguish without the occurrence of rain. While the Intergovernmental Panel on Climate Change identified an increase in bushfire risk, this increase is occurring at the higher end of anticipated rates (Allen et al., 2011). The numbers of bushfires in Australia have increased 40% over the five years to 2016 (Dutta, Das, & Aryal, 2016). In 2013, this represented about 238,940 wildfires for the year, as measured through satellite imagery, although many of these represent small fires (SCRGSP, 2016). Given the current climate change mitigation policies, the world is on track to have a temperature rise of 3.1°C to 3.7°C (Climate Action

J. Stanley (✉)
Melbourne Sustainable Society Institute, School of Design,
The University of Melbourne, Parkville, VIC, Australia
e-mail: janet.stanley@unimelb.edu.au

Tracker, 2018). The rise in occurrence and intensity of bushfires in Australia can be observed even with the current temperature increase of 1.0°C since 1910. This is not only an Australian problem, with many catastrophic fires occurring internationally in the past few years and in countries that have never experienced bad fires previously, such as Sweden and the United Kingdom, and even the Arctic Circle and Alaska.

An important consideration for a fair and well-functioning society should be to understand and respond to the variable impacts of extreme events, including bushfire, on different people (March et al., 2018). This should include attention to equity considerations in relation to ability to avoid exposure to bushfire, respond to a bushfire and recover from a bushfire. These issues could be considered on a geographic, group or individual basis, and may refer to equalising competition in a market approach, equalising abilities to respond, or equalising outcomes after an extreme event. It is argued in this chapter that there is a need for equity for all people who will be impacted, including those who are alive now, and future generations, in that people should be provided with the capability to be what they wish, supported by a society that provides conditions to make this possible (Sen, 1992). This value position holds when considering the potential environmental, social and economic impacts of a bushfire on exposed people.

However, considerations of equity could also be applied in a different way. This chapter argues that a social justice failure for one group of people is creating significant problems for other members of society. It argues that government policy failures, leading to poor outcomes for some youth living in rural/urban interface areas, increase the risk of the occurrence of bushfire arson. This issue is illustrated using the city of Melbourne and regional settlements in Victoria. Such a risk is further increased by policy failures to reduce the occurrence of climate change, an issue that could further be viewed as a social justice failure, as bushfire risk is not equal for all people, it being dependent on geographical location.

11.2 The Cause of Ignition of Bushfires in Australia

A bushfire or wildfire is defined as a fire that occurs in vegetation rather than confined to a building or structure. However, the fire may move from the original source of ignition, such as in a dumped car, to become a bushfire, leading to some variability as to how the fire is officially categorised. The cause of ignition of bushfires in Australia is commonly understood as being accidental, suspicious, deliberate, natural, re-ignition or embers from another bushfire, a prescribed burn that escapes, other and unknown (Bryant, 2008). However, many bushfires are not officially recorded and/or the cause of the fire investigated. Where a record is made, the cause of the fire can vary according to how the ignition categories are defined and the cause is understood. Thus, the causes of ignition are often uncertain.

In 2008, Bryant undertook an analysis of 280,000 bushfires from 18 major fire and land management agencies from all Australian states and territories, covering the years 1997/98 to 2001/02. To account for variation in reporting, the causes of ignition were re-categorised into consistent descriptions. She found the biggest category, covering 37% of bushfires, was recorded as 'suspicious', with another 13% 'deliberately lit'. A further 35% of fires were 'accidentally lit', 6% due to 'natural causes', 5% 're-ignition' and 4% due to 'other' causes. Thus, 85% of bushfires were likely to have been lit by people.

11.3 Who Lights Fires and Why

Bushfires lit by people can be viewed as due to a malicious intent or an accidental outcome. Accidental or reckless bushfires can occur due to a range of activities, such as leaving a campfire unattended, burning off rubbish or timber windrows, and using a combustion engine vehicle or machine on a high fire danger day, with the risk that it may throw out sparks that start a fire. At times it may be difficult to identify the thinking behind the ignition, although in Australia the legal response to an intentional or accidental fire occurrence is very different.

The largest group of people who maliciously light fires are youth aged about 14 to 20 years of age. They are thought to account for about 40% of bushfires. Around further 14% of bushfires are lit by children under this age. Fire lighting activity appears to increase again for those over 30 years of age, especially with unemployed males. The chances of a bushfire resulting from accidental or malicious ignition is likely to increase as the fire danger rating is judged to be a 'severe', 'extreme' or 'catastrophic or code red' bushfire danger day (Bradstock, Penman, Boer, Price, & Clarke, 2014).

Youth who commit arson are predominantly males, who may engage in substance abuse and other criminal behaviours, including theft, criminal damage and vandalism (Dolan & Stanley, 2010). They are likely to have absent fathers, little home supervision or discipline, have experienced child abuse and neglect, and associate with an anti-social peer group. They often experience learning difficulties with below average academic achievement. Lighting fires may give a feeling of excitement, defiance and power, or it may be an expression of displaced anger. There may be no attempt to extinguish the fire, little consideration of consequences, feelings of remorse and fear of punishment. Fortunately fire-lighting behaviour diminishes as the young person ages. Older males who engage in fire-lighting also have a history of social and educational disadvantage, poor family functioning in childhood, and low self-esteem along with an interest in fire (Doley, Dickens, & Gannon, 2016).

11.4 Where Bushfires Occur

Many bushfires occur near urban areas that border with rural or forested land (Bryant, 2008). Often these places are new housing developments that create urban sprawl on the fringe of Melbourne, Sydney and regional centres, as well peri-urban development a short distance from the main urban settlement. Families tend to be attracted to these areas as they offer more affordable housing. However, this saving often comes with other costs, the housing price reflecting a lack of accessibility to jobs and services, particularly reflecting a lack of public transport (Bryant, 2008; Price, 2013; Stanley, Stanley, & Hansen, 2017). Movement into forested areas not far from a major urban area is also taking place in many countries, with a pursuit of a particular lifestyle where people permanently live or have a holiday house close to nature, thus in areas more vulnerable to bushfire (Bryant, 2008; Price, 2013). Melbourne's rural/urban interface area is said to be the most vulnerable for bushfire in the world (Buxton, Haynes, Mercer, & Butt, 2011). This risk is increasing as the population numbers increase in these interface areas (Collins, Owen, Price, & Penman, 2015).

11.5 Why Bushfires Occur in This Interface Area

Thus, two issues tend to coincide, leading to a higher risk of bushfires in the rural/urban interface areas. These are the penetration of urban areas, such as new housing developments, into locations where the environment is vulnerable to bushfires. These new developments are often associated with poor infrastructure provisions, leaving some youth disadvantaged, with lower levels of social inclusion and opportunities in life, thus vulnerable to committing crime, including malicious fire lighting. The association between disadvantage and crime is recognised in the environmental criminology and psychology literature (see e.g. Baumeister, DeWall, Ciarocco, & Twenge, 2005; Grubb & Nobles, 2016). These associations are discussed more fully below.

The drivers of social outcomes can be understood as occurring in four broad clusters (Stanley, 2011a). These are societal impacts, climate change being one of these, but also population growth, the world economy, conflict, ecosystem services and so on. The second cluster relates to government policy, infrastructure, services, transport and capital resources, as well as land use decisions. Community dynamics make up the third cluster, encompassing issues such as community support, sense of community, social capital and sense of place. The fourth cluster relates to personal characteristics, such as age, ethnicity, gender and personality, all largely fixed and not directly subject to policy changes, but personality and character may arise due to adverse experiences. Examples could be anger and feelings of helplessness, leading to a faulty perception of how to reveal power through fire. The reasons for fire in interface areas can be understood in relation to these four clusters.

Climate change and population growth are particularly important components of societal impacts in relation to bushfire. Australia is one of the highest per capita emitters of greenhouse gases in the world (Flanagan, 2019). The failure to put in place policy to adequately address climate change is greatly increasing bushfire risk. The changes in the climate, particularly rises in temperature, and changes in rainfall patterns increase the likelihood that a bushfire, once lit, will have the fuel conditions for the fire to grow in size and intensity. Thus, even if the same numbers of people are lighting bushfires as in the past, the risk of a serious bushfire occurring is higher and growing, as the climate becomes more extreme.

Australia is experiencing rapid population growth, with overseas migration the major contributor. The total population of Greater Melbourne, for example, increased by a quarter over the 2006–16 decade, averaging 2.3% annually between 2011 and 2016. Population growth rates of some local government areas in Melbourne are well above these rates, particularly those on the outer urban fringe. At the same time, there has been a failure to adequately support this population growth on the fringe of the city with essential infrastructure, particularly in relation to the provision of public transport (Brain, Stanley, & Stanley, 2019). If this trend continues, then approximately $A376 billion will need to be spent to remedy this shortfall by 2031; thus, the problem is far from insignificant (Brain et al., 2019).

The impact of this infrastructure shortfall in Melbourne can be seen in the gross regional product per capita of working age population that grew more slowly where the population growth in the Local Government Area was faster than about 2% annually, on average, over the 1992–2017 period. This population growth was greatest in fringe suburbs, whose residents went backwards relative to the state as a whole in terms of capturing income from economic activity (Brain et al., 2019). Poor land use planning has led to urban sprawl, rather than facilitating an increase in population density, an outcome that leads to higher costs associated with the provision of infrastructure and public transport (SGS Economics & Planning, 2016). This is due to the greater distances to reach the same numbers of residences, and the absence of pre-existing infrastructure on which to build new additions.

There is a considerable interplay between infrastructure and community dynamics, particularly in relation to achieving accessibility, a vital component of realising social inclusion and personal wellbeing. A major study examined the particular drivers needed to achieve social inclusion and wellbeing, based on a large Victorian sample (see e.g. Stanley, 2011b; Stanley et al., 2017). The definition of social inclusion was adapted from the work of Burchardt, LeGrand, and Piachaud (2002), while wellbeing was measured using two prominent instruments in the psychological literature, subjective wellbeing and psychological wellbeing (Ryan & Deci, 2001). A model based on 1% statistical associations identified the major drivers as:

- income (by implication, education and work);
- accessibility (transport);
- social capital, especially bridging social capital and connections with the community (personal relationships and connections);
- self-esteem, confidence; and
- control over your personal environment (such as capabilities to make choices, problem solve).

The examination of socio-economic outcomes for people in the fringe suburbs in this study suggests that they often have poorer outcomes, when a comparison is made with those living in the more established suburbs and inner Melbourne. Most of the drivers that lead to social inclusion and wellbeing are lower on the urban fringe, compared to middle and inner Melbourne areas. In particular, the ability to be mobile is a critical factor in realising the conditions needed to achieve the other drivers of social inclusion and wellbeing. Without transport it is hard to reach education and jobs, interact with others and engage with the local community, all these associated with confidence and self-esteem. The outer suburbs of Melbourne, along with the edges of rural settlements, have a scarcity of public transport, especially local public transport, the urban sprawl and low housing density increasing the distance to obtain services and meet other needs, such as recreation. The research found that those people with the lowest levels of social inclusion also made fewer trips than those showing higher levels of inclusion (1% significance).

Examining the detail a little further, a broad range of socio-economic variables suggests that higher levels of disadvantage are found in the outer Melbourne areas and in some rural settlements in comparison with other urban areas. Some examples are given. If Greater Melbourne is divided into the inner, middle and outer areas, each area holding approximately one-third of available jobs in Melbourne, then outer Melbourne has about three times less the number of jobs per 1000 population than inner Melbourne (Brain et al., 2019). Yet, at the same time, population growth in outer Melbourne, between 2011 and 2016, was 57.5%, compared with 20.2% in inner Melbourne. Thus, population growth is highest where there are fewer jobs, necessitating longer travel distances, from areas with a poorer public transport service.

Educational levels are lower in the outer suburbs of Melbourne, with fewer people with a higher degree than can be found in middle and inner Melbourne (Brain et al., 2019). The educational disadvantage begins right from pre-school, with a tendency for both the rapidly growing outer growth areas of Melbourne and many rural settlements to have a lower than state average for developmental milestones needed for school entry at five or six years of age (Australian Government, 2016). The impact can be seen in the data that shows a strong negative correlation between achievement in literacy and numeracy standards in year 9 and the proportion of children who are vulnerable on one or more developmental domains on school entry ($r = -0.676$; $p = 0.000$ on literacy; $r = -0.602$; $p = 0.000$ on numeracy) (Brain et al., 2019).

Social capital is where individuals can use membership in groups and networks to secure benefits relating to social connections, and economic and cultural resources (Bourdieu, 1985). Bonding social capital refers to networks between close family members and neighbours, while bridging social capital refers to broader networks, such as work colleagues, which lead to wider resource community connecting opportunities (Stone, Gray, & Hughes, 2003). Thus, achievement of bridging capital is likely to involve further distances of travel. Indicators of social capital (trust in others and size of networks) suggest that social capital is lower in faster growing suburbs and increases the closer the local government area is to Melbourne CBD.

In the newly developed urban/rural fringe areas community dynamics are commonly characterised by a relatively low median age and/or a high proportion of young persons, who as already noted, are often socio-economically disadvantaged (Nicolopoulos, Murphy, & Sandinata, 1997). With the current high housing prices, young families move to outer areas to access more affordable housing, ironically, the price reflecting the poor accessibility. As these communities are new with people arriving from many other places, there is a risk of disjointed communities associated with disconnect from previous community and support groups. The absence of public transport to enable youth to undertake social connections and activities further isolates those who move to fringe suburbs (Stanley et al., 2017). As a result, a greater concentration of other problematic and antisocial behaviours can occur (Pease, 1998).

These issues are likely to be exacerbated by the high levels of youth unemployment, which sits between 15% and 20% of youth in many of the outer fringe suburbs of Melbourne. This measure of youth unemployment does not include under-employment, nor those disengaged from education and searching for work. In February 2017, the under-employment rate for youth aged 15 to 24 years of age sat at about 18% (Vandenbroek, 2017). For example in the outer suburb of Frankston, about half of the males aged 15 to 19 have left school, and are not fully occupied, with approximately

- 11.8% disengaged (not registered as unemployed or in education),
- 22% registered as unemployed and looking for work, and
- 15% under-employed.

Youth disadvantage is also present outside Melbourne, in regional settlements in Victoria, where bushfire ignitions are also higher where urban meets rural areas, when compared with more isolated rural areas (Llausàs, Buxton, & Bellin, 2016). A study of rural transport found that youth in

South Western Victoria experience the highest levels of transport disadvantage; and wellbeing levels are lower than those found in urban Melbourne as a whole (Stanley & Banks, 2012, unpublished findings from ARC study). Of greatest concern is the discrepancy between self-assessed perception of their future, where the average score on a 10-point wellbeing scale was 7.2 in urban Melbourne and 5.6 in rural Victoria. Rural youth were often not able to take advantage of education initiatives designed to keep youth at school, such as the VET and VCAL (the Victorian Certificate of Applied Learning) schemes, due to an absence of transport to access these opportunities. Similarly, job opportunities on rural farms could not be taken up. Youth who have never experienced being in the workforce full time and are unable to get work risk longer term disadvantage through loss of motivation and a reduction in 'employability'.

11.6 Connecting Disadvantage and Bushfires

As noted above, hot spots for bushfires are commonly located on the outer fringes of metropolitan and regional settlements, areas characterised by recent urban expansion, rapid population growth or poor infrastructure provisions (Bryant, 2008; Muller, 2009). The structural and community characteristics associated with this growth are likely to increase the risk of offending and anti-social behaviours, particularly amongst young people (Pease, 1998).

An examination of the times that bushfire most commonly occurs also suggests that youth may be a significant proportion of the offenders. Up to half of maliciously lit fires occur on a Saturday or Sunday, while on a weekday, the most common time of occurrence is between 3 pm and 6 pm (Bryant, 2008). The window between 3 pm and 6 pm on weekdays reflects the time in which young persons may travel from school unaccompanied by an adult. The distance between the location of the offence and the offender's home tends to be small, the literature reporting studies finding this to be from half a kilometre to ten kilometres away (Catry, Damasceno, Silva, Miguel Galante, & Francisco Moreira, 2007; Davidson, 2006; Price, 2013). Thus, the distance suggests, while some offenders will travel by car, the distance is markedly a walk or comfortable bicycle ride. This is supported by research on accessibility in a rural area, the LaTrobe Valley, in Victoria, an area known as experiencing frequent bushfires (Stanley, Stanley, Balbontin, & Hensher, 2018). Youth were found to be highly transport disadvantaged. While their frequency of trip making was quite high when compared with other population groups, it comprised short distances only, achieved by active travel. This also increased their lack of opportunities around education and employment, compounding social exclusion.

There are suggestions that increased fire frequencies in interface areas are facilitated by the rapid growth of urban sprawl and its increased proximity to rural and natural environmental areas, that is associated with factors that may arise that increase the likelihood of offending, particularly among the young. These include greater physical and social separation from previous and/or available social networks, limited or non-existent recreational facilities, and at least initially, disjointed communities at a social level, where many people have arrived from other areas. In Australia, the absence of public transport to enable youth to undertake social connections has also been discussed (Stanley et al., 2017).

Interestingly, open and natural areas are increasingly being understood as important areas that lead to improvements in health and wellbeing, also offering a sense of place and belonging to local residents (European Union, 2011; Gill, 2011; Marselle, Irvine, & Warber, 2013). However, when it comes to the risk of bushfire, the presence of open space and particularly natural environmental areas increases the risk of bushfire. Many of the newer outer suburbs of Melbourne, and the outer areas of rural settlements have proportionately higher levels of open space and natural areas than is found in middle and inner areas of Melbourne (Victorian Planning Authority, 2017). While some of this open space may be flat open ground, for example, in the northern areas of Melbourne, the presence of grassland presents a bushfire risk and much of eastern Melbourne has forested areas mixed with farming land and the intrusion of urban development. These lead to opportunities for ignition of bushfires, offering a source of fuel and commonly privacy for fire lighting, away from the more populated urban structures.

Thus, it could be argued that characteristics of the urban interface across each of the four clusters, outlined above, may be contributing to a failure in social justice. Regarding societal trends, the failure to address climate change is increasing the bushfire risk through increasing the presence of a dry environment more likely to burn once ignited. The fringe areas of urban settlements tend to have the poorest infrastructure, particularly transport, services and job opportunities, with high levels of youth disadvantage. Such disadvantage risks a range of negative emotions, such as lowered self-esteem, anger, frustration and cutting off emotions, thus leading to impacts on personal characteristics (Twenge & Baumeister, 2005). Those with lower levels of social inclusion have a high risk of experiencing negative affect (Stanley et al., 2017). Measured using the Psychological Wellbeing Scale (Ryff, 1989), on a scale from 1 to 5, with 5 being the highest level of negative emotions, the average of those who have no social exclusion risks is 1.7 in Victoria, while those at high risk of social exclusion average 4.8 on negative affect. The (mainly male) youth who commit arson are thought to have a low socio-economic background, are likely to have experienced child abuse and/or neglect, may have poor impulse control, and engage in other criminal/anti-social behaviour and have low academic achievement (Dolan & Stanley, 2010). All this is not helped by the way unemployed youth are labelled and stigmatised (Whiteford, 2019). Schemes that offer hardly a living wage for unemployed youth (Newstart), drug-testing, online compliance schemes and terms such as 'leaners not lifters' give the perception of unworthiness.

Thus, the way we are shaping cities and urban development impacts on personal life chances, creating social exclusion and higher levels of disadvantage. Yet the cost of these failures is high, in terms of the potential cost of bushfires, particularly where associated with loss of life and injury, loss of businesses and property, and losses to the environment. The social cost of injustice is also high, to the individuals concerned, their families and society generally. Work by Wilkinson and Pickett (2009) has shown that the greater the increase in inequality within countries, the greater the social problems faced by that country, covering issues such as crime, single parenthood, life expectancy, trust, mental illness, imprisonment and literacy. Australia sits at about the middle of a scale showing inequality in 21 countries.

11.7 What Is the Role of Prevention?

Bushfire prevention rests around action within each of the four domains. At present the prevention of bushfires in Australia and internationally is heavily reliant on fuel reduction through prescribed burning, largely undertaken away from housing in more forested areas, for reasons of safety (Llausàs et al., 2016). The wide range of options to prevent bushfires is largely overlooked, especially in relation to malicious fires lit by youth in rural/urban interface areas (Stanley, March, Ogloff, & Thompson, 2020). Indeed, programs that target the reduction of malicious fire lighting are almost non-existent in Australia and internationally.

It is critical for Australia and the rest of the world to reduce greenhouse gas emissions. Many countries are experiencing a growth in severe bushfires, the growth trend increasing over time. The cost of improving transport and infrastructure in the outer areas of Melbourne and in rural towns is high. However, the cost of not improving the wellbeing of disadvantaged youth is also very high, and only likely to increase given population growth and climate change. At present the emphasis in transport in Melbourne is to build major car-based freeways, with the faulty perception that building more roads will reduce congestion. The cost of these freeways is at the opportunity cost of public transport and local transport, as well as the provision of other local infrastructure to improve the wellbeing of youth and other residents in the rural/urban interface areas of Melbourne and other regional settlements.

Policies that address both the risk of bushfires and the growth in inequity in relation to youth, such as complementary long-term land use planning, urban planning policies, and transport strategies, alongside emergency service planning, are needed. At present there are almost no policy or program links between planning and emergency services, although a major three-year study is presently being undertaken to understand how connections can be made (March et al., 2018). A strong case could be put for moving towards the establishment of 20-minute neighbourhoods, where most people can reach most of the services they need most of the time by a 20-minute trip by walking, cycling or public transport. Issues like quality of life, third places, such as cafes, community houses, open schools, high-quality open areas and playgrounds, where people can meet and establish a sense of community, need to be part of this approach. Lowering population levels to the point where their needs can be adequately served is important. Social policy around preventing child abuse and neglect, much greater support for educational attainment and better meeting the needs around youth are all part of addressing the bushfire risk. It is perhaps no coincidence that many of the countries currently experiencing severe wildfires are those with high levels of disadvantaged youth and youth unemployment.

The issue is well put by Flanagan (2019), who points out that the Federal Government spends almost four times as much on fossil fuel subsidies, 1.96% of gross domestic product, than it spends on early childhood education and care, 0.5% of gross domestic product. *"In other words, for every public dollar we spend building a future for our youngest citizens, we spend nearly four dismantling that future"*. This is surely an issue of social justice in itself, but also one of disaster justice for many people vulnerable to bushfires on the urban boundaries.

References

Allen, S. K., Barros, V., Burton, I., Campbell-Lendrum, D., Cardona, O., Cutter, S., ... Wilbanks, T. (2011). Managing the Risks of Extreme Events and Disasters to Advance Climate Change Adaptation: Summary for Policy Makers, Special Report of the Intergovernmental Panel on Climate Change. Retrieved from https://www.ipcc.ch/pdf/special-reports/srex/SREX_FD_SPM_final.pdf

Australian Government. (2016). Australian Early Development Census 2015. Retrieved from https://www.aedc.gov.au/

Baumeister, R., DeWall, C., Ciarocco, N., & Twenge, J. (2005). Social Exclusion Impairs Self-Regulation. *Journal of Personality and Social Psychology*, 88(4), 589–604.

Bourdieu, P. (1985). The Forms of Capital. In J. Richardson (Ed.), *Handbook of Theory and Research for the Sociology of Education* (pp. 241–258). New York: Greenwood.

Bradstock, R., Penman, T., Boer, M., Price, O., & Clarke, H. (2014). Divergent Responses of Fire to Recent Warming and Drying Across South-Eastern Australia. *Global Change Biology*, 20, 1412–1428.

Brain, P., Stanley, J. K., & Stanley, J. R. (2019). *Making the Most of Our Opportunities: First Report to the Municipal Association of Victoria*. Melbourne: NIEIR and Stanley and Co.

Bryant, C. (2008). *Understanding Bushfire: Trends in Deliberate Vegetation Fires in Australia*. Technical and Background Paper No. 27. Canberra: Australian Institute of Criminology.

Burchardt, T., LeGrand, J., & Piachaud, D. (2002). Degrees of Exclusion: Developing a Dynamic, Multidimensional Measure. In J. Hills, J. LeGrand, & D. Piachaud (Eds.), *Understanding Social Exclusion* (pp. 30–43). Oxford: Oxford University Press.

Buxton, M., Haynes, R., Mercer, D., & Butt, A. (2011). Vulnerability to Bushfire Risk at Melbourne's Urban Fringe: The Failure of Regulatory Land Use Planning. *Geographical Research*, 49(1), 1–12.

Catry, F., Damasceno, P., Silva, J., Miguel Galante, M., & Francisco Moreira, F. (2007). Spatial Distribution Patterns of Wildfire Ignitions in Portugal, Conference Paper, Wildfire, January, Seville, Spain.

Climate Action Tracker. (2018, April). Retrieved from https://climateactiontracker.org/countries/australia/

Collins, K., Owen, A., Price, F., & Penman, T. (2015). Spatial Patterns of Wildfire Ignitions in South-Eastern Australia. *International Journal of Wildland Fire*, 24, 1098–1108.

Davidson A. M. (2006). Key Determinants of Fire Frequency in the Sydney Basin. Unpublished Honours Thesis, Australian National University, Canberra.

Dolan, M., & Stanley, J. (2010). Risk Factors for Juvenile Firesetting, Advancing Bushfire Arson Prevention in Australia. *Report from Collaborating for Change: Symposium Advancing Bushfire Arson Prevention in Australia*, held in Melbourne, 25–26 March, 2010, Monash University and Australian Institute of Criminology (pp. 31–32).

Doley, R., Dickens, G., & Gannon, T. (2016). Deliberate Firesetting – An Overview. In R. Doley, G. Dickens, & T. Gannon (Eds.), *The Psychology of Arson: A Practical Guide to Understanding and Managing Deliberate Firesetters* (pp. 1–10). New York: Routledge.

Dutta, R., Das, A., & Aryal, J. (2016). Big Data Integration Shows Australian Bush-Fire Frequency Is Increasing Significantly. *Royal Society Open Science*. Retrieved from http://rsos.royalsocietypublishing.org

European Union. (2011). *Cities of Tomorrow: Challenges, Visions, Ways Forward*, Brussels.

Flanagan, F. (2019). Climate Change and the New Work Order. *Inside Story*, February 28. Retrieved from https://insidestory.org.au/climate-change-and-the-new-work-order/

Gill, T. (2011). *Children and Nature: A Quasi-Systematic Review of the Empirical Evidence*. London: London Sustainable Development Commission, Greater London Authority.

Grubb, J., & Nobles, M. (2016). A Spatiotemporal Analysis of Arson. *Journal of Research in Crime and Delinquency, 53*(1), 66–92.

Llausàs, A., Buxton, M., & Bellin, R. (2016). Spatial Planning and Changing Landscapes: A Failure of Policy in Peri-Urban Victoria, Australia. *Journal of Environmental Planning and Management, 59*(7), 1304–1322.

March, A., Nogueira de Moraes, L., Riddell, G., Dovers, S., Stanley, J., van Delden, H. ... Maier, H. (2018). Australian Inquiries into Natural Hazard Events: Recommendations Relating to Urban Planning for Natural Hazard Mitigation (2009–2017), University of Melbourne, The University of Adelaide, Australian National University.

Marselle, M., Irvine, K., & Warber, S. (2013). Walking for Well-Being: Are Group Walks in Certain Types of Natural Environments Better for Well-Being Than Group Walks in Urban Environments? *International Journal of Environmental Research and Public Health, 10*(11), 5603–5628.

Muller, D. (2009). *Using Crime Prevention to Reduce Deliberate Bushfire in Australia*. Research in Public Policy Series No. 98. Canberra: Australian Institute of Criminology.

Nicolopoulos, N., Murphy, M., & Sandinata, V. (1997). *Socio-economic Characteristics of Communities and Fires*. NSW Fire Brigades Statistical Research Paper 4/97. Sydney, NSW: NSW Fire Brigades.

Pease, K. (1998). *Repeat Victimization: Taking Stock*. Home Office Police Research Group, Crime Detection and Prevention Series, Paper 90. London: Home Office.

Price, O. (2013). Reducing Bushfire Risk: Don't Forget the Science. *The Conversation*, October 11. Retrieved from http://theconversation.com/reducing-bushfire-risk-don't-forget-the-science-19065

Ryan, R., & Deci, E. (2001). On Happiness and Human Potentials: A Review of Research on Hedonic and Eudaimonic Well-Being. *Annual Review of Psychology*, 52, 141–166.

Ryff, C. (1989). Happiness Is Everything, Or Is It? Exploration on the Meaning of Psychological Well-Being. *Journal of Personality and Social Psychology*, 57, 1069–1081.

SCRGSP (Steering Committee for the Review of Government Service Provision). (2016). *Report on Government Services 2016*, Vol. D, Emergency Management, Productivity Commission, Canberra, Australia.

Sen, A. (1992). Capability and Wellbeing. In M. Nussbaum & A. Sen (Eds.), *The Quality of Life* (pp. 30–53). Oxford: Clarendon Press.

SGS Economics & Planning. (2016). *Comparative Costs of Urban Development: A Literature Review: Final Report*, Infrastructure Victoria, July.

Stanley, J. (2011a). *Social Exclusion, New Perspective and Methods in Transport and Social Exclusion Research* (pp. 27–44). Emerald Press, UK.

Stanley, J. (2011b). Measuring Social Exclusion. In G. Currie (Ed.), *New Perspective and Methods in Transport and Social Exclusion Research* (pp. 77–90). Bingley, UK: Emerald Press.

Stanley, J., & Banks, M. (2012). *Transport Needs Analysis for Getting There and Back: Report for Transport Connections*: Shires of Moyne and Corangamite, June, Monash University, Melbourne.

Stanley, J., March, A., Ogloff, J., & Thompson, J. (2020). *The Prevention of Human-Lit Wildfire: International Best Practice*. Delaware, USA: Vernon Press.

Stanley, J., Stanley, J., Balbontin, C., & Hensher, D. (2018). Social Exclusion: The Roles of Mobility and Bridging Social Capital in Regional Australia, *Transportation Research Part A: Policy and Practice*. Retrieved May 29, 2018, from http://web.education.unimelb.edu.au/assets/pospsych/Social%20Exclusion%20and%20the%20Value%20of%20Mobility.pdf

Stanley, J., Stanley, J., & Hansen, R. (2017). *How Great Cities Happen*. Cheltenham, UK: Edward Elgar Publishing.

Stone, W., Gray, M., & Hughes, J. (2003). *Social Capital at Work: Towards a Theoretically Informed Measurement Framework for Researching Social Capital in Family and Community Life*. Research Paper 24. Melbourne: Australian Institute of Family Studies.

Twenge, J., & Baumeister, R. (2005). Social Exclusion Increases Aggression and Self-Defeating Behaviour While Reducing Intelligent Thought and Prosocial Behaviour. In D. Abrams, M. Hogg, & J. Marques (Eds.), *The Social Psychology of Inclusion and Exclusion* (pp. 27–46). East Sussex, UK: Psychology Press; Taylor and Francis.

Vandenbroek. (2017). *Underemployment Statistics: A Quick Guide*. Research Paper Series 2016–2017, Parliamentary Library, Parliament of Australia.

Victorian Planning Authority. (2017). *Metropolitan Open Space Network: Provision and Distribution*, Melbourne.

Whiteford, P. (2019). Future Budgets Are Going to Have to Spend More on Welfare, Which Is Fine. It's Spending on Us. *The Conversation*, March 7. Retrieved from https://theconversation.com/future-budgets-are-going-to-have-to-spend-more-on-welfare-which-is-fine-its-spending-on-us-111498

Wilkinson, R., & Pickett, K. (2009). *The Spirit Level: Why Equality Is Better for Everyone*. London: Penguin.

CHAPTER 12

Issues of Disaster Justice Confronting Local Community Leaders in Disaster Recovery

Valerie Ingham, Mir Rabiul Islam, John Hicks, and Oliver Burmeister

12.1 Introduction

In Australia, a paradigm shift is currently underway in emergency management. Federal government policy, notwithstanding the fact that states have primary responsibility for emergency management, is seeking to guide emergency management away from its current dominant reliance on decision-making based on command and control by government-authorised responders from the emergency services sector to a position of

---

V. Ingham (✉)
Australian Graduate School of Policing & Security, Charles Sturt University, Bathurst, NSW, Australia
e-mail: vingham@csu.edu.au

M. R. Islam
School of Psychology, Charles Sturt University, Bathurst, NSW, Australia
e-mail: rislam@csu.edu.au

J. Hicks
School of Accounting & Finance, Charles Sturt University, Bathurst, NSW, Australia
e-mail: jhicks@csu.edu.au

© The Author(s) 2020
A. Lukasiewicz, C. Baldwin (eds.), *Natural Hazards and Disaster Justice*, https://doi.org/10.1007/978-981-15-0466-2_12

shared responsibility, where local community organisations (the community services sector) are incorporated into the disaster planning and recovery of their own community. At the heart of the paradigm shift of 'who is responsible' for community disaster resilience is the National Strategy for Disaster Resilience (COAG, 2011) and Emergency Management Australia Handbook 6: Community Engagement Framework (2013), both of which place a strong emphasis on the concept of 'shared responsibility' and instruct local community and emergency service organisations to cooperate in building local disaster resilience. The motivation for the shift—ostensibly building more disaster-resilient communities—may be traced back to a broader concern for the increasing frequency of large disaster events and the potential for reducing the cost of enhancing resilience by engaging largely underutilised or unused social resources (Islam, Manock, Sappey, Hicks, & Ingham, 2012). By devolving responsibility for their own wellbeing and resilience to individual local communities (i.e. activating local organisations and volunteers from various walks of life) it is proposed that an enhanced community recovery can be expected (Gil-Rivas & Kilmer, 2016; Madsen & O'Mullan, 2016) and public savings can be made (Olivia, 2018).

Justice is described in the literature as a value which emphasises fairness and equality among individuals (Beauchamp & Childress, 2009; Pakrasi, Burmeister, McCallum, Coppola, & Loeb, 2015). For Slote (2010), justice is an ideal which sees our historical and personal connections with others as the basis for positive and caring responses to crisis situations in the community. Justice, within the context of a community preparing for, responding to and recovering from a disaster is about recognising the community as a fully participating entity, rather than something upon which the emergency services 'work' and then go away. In this chapter we highlight the ways in which community leaders, in particular, are able to contribute and what happens when they do this without fair and just resourcing. Considerations include the costs to community services sector when responding to government calls for assistance in the management of disaster events. Further, we argue that those costs should not be exacerbated by government failure to ensure the efficient integration of their activity with that of the emergency services sector, nor by the inadequacy

---

O. Burmeister
School of Computing & Mathematics, Charles Sturt University,
Bathurst, NSW, Australia
e-mail: oburmeister@csu.edu.au

of a funding provision which prevents them from achieving the outcomes of which they are capable. In this way we understand issues of disaster justice to include "*distributive inequity, lack of recognition, disenfranchisement and exclusion, and, more broadly, an undermining of the basic needs, capabilities, and functioning of individuals and communities*" (Schlosberg & Collins, 2014, p. 361).

This chapter focusses on the Blue Mountains community as a case study. It examines the extent to which the call by government for the community services sector to participate in responding cooperatively with the emergency services sector in the building of local disaster resilience has been accompanied by an appropriate level of resource provision. It also looks at what can be done so that the government-guided integration of the activities of both sectors is achieved in a manner which ensures a just treatment of the community services sector in the process. Section 12.2 places our study within the context of the Blue Mountains community which was exposed to disastrous fires in the spring of 2013. Section 12.3 reviews the literature on emergency service sector and community services sector cooperation and Sect. 12.4 discusses our methodology. The data collected during our research is analysed in Sect. 12.5 and conclusions based on this analysis are presented in Sect. 12.6. Finally, Sect. 12.7 provides two recommendations that we believe will enhance justice in the management of disasters in Australia.

12.2 2 October 2013, Blue Mountains Fires

On Wednesday, 16 October 2013, a military exercise sparked the State Mine Fire near Lithgow, NSW. Due to the hot and windy conditions this fire quickly escalated into a threat for Upper Mountains' residents. The following day the Mt York fire and Mt Victoria fires started, crossed containment lines and destroyed ten houses. Most emergency service crews and resources were deployed to fire fighting and containment activities in the upper Blue Mountains. During this period a third fire occurred in the lower Blue Mountains, quickly named the Linksview Road Fire.

The Linksview Road Fire was caused by wind toppling a tree onto electricity wires and over 200 homes were lost and a further 100 damaged (Wylie, 2018). Community meetings were held throughout the region during this time and on 20 October the then NSW Premier, Barry O'Farrell, declared a state of emergency. This declaration hands the control of a disaster event from local authorities to the State and thus enables a higher level of immediate fire-fighting resourcing and later, recovery aid and assistance. On Wednesday, 23 October, as the weather conditions intensified, the then Minister for Emergency Services advised, "*residents who did not have a bush fire survival plan*" to leave the Blue Mountains (Levy, 2013). The fires were declared extinguished on 13 November 2013. The State-led recovery operations lasted five months before handover to the Blue Mountains City Council.

The engagement of the local community services sector in the immediate recovery period was facilitated by a combined interagency meeting called by the community peak body—Mountains Community Resource Network—within ten days of the Linksview fires. The meeting attracted 55–60 community leaders, including a representative from the Ministry for Policing and Emergency Services (now Office of Emergency Management) and the Rural Fire Service. At the initial interagency meetings a work plan was created in addition to a number of sub-committees with short, medium and long-term planning blocked in. The first objective was the coordination of immediate relief and recovery work. The interagency meetings proved to be a major burden for community organisation staff as they involved a large time commitment with various sub-committee meetings requiring attendance for a couple of hours once or twice a week, and Recovery Committee meetings two or three hours twice a week, that is, much 'busy work' for the middle tier. The community sector took a community development approach to psychosocial recovery, empowering people to manage their own recovery. By their own account, there was a lot of 'touchy feely' caring and sharing with cups of tea, as well as assistance with locating information and filling in the inevitable forms.

Meanwhile the local emergency services sectors were in the response and immediate recovery phases of emergency management. The emergency services are comprised of NSW Fire & Rescue who attend structural fires; the Rural Fire Service comprised in the main of local volunteers and mandated to focus on rural fires in relation to fire suppression, that is, their core business is not education; and the State Emergency Service, whose core business is flood and storm response, that is, chainsaws and boats. The recovery activities for local emergency services focused on the operational aspects of mopping up fires and debris removal, cordoning off streets to assess damage and mark up asbestos contamination, communicating to the community regarding what they could and couldn't do, and in the eyes of some residents, obstructing their access to the remains of their home.

12.3 The Intended Outcome of Policy Intervention

In this section we highlight the policy outcomes being targeted by the relevant policies for disaster planning and recovery.

12.3.1 Shared Responsibility and Resilience Policy

The Hyogo Framework for Action 2010–2015: Building the Resilience of Nations and Communities to Disasters was issued by the United Nations Office for Disaster Risk Reduction (UNISDR) and advocated the adoption of goals and priorities which although broad, were designed to have influence from a national to local level. The subsequent Sendai Framework for Disaster Risk Reduction 2015–2030 (UNISDR) similarly provided strategies for the reduction of new and existing disaster risks. One of its catch phrases is 'build back better'. Critics of the Sendai Framework point out that reliance on existing governmental structures predisposes it to have little influence at the local level. According to Burnside-Lawry and Carvalho (2016, p. 4) this could be due to the need for "*strong political leadership and inter-departmental coordination*" which may or may not be present.

Within the same genre of advocated policy, the Council of Australian Governments, in a document titled 'National Strategy for Disaster Resilience' (COAG, 2011), calls for all sections of Australian society to be involved in building disaster resilience through taking a 'shared responsibility' approach to planning and preparation for disaster. 'Shared responsibility' is introduced by COAG in the National Strategy for Disaster Resilience in this way:

> *There is a need for a new focus on shared responsibility; one where political leaders, governments, business and community leaders, and the not-for-profit sector all adopt increased or improved emergency management and advisory roles and contribute to achieving integrated and coordinated disaster resilience.* (COAG, 2011, p. 2)

The COAG call envisages that all levels of government, business and the non-government sector will be actively involved in disaster and business continuity planning.

12.3.2 Emergency Management at State and Municipal Level

The formal relationship between local emergency services and community organisations in relation to disaster arrangements is legislated by the *State Emergency and Rescue Management Act 1989* (NSW) (SERM Act, 1989). This Act establishes State, Regional and Local Emergency Management Committees, overseen by the Minister for Police, and Minister for Emergency Services in NSW. At the local level, the Local Emergency Management Committee (LEMC), funded by the State government, is "*responsible for the preparation of plans in relation to the prevention of, preparation for, response to and recovery from emergencies in the local government area for which it is constituted*" (SERM Act, 1989: 29/1).

In NSW the LEMC is the operational heart of local disaster planning and the decision makers are located within the emergency services. An 'observer' group has right of attendance but no voting power, and usually the observers are invited representatives from nationally recognised non-government organisations (NGOs), most commonly the Red Cross and Salvation Army.

Within the Local Emergency Management Plan (EMPLAN) there are three parts for the LEMC to complete: Administration, Community Context and Restricted Operational Information. The EMPLAN requires the LEMC to devise a community profile, stating that "*[t]he community profile assists the LEMC to understand the diverse needs, values and priorities of local community and its characteristics within the broader environment. This information is critical to inform planning and emergency operations*". The EMPLAN guidelines indicate hegemony in that the current disaster management policy provisions in NSW are framed within an emergency service narrative which places the community as a group to be 'consulted' and possibly 'engaged', but not actually participate in the formulation of disaster plans.

12.3.3 Community Sector Leadership

Under the National Strategy for Disaster Resilience there is an increased burden of responsibility for community leaders to build capacity within their organisation and in their community in order to deal with disaster (Redshaw, Ingham, McCutcheon, Hicks, & Burmeister, 2018). From our research we have drawn the conclusion that leaders in the community sector feel that they are often expected to be multi-skilled facilitators when it comes to problem solving. In addition to being able to mobilise a largely volunteer workforce, provide services for the more marginalised people in their community and secure funding for these activities, they are expected to have *"skills in community planning, facilitation, team building and conflict resolution, and importantly as being able to move from project to project, dealing with a range of issues and implementing a range of solution strategies"* (Davies, 2009, p. 382). Local community organisations are typically able to draw on a pool of volunteers and as Fitzpatrick and Molloy (2014) state they are *"ideally placed for promoting the messages of disaster management and building resilience as they exist to support the communities they service, and they are already embedded and connected at a grassroots level"*. Another researcher, Rivera (2016, p. 173) observes that *"[v]oluntary organizations are involved in EM planning more extensively when the usage of volunteers in disaster assistance activities is integrated into disaster plans"*. It is clear that participation is key to gaining an effective approach to shared responsibility.

12.3.4 Summary

The aspirations of policy makers at all levels—international, national, state and local—are common in that they desire to see a shared responsibility for the building of disaster resilience in the community. However, at the international level, and to a lesser extent, the national level, the ability to act on these ambitions is limited largely to advocacy. Unfortunately, advocacy can often run ahead of implementation—especially when the advocates are not, themselves, in a position to determine policy or to implement it. At the state level, where policy is both determined and implemented, there are often competing needs which obfuscate the focus on what needs to be done and the urgency with which it needs to be undertaken. Where there are groups at the local level which have embraced the call for shifts in policy direction, the apparent lag in reaction at the state level can become exceedingly frustrating. Whilst shared responsibility is a common ideal, it faces many challenges emanating from the fact that there is a multitude of people and organisations involved—and usually at a time of unpredictable and violent events which are creating trauma within local communities. Nevertheless, a way needs to be found to appropriately recognise and incorporate community organisations and the skills they can bring to combating a common threat to the community.

12.4 Approach to the Study

It is vital that research professionals engage with participating communities on a foundation of equality of voice and contribution. In order to test policy, highlight challenges and open opportunities to provide contextual feedback, a foundation of mutual respect and trust in the iterative process must be built, rather than a hegemonic imposition of researcher upon community.

The research partnership between the Disaster & Community Resilience research group, Blue Mountains City Council, and two neighbourhood centres received ethical approval from Charles Sturt University in 2014 to research community connections within the Blue Mountains. The project employed a participatory action framework and involved a Council-distributed survey, interviews and focus groups with community members, including vulnerable people and their carers, as well as community leaders. This chapter explores the community leader interviews of 2014 and 2018 and reports only on the disaster justice issues identified.

Interviews were conducted with seven community leaders in 2018, five of whom also participated in 2014, making a total of nine individuals overall. The five-year interval since the 2013 fires was selected as enough time had elapsed for participants to reflect in hindsight on their experience. The participants work and reside in a location of relatively small population and to protect their identity we have coded them by number rather than profession. The participants include a fire services officer, a school principal, leaders from large NGOs and managers of neighbourhood centres. The interviews were taped, transcribed by an external service and de-identified.

The 2018 interview schedule was based on the same semi-structured and open-ended discussion starters of 2014, adjusted for 2018 with the addition of a few clarifying points of introduction prior to each question. The interviews sought information regarding organisational community involvement prior to and after the fires, sources of information and community connections with other organisations, dealing with traumatised people (both their own workers and the community they serve) and their own personal wellbeing and coping strategies.

Research group members collaborated on the identification of themes pertinent to community leaders, specifically in relation to participant capacity and resourcing for operating in community recovery.

12.5 What *Is* Happening

Prior to the October 2013 Blue Mountains fires, neighbourhood centres and other community service organisations such as schools and childcare centres had very little knowledge of local emergency management plans or evacuation procedures. They were not represented on the LEMC and candidly admitted to being unaware of its existence. Eight of the nine local community leaders had little more than ad hoc connections with local emergency services prior to the fire—fortuitously made through their informal friendships. Thus, there were no formal connections between local community organisations and the LEMC, or emergency services in general.

Likewise, prior to the October 2013 fires, local emergency services had very little knowledge of how closely the community sector worked with vulnerable and at-risk residents, the very people they needed to oversee the evacuation of. Plenty of information in the form of brochures and educational materials was produced by the fire services; however, there was no systematic plan to tap local community services for its dissemination. Education was not considered as core business for fire services, although the shared responsibility mandate has now moved public education into view.

In this section we present the disaster justice issues faced by local community leaders in the recovery period. The first adverse situation which stymied community leader recovery efforts was the inequality they experienced as leaders when operating in a multiagency environment. We have termed this 'inequality of voice'. The three other adverse situations which follow are inequality of funding vs. FACS and Office of Emergency Management expectations; community leader wellbeing—burn out, guilt and health issues; and blurring of professional and personal boundaries.

12.5.1 Inequality of Voice in Disaster Planning and Recovery

Six of the community leaders experienced marginalisation in various leadership forums, in that they were not included in briefing sessions, consulted as to their expert knowledge of vulnerable people and for some they were not invited to networking events:

> *You see outside people rock up and just make decisions and do things and talk about the community in front of the community… it just annoyed me and I thought 'People here need a voice.'* Participant 9: 2018, p. 4

> *There was a complete disconnect—he* [initial officially appointed recovery manager] *had utter contempt for the community sector, for the workers in it, for the work that we do. He didn't understand it—he thought it was women's work—should be done for free...'Real' work is putting out fires and wearing a hard hat.* Participant 7: 2018, p. 10

> *We don't have a formal place at the Local Emergency Management Committee for the Blue Mountains. We are saying we need a voice at the table here, we need some kind of representation here, we need ongoing and sustainable structures that actually put us in joint planning and joint discussions.* Participant 5: 2014, p. 16

Knowledge of local vulnerable people was not sought by the initial recovery team beyond creating an ad hoc list, despite the strong community leader focus on the aged and the isolated:

> *They failed to understand that we have a captive audience for many hours in the day and that we were a great port of call... I'm in charge of 1100 people right in the middle of the area and they didn't understand that we are the services and I manage the staff.* Participant 6: 2014, p. 12, 13

The media communications of six community leaders, written for dissemination to the broader community immediately after the fires, were discarded by the official recovery media manager:

> *We had enormous difficulties, in fact it was impossible to get any of our messages about what's normal and what's an abnormal reaction, what are the things that you might look out for, what's happening with your kids, are they bed wetting, whatever it might be.* Participant 7: 2014, p. 14

> *We're already a communication system, provide what you need to be communicated and we can do it immediately to all our* (members) *across the mountains...they didn't think about contacting us.* Participant 6: 2018, p. 12

Public meeting protocols were ignored, causing confusion for some people and the silencing of others:

> *There are guidelines that MPES has about how you run a public meeting in an emergency, and the sorts of calm that you need, for example what are the key messages that you're trying to get out, the need for an independent MC—so not the person who's answering the questions also emceeing.* Participant 7: 2014, p. 15

12.5.2 Inequality of Funding vs. FACS and Office of Emergency Management Expectations

Fire services, whether volunteer or career, organise support for major fires by responding appliances and crews from surrounding stations, whilst simultaneously organising backup for the depleted station. On the other hand, the community sector received no extra funding or resourcing for their own backup, despite one neighbourhood centre being ordered by Ministry of Police & Emergency Services (formerly MPES, now Office of Emergency Management) to send a worker to another neighbourhood centre. No funding was provided to backfill the vacant position, and the message the community sector 'heard' was that the day-to-day services delivered by the seconded person were of no consequence and could just be halted in an instant.

The funding of back-filled positions for community leaders would have assisted enormously and during the interviews formed a major focus of attention. As it was, community leaders worked two jobs, became run-down and sick, missed important family events and generally felt torn between the demands of the recovery and the lack of resourcing, as illustrated in the following interview excerpts:

> *What does it actually cost an organisation over and above the budget? There's a financial cost to maintaining service to a community for an organisation that's impacted by fire.* Participant 5: 2014, p. 28

> *I've said, as a bottom line as a system, if we believe education's critical and we make laws to say it is, then we need to show that through our funding after natural disasters.* Participant 6: 2014, p. 23

For some there was a lack of support from their own organisations:

> *I didn't get a call from my supervisors for two weeks … and I said 'Hello', I said 'You know you're 12 days too late for me', and he went, he didn't know what to say, 'I said this phone call is twelve days too late.'* Participant 6: 2018, p. 16

> *Somebody even came from central office who couldn't pronounce the names of towns up here, who was one of the key speakers. A lot of people got very terrified.* Participant 9: 2014, p. 8

12.5.3 Who Leads the Recovery?

Community leaders were swung into the recovery space by virtue of the geographic location of their service and thus their local knowledge. They had little choice but to hop-to and do what they could. On the other hand, the official recovery team was assembled from multiple resources, brought into the area and charged to lead the recovery, almost totally ignoring the contribution and sacrifices of the local community leaders and their organisations.

One of the issues identified in the recovery space was role confusion. It became clear that the official appointment of a recovery manager with a strong history and public identification with a particular emergency service resulted in the public identifying *that* emergency service as being in charge of the recovery. The recovery manager was aided by other official appointments from the same previous service, which served to compound the situation. It was also evident that a recovery manager with an emergency service background brought certain cultural perspectives to the recovery when engaging with the community sector:

> *And that zone was just a life in itself and bore virtually no relationship to people on the ground I would say... very early on I said we need to be involved in, we need a seat at the Wellbeing table, because we basically were going to be the ones ending up carrying the bag.* Participant 9: 2014, p. 12

> *We didn't have access to the power brokers.* Participant 5: 2018, p. 3

The lack of recognition of community leader local knowledge and expertise by the officially appointed recovery team distressed local community leaders:

> *We're not seen as valid in terms of the disaster hierarchies ... legislated for instruments—community sectors aren't there yet. We're the people who are in communities, contactable by communities and seen to be responsive and there for communities. So, a whole capacity around preparedness and response was never tapped.* Participant 5: 2018, p. 2

> *I felt very slighted and I just thought, I don't get that people don't get our role.* Participant 6: 2014, p. 14

> *We do have a role, and even if it's not validated by the combat and command agencies, we are saying 'we have a role', this is around how we enhance our com-*

munity's capacity to work in natural disaster and emergency. Participant 5: 2014, p. 26

I wish I'd had someone come and say, "Right, here's your local services, here's the things they can provide." We had to source it all. That peeved me off. By the time they started bringing stuff to us, it was too late. Beyond Blue is just being rolled out now, can you believe it? Participant 6: 2014, p. 31

12.5.4 Community Leader Wellbeing: Burn Out, Guilt and Health Issues

Unlike their well-resourced emergency service counterparts who are governed by statutes, protocols and guidelines, the community services sector is staffed by locals who are generally living in the disaster zone for years to come. They drive to work, shop and attend recreational facilities along with their clientele. Some took to shopping at midnight to avoid meeting clients at the end of a long hard day. We found six leaders in 2018 were feeling burned out, unacknowledged and considering early retirement:

I'm buggered, I haven't got anything left in the tank, people think I'm functioning well and I can talk like I'm functioning well, but I'm not you know, I'm not. Participant 6: 2018, p. 15

So as opposed to say sleeping pills or something like that I am prepared to accept whatever the health consequences are because what's the alternative? I need to go to sleep and it's my reward at the end of the day—a glass of wine. Participant 7: 2018, p. 21

Look in amongst it all, I thought my job was to remain relatively composed but I found myself crying at the drop of a hat. Participant 5: 2018, p. 14

The issue of time and the increase in workload and subsequent emotional and physical effect was also major theme:

[A]ction plan templates and reporting templates and a needs analysis template, paperwork that you are obliged to fill in that were designed by someone who is a bureaucrat and it makes perfect sense to them, but when you're on the ground and have no time at all and very little attention to spare then it becomes an incredibly onerous thing to do. Participant 7: 2018, p. 5

Hours and hours and hours. So, it was hours in meeting times, hours in preparation, hours in submission writing…So, then you have to balance what gets dropped. Participant 5: 2018, p. 12

12.5.5 Blurring of Professional and Personal Boundaries

The blurring of personal and professional boundaries was keenly felt as a consequence of residing and working in a small community. On the day of the fires the following participant was caught in a desperate situation, wanting to keep their service open for fire-affected residents but receiving desperate phone calls from home:

They're ringing me up saying "You've got to come home, you've got to come home! I'm panicking I'm panicking, you have to come home, you can't be there at work!" And so finally at two o'clock in the afternoon I just said "Alright I'll come home." Participant 4: 2014, p. 9

Eight of the nine participants felt the burden of living and working in the disaster-affected area very keenly. Leaving their personal destruction issues and driving past that of others on their way to work, spending the day with fire-affected residents and then driving past it all again to go home proved a challenge:

I did feel a very big burden because this is my community and I knew a lot of these people before, like I'd grown up with these people, I've known them all my life—I have this role, I have this role, but I've known you for 30 years or something, do you know what I mean?... My stress was just through the roof at the end just because, you know I didn't want to fail, I didn't want the community to look back and go 'you know that was a stuff up?' Participant 9: 2018, p. 19, 20

And when you live in this community you live and breathe everyone else's story every day of your life. Participant 6: 2014, p. 28

12.6 RECOGNISING THE DISASTER JUSTICE ISSUES

The research allowed us to identify and draw out the disaster justice issues experienced by nine Blue Mountains community leaders. Our findings correlate with the research literature in which recent case studies highlight disaster justice issues relating to the chronic under-resourcing and lack of acceptance of the community services sector by the officially appointed recovery team—despite the increased demands made on them by the 'shared responsibility' policy. These impacting pressures originate from the needs of the vulnerable within the community, the demand of the shared responsibility policy and the community service leaders not being included in emergency management decision-making processes. This returns the discussion to the value of justice. To avoid simply reinforcing existing power structures we need a better connection between government and community organisations which "*aids in the arrangement of power so that the ideals of democracy, freedom and justice are attained as well as possible*" (Brey, 2007, p. 73). As Brey has put it, "*power relations are both established by the actions of agents and by the workings of social structures*" and "although power relations do not require intentionality, the exercise of power always does" (p. 74).

The overarching argument of this chapter is that devolving responsibility for building local disaster resilience and encouraging communities to participate in their own disaster recovery, must be carefully planned and appropriately resourced. Otherwise, as our research demonstrates, the cost is born by community leaders who burn out, experience guilt, grief and distress at what they can see needs to be done and yet cannot accomplish due to the stark lack of resourcing in the community sector and the reluctance of official recovery teams to accept community leaders as equal partners in the process.

In this chapter we have explored the difference between what *ought* to be happening (Sect. 12.3) and what *is* happening (Sect. 12.5) to demonstrate the disaster justice issues facing Blue Mountains community leaders. Encapsulated, through the National Strategy for Disaster Resilience and contingent concept of 'shared responsibility' (COAG, 2011), the government is instructing the community sector to do something without costing it in the same way they cost and resource the emergency services sector.

The underlying assumption of government in the National Strategy for Disaster Resilience is an understanding that resilience is not a static state, but rather it is to be grown and developed. We agree, however, that the National Strategy for Disaster Resilience misses the next vital step in the process, which is the nuts and bolts of the 'how'. For instance, how are local councils, the non-government sector and emergency services going to cooperate and share responsibility when there is a gross imbalance of decision making authority perpetuated by excluding NGOs from the LEMC, a lack of resourcing to backfill paid NGO staff and a reliance on the goodwill of community leaders and their volunteers while looking over their shoulders they see emergency service volunteers provided with uniforms, refreshments and carefully regulated working hours and conditions?

The results demonstrate that there is currently a serious disaster justice situation. Shared responsibility as a concept is consistent with the current narrative of participation and inclusion; however, in reality, there is a hegemonic situation whereby the emergency services hold the power largely due to heavy government resourcing and legislated authority. Therefore, shared responsibility as a concept is plausible and commendable, but in reality it is not possible until the situation is equalised and the community sector is officially recognised through policy support and government resourcing.

12.7 Recommendations

The same moral duty applied to the local emergency services sector should be applied to the community sector. To achieve a truly equitable platform for shared responsibility we recommend:

12.7.1 Funding Policy to Backfill Community Leader Positions

Government funding to advance community leaders from unaffected locations into disaster affected community organisations to assume the daily managerial functions and thus release the local community leader to work in the recovery space.

12.7.2 Structural Changes to the Composition of the Local Emergency Management Committee

The participation of local community leaders in all phases of disaster management should be reflected in practice as well as in policy. The structure of the LEMC is challenged by the findings of our research. There needs to be an incorporation of community leaders into all aspects of disaster planning, a seat at the table, the sense that others 'have their back'. Simply clarifying roles and responsibilities of the various actors (emergency service and community services) is not enough to facilitate equality.

References

Beauchamp, T. L., & Childress, J. F. (2009). *Principles of Biomedical Ethics*. Oxford: Oxford University Press.

Brey, P. (2007). The Technological Construction of Social Power. *Social Epistemology, 22*(1), 71–95.

Burnside-Lawry, J., & Carvalho, L. (2016). A Stakeholder Approach to Building Community Resilience: Awareness to Implementation. *International Journal of Disaster Resilience in the Built Environment., 7*(1), 4–25.

COAG. (2011). *National Strategy for Disaster Resilience*. Barton, ACT: Council of Australian Governments.

Davies, A. (2009). Understanding Local Leadership in Building the Capacity of Rural Communities in Australia. *Geographical Research, 47*(4), 382.

Emergency Management Australia Handbook 6: Community Engagement Framework. (2013). Retrieved from https://knowledge.aidr.org.au/resources/handbook-6-community-engagement-framework/

Fitzpatrick, T., & Molloy, J. (2014). The Role of NGOs in Building Sustainable Community Resilience. *International Journal of Disaster Resilience in the Built Environment, 5*(3), 292–304.

Gil-Rivas, V., & Kilmer, R. P. (2016). Building Community Capacity and Fostering Disaster Resilience. *Journal of Clinical Psychology, 72*, 1318–1332.

Hyogo Framework for Action 2010–2015: Building the Resilience of Nations and Communities to Disasters. Retrieved from https://www.unisdr.org/we/coordinate/hfa

Islam, R., Manock, I., Sappey, R., Hicks, J., & Ingham, V. (2012). Flooding in Bangladesh and Australia: Applying an Interdisciplinary Model. *International Journal of Interdisciplinary Social Sciences, 6*(8), 81–92. Retrieved from http://iji.cgpublisher.com/product/pub.88/prod.1515/m.2

Levy, M. (2013). NSW Bushfire: Minister Urges Residents to Leave Blue Mountains. *Sydney Morning Herald*. Retrieved February 26, 2019, from https://www.smh.com.au/environment/weather/nsw-bushfires-minister-urges-residents-to-leave-blue-mountains-20131023-2vzz3.html

Madsen, W., & O'Mullan, C. (2016). Perceptions of Community Resilience After Natural Disaster in a Rural Australian Town. *Journal of Community Psychology, 44*, 277–292.

NSW EMPLAN. Retrieved from https://www.emergency.nsw.gov.au/Pages/publications/plans/EMPLAN.aspx

Olivia, S. (2018). Measuring the Economic Resilience of Natural Disasters: An Analysis of Major Earthquakes in Japan. *City, Culture and Society, 15*, 53–59. https://doi.org/10.1016/j.ccs.2018.05.005

Pakrasi, S., Burmeister, O. K., McCallum, T. J., Coppola, J. F., & Loeb, G. (2015). Ethical Telehealth Design for Users with Dementia. *Gerontechnology, 13*(4), 383–387. https://doi.org/10.4017/gt.2015.13.4.002.00

Redshaw, S., Ingham, V., McCutcheon, M., Hicks, J., & Burmeister, O. (2018). Assessing the Impact of Vulnerability on Perceptions of Social Cohesion in the Context of Community Resilience to Disaster in the Blue Mountains. *Australian Journal of Rural Health, 6*(2), 14–19.

Rivera, J. D. (2016). Organizational Structure and Collaboration: Emergency Management Agencies and Their Choice to Work with Voluntary Organizations in Planning. *Risk, Hazards & Crisis in Public Policy*, 173. Retrieved from https://onlinelibrary-wiley-com.ezproxy.csu.edu.au/doi/epdf/10.1002/rhc3.12105

Schlosberg, D., & Collins, L. B. (2014). From Environmental to Climate Justice: Climate Change and the Discourse of Environmental Justice. *WIREs Climate Change, 5*, 359–374.

Sendai Framework for Disaster Risk Reduction 2015–2030: UNISDR. Retrieved from https://www.unisdr.org/we/coordinate/sendai-framework

Slote, M. (2010). Justice as a Virtue. In E. N. Zalta (Ed.), *Stanford Encyclopaedia of Philosophy*. Stanford, CA: Stanford University.

State Emergency Management Legislation: SERM Act 1989. Retrieved from https://www.legislation.nsw.gov.au/acts/1989-165.pdf

Wylie, B. (2018). Blue Mountains Blazes: Arcing Powerlines Sparked Fires That Destroyed Almost 200 Homes. Retrieved February 26, 2019, from https://www.abc.net.au/news/2018-05-23/falling-trees-caused-devastating-blue-mountains-fires/9790740

CHAPTER 13

Disaster, Place, and Justice: Experiencing the Disruption of Shock Events

David Schlosberg, Hannah Della Bosca, and Luke Craven

13.1 Introduction

Unprecedented global processes such as climate change and urbanisation are likely to not only change and transform urban areas but also metamorphosise and disrupt the concepts and certainties that support everyday life to make what was *"unthinkable yesterday ... real and possible today"* (Beck, 2016). Our interest lies in this distinction between events that change or transform urban and peri-urban places, and events that disrupt or undermine their very meaning and value for residents and communities. Specifically, we explore how this distinction between impact and metamorphosis of place is illustrated by the experiences of loss reported as a result of shock events—in this case a peri-urban bushfire.

D. Schlosberg (✉) • H. Della Bosca
Sydney Environment Institute and the Department of Government and International Relations, University of Sydney, Sydney, NSW, Australia
e-mail: david.schlosberg@sydney.edu.au

L. Craven
First Person Consulting, Melbourne, VIC, Australia

© The Author(s) 2020
A. Lukasiewicz, C. Baldwin (eds.), *Natural Hazards and Disaster Justice*, https://doi.org/10.1007/978-981-15-0466-2_13

This approach requires engaging with and beyond empirical, physical, or infrastructural impacts of such events to examine what Carpenter, Folke, Scheffer, and Westley (2009) would term "noncomputable" factors of resilience—the emotional, personal, and innately subjective individual experiences of those whose lives are not only simply impacted, but thoroughly and conceptually redefined. Almedom and Tumwine (2008) define resilience, in part, as the ability of society to actively extract meaning from a shock event with the goal of maintaining normal function without fundamental loss of identity. This recognition of ontological factors of resilience is critical considering the increasing evidence that the emotional, lived experiences of place disruption have the potential to harm individual and community well-being through time (Askland & Bunn, 2018; Tschakert, Ellis, Anderson, Kelly, & Obeng, 2019). In the resilience literature, calling for increased awareness of intangible human loss is not only important, but also a means of justice in the face of coming shock events (Glandon, 2015; Magee, Handmer, Neale, & Ladds, 2016). These arguments suggest that the focus of resilience research must increasingly be turned towards experiences and meaning, rather than physical impacts of shock events alone, because ultimately *"resilience is experienced as a social narrative, not as a set of numbers"* (Glandon, 2015, p. 27); or, more specifically, the costs of property damage. Identity and attachment to place and community are key to such an expansive understanding of resilience. Indeed, placing meaning and identity at the heart of resilient communities becomes critical in a metamorphosing world in which relationships, experiences, and attachments are at risk.

In order to engage with the disruption of lived experiences, we draw on the concept of place attachment (Manzo & Devine-Wright, 2013). As attachment is about the bond between people and the places they live, a focus on the concept can bridge an exploration of changes in the physical world and subjective experiences and meanings of these changes. Place attachment is a dynamic, two-way relationship between specific environments and individual and community identity, and is emotionally constituted through place-based experiences, memories, and understandings through time (Devine-Wright, 2015). Events that alter the lived environment may require people to remake or redefine their emotional connections to place (Fullilove, 1996) and this is particularly true of shock events (Morrice, 2013). We use this conceptual approach in order to analyse and compare the experiences of community members after a major bushfire in the Blue Mountains outside of Sydney, Australia, in 2013. We draw on resident and service provider focus group discussions to examine the relationship between shock events, conceptual disruption, and loss of place.

Our aim in this chapter is to consider the experiences of loss associated with this shock event in order to better understand the meaning of loss of place, its experience over time, and the relationship between that loss of place and other impacts to individuals and communities. We start with a discussion of the meaning of place attachment, and its relationship to a concept of environment and climate justice in the context of resident identity. We then illustrate the relationship between this idea of place attachment and the threat of shock events with reference to the bushfire. And we explore why and how this discussion pushes against more traditional infrastructure-focused understandings of approaches to resilience, closing with a suggestion for more place-aware resilience policymaking.

13.2 Place Attachment and Resilience: Identity, Emotion, and Imagined Futures

Physical places give people ontological meaning in their day-to-day material lives. Important here is Graham et al.'s (2013, p. 49) concept of "lived values", or the everyday practices or articulations through which communities and individuals express the values that define "*what is important in their lives and the places they live*". Events that change that relationship between experiences of the physical world and their meaning can significantly disrupt these values and change how people relate to their environment (Rawluk et al., 2017). As Tweed and Walker (2011) and Amundsen (2015) highlight, the increasing intersection of events that disrupt the habitual functions of urban areas suggests that engaging productively with questions of community resilience requires understanding processes of disruption.

Askland and Bunn (2018) highlight the ontological nature of place attachment, and that its interruption or discontinuation may create not simply actor-centric psychological distress, but rather an ontological concern. In other words, disrupting this connection to place may fundamentally challenge the way a person understands the world and their place in it. Real or perceived changes to the values that make a place special may trigger unwanted community conflicts (Gee, 2010), or it may trigger feelings of dissatisfaction and loss in individuals and communities, especially when that change challenges multi-generational constructions of home and identity (Nicolosi & Corbett, 2018). The loss of personal agency and identity as a result of changes in day to day life can lead to "*cascading social and environmental problems*" above and beyond the original disruption (Barnett, Tschakert, Head, & Adger, 2016, p. 977). There is an emerging literature on the emotional impact of climate displacement in the form of place disruption. Tschakert et al.'s (2017, p. e476) work finds that climate change related disruption to daily life may manifest as stress, loss, grief, hopelessness, and alienation if it impedes identity and agency, ultimately affecting the "mental, emotional, and spiritual health and well-being" of individuals and communities. One recent paper on the emotional impacts of the damage occurring on the Great Barrier Reef in Australia coins the phrase 'Reef Grief' to describe feelings of loss (Marshall et al., 2019). Ultimately, such an experience can be described by the concept of "solastalgia", the existential distress experienced by people when they feel that their identity is no longer supported or reflected in their home places due to profound environmental change (Albrecht, 2005). The Guardian (Macfarlane, 2016) has highlighted solastalgia as an important part of the necessary language of the Anthropocene, "*a modern uncanny, in which a familiar place is rendered unrecognisable by climate change or corporate action: the home becomes suddenly unhomely around its inhabitants*".

Much of the recent work on place attachment and loss looks at experiences of climate change. One key argument is that it is this unique ability of place attachment analysis to place lived experiences of loss at the forefront of analysis that makes it "*a better starting point for climate change adaptation than an emphasis on climate change impacts*" (Amundsen, 2015, p. 257). Again, this recognises that it is not change events themselves that are significant in isolation, but how such events disrupt the qualities that make a place special to its inhabitants (Gee, 2010; Morrice, 2013) to an extent that diminishes the capability of individuals and communities to live the lives they desire or expect, in the way that offers connection and continuity to place and each other.

Overall, there is a growing body of literature to suggest that when people are unable to remake or redefine these connections to place due to shock event disruptions, they are likely to experience feelings of loss. Loss here is defined as occurring when people "*are dispossessed of things that they value, and for which there are no commensurable substitutes*"

(Barnett et al., 2016, p. 976). In this, feelings and experiences around identity, safety, and belonging (Manzo & Devine-Wright, 2013) are implicated in emotional experiences of physical change that disrupt relationships to, and experiences of, place (Barnett et al., 2016; Carrus, Scopelliti, Fornara, Bonnes, & Bonaiuto, 2014).

The experience of place-based change and ontological loss is linked with temporal meaning and evaluations. This is emphasised in Manzo's metaphor of place as a *"bridge to the past"* (2005) in which specific places embody memories or values that act as a connection to the past, providing a point of access to and continuity with significant emotional and ontological experiences. Again, the emotional implications of place attachment and its disruption are most evident in personal perceptions of feelings of safety, threat, or belonging, particularly where place intersects with ongoing communal social identity (Manzo, 2005). Thus we can see that place attachment is not solely a result of present associations, but importantly, associations through time in such a way that place-based disruption "*due to spatial and temporal dissonance underpins ontological anxiety*" (Askland & Bunn, 2018, p. 16). Manzo's bridge metaphor can be extended here to incorporate constructions of the future, long established as a critical aspect of place attachment (see Askland & Bunn, 2018; Milligan, 1998), particularly with respect to the desirability of imagined futures (Della Bosca & Gillespie, 2018; Schlosberg, Craven et al., 2018). Changes to local environments may enable or challenge the continuation of valued constructions of the past, as well as the desirability of specific futures.

Disruption to longstanding connections and meanings of place have also long been understood as an issue of social and environmental justice, relating to power, inequity and vulnerability, recognition, procedural justice, and capabilities (Glandon, 2015; Schlosberg, Craven et al., 2018; Stanley, 2009). Issues of power and agency have become apparent through the increased recognition that day-to-day experiences are shaped by emotional and values-based constructions of place and time (Stanley, 2009). The inclusive recognition of the emotional implications of place-based identity threats is key in procedurally acknowledging and representing contested or conflicting community experiences in decision-making processes (Ojala & Lidskog, 2017). Both Barnett et al. (2016) and Tschakert et al. (2017) call for procedural acknowledgement and integration into people-based climate resilience approaches that account for the meaning and values embodied in place. It has led Groves (2015), Agyeman, Schlosberg, Craven, and Matthews (2016), and more recently Schlosberg, Rickards, and Byrne (2018) to argue that the disruption of the day-to-day capabilities that enable individuals and communities to live as they want to live is an aspect of environmental injustice.

Such an understanding of the importance of relationality and place in the concept of environmental justice is growing. As Delaney (2016, p. 3) notes, "*(in)justice is intrinsically social and relational in the sense that claims of injustice necessarily call into account inherently social states of affairs concerning contingent social arrangements—including socio-spatial arrangements*". In addition, Groves (2015) has argued that the colonisation of attachment to place, driven by the diverse impacts of a changing climate, can be conceptualised as a distinct form of environmental injustice. According to his argument, if attachment is a constitutive part of how people inhabit particular environments, then disrupting those attachments can do damage to both individual and collective well-being. Harms to attachment erode "*forms of agency embedded in attachments to place and collectives*" (p. 870), resulting in people losing "*a sense of themselves as doers and actors*" (p. 858). This rupture of residents' identities has marked effects, essentially taking away their capacity to "*negotiat[e] a future for themselves and their children*" (p. 859). As Schlosberg, Rickards, and Byrne (2018, p. 593) argue,

> how we live in an environment [and] people's experience of and relationship to places is an important element of broader questions about environmental justice (or injustice). Here, justice hinges on a sense of a positive place attachment—and avoidance of negative impacts on place, such as pollution and threats to environmentally based, culturally valued practices.

In addition, and from a capabilities approach to justice and environmental justice, the undermining of a broad range of capabilities is related to loss of place. Disruptions to place attachment can bring loss of cultural attachments to place as well as harms to a range of basic needs, including water, shelter, food, health care, and more—and do so in ways that are inequitably distributed. The links between place disruption and environmental injustice are numerous.

The significance of emotional and personal constructions of place is increasingly recognised in the resilience literature, particularly through a lens of hope (Head, 2016) or—more commonly—grief (Cunsolo & Ellis, 2018; Marshall et al., 2019). In a review paper of the Rockefeller Foundation's Asian Cities Climate Change Resilience Network (ACCCRN) program, Friend and Moench (2013) highlight that resilience is experienced by urban residents as the comparative difference in day-to-day life before and after a shock event. This reinforces Pelling and Manuel-Navarrete's (2011) observation regarding the innate subjectivity of resilience. The increasing attention to place attachment in resilience literature speaks to "*a spatially and socially situated approach to assessing vulnerabilities and resultant strategies*" associated with local experiences of change and transition (Barr & Devine-Wright, 2012, p. 527). Beyond resilience as a theory, a set of capacities, or a strategy, we can see that resilience is an emergent property of interwoven capabilities undermined by this type of place-based loss. As we will go on to show, for example, we can identify place attachment or its absence as a 'fertile functioning' or 'corrosive disadvantage', respectively (Wolff & de-Shalit, 2007)—they can lead to a cascading relationship, positive or negative, across complex phenomena that support or undermine the capabilities necessary for justice. Resilience, then, can respond to the complex impacts and interactions surrounding shock events that can provide a 'fertile' and just policy response to the impacts on place, identity, and well-being.

All of this marks an important departure from the scientific and technical detachment of conventional resilience approaches which were understood to produce "*objective knowledge that is unmarked and disembodied and thus epistemologically and normatively superior*" (Stanley, 2009, p. 1008). The academic realm of resilience literature contains an abundance of research that focuses primarily on physical disruptions to place, neglecting the "*psychological, symbolic, and particularly emotional* aspects *of healthy human habitats*" (Agyeman, Devine-Wright, & Prange, 2009, p. 509). Conceptions of resilience that privilege infrastructure and services can devalue or exclude 'noncomputable' (Carpenter et al., 2009) lived experiences, and are unable to capture or reflect important factors of individual and community well-being, and justice, through time. This is not to discount the role of the physical damage of shock events plays in the physical, mental, and emotional recovery of individuals and communities, but rather to highlight that exclusively focusing on physical impacts renders the intangible experiences of loss invisible.

And yet, even with the evolution of the literature, the criticality of the emotional, social, and cultural impact on identity, agency, and justice has not often been translated into official strategies or approaches to minimise experiences of loss. Such intangible impacts of shock events tend to be overlooked in state assessment models of shock events (Magee et al., 2016).

While there are inevitable challenges in making room for place attachment's intangibilities in policy fields that respond best to quantitative-centric datasets, this mounting evidence suggests that if we, as researchers, want to inform holistic and just resilience applications, we must better represent these less obviously visible yet critical realities of change, disruption, and loss.

Our contribution to emerging work on the intersection of urban resilience and experiences of place attachment disruption is twofold. First, we offer an empirical examination of lived experiences in peri-urban Sydney, Australia, of an important shock event—one that will inevitably repeat across the country in a climate-challenged environment. Second, in engaging with both residents and a range of community service providers, we respond to calls in the resilience literature (see Hassler & Kohler, 2014) for wider stakeholder input in understanding these processes of disruption, specifically through engaging with the lived experiences of local individuals and communities. Our approach and analysis aim to draw explicit policy focus to the processes of place attachment disruption underpinning trauma in experiences of shock events. In so doing, we provide an evidence base for place-sensitive resilience approaches that more effectively support and enhance community well-being.

13.3 Methodological Context

In this chapter we examine community experiences of a key shock event in the greater Sydney region—the 2013 Blue Mountains Bushfires. In October 2013, hot and windy conditions triggered widespread bushfires across NSW. On 16 and 17 October, three separate fire fronts broke out in the Blue Mountains, originating in Lithgow, Springwood, and Mount Victoria. 196 homes were destroyed, and significant damage done to 132 others, totalling $180 million in losses. 65,000 hectares of mostly national parkland were burnt, taking a dramatic toll on wildlife (Milman, 2013). The State government declared the bushfire areas a natural disaster zone, with the Blue Mountains Council stating that the bushfires were the worst disaster in Blue Mountains history.

This event is one of a number identified in Resilient Sydney's 2016 Preliminary Resilience Assessment (PRA) of major shock events (Resilient Sydney, 2016). While the scale of participation for this single case study inhibits more broadly representative conclusions being drawn, we believe it provides a valuable entry point into lived community experiences, invites comparative research, and is sufficient for the narrative exploration taken in this paper. Below, we provide an overview of the methodology and recruitment strategy used in this study, before outlining the case study and participant groups.

The study employed System Effects, a mixed method framework that aims to capture the lived experience of complex phenomena (Craven, 2017) and enables a high level of individual participant detail to be reflected within aggregated findings. The System Effects method requires individual participants to create a series of impact maps to capture their individual experience, followed by small group discussions that allow reflection and further elaboration of individual and communal experiences. The results and discussion of this paper are based entirely on the participant transcripts generated through these focus group discussions, as they provide the level of narrative detail required to enable place attachment analysis. We found that the holistic thinking demanded of participants for the System Effects mapping exercises enabled them to think about the rich and layered interconnections between the different aspects of resilience and individual and community well-being.

This work involved recruiting Sydney participants with personal experience of the 2013 bushfires. Participants were invited to attend a three-hour focus group (held within the community of the target event), between August and December 2017. We identified two participant group types to target for recruitment—local residents and professional/voluntary service providers attending the area at the time of the event. Both groups were identified as having distinct lived experiences, and as groups with a high level of policy relevance. An important aspect of this research was to broaden the definition of who and how people are 'impacted' in these types of shock events, and the experiences of these two different groups were central to achieving a more comprehensive understanding of stakeholder impacts.

All recruitment was conducted by the office of Resilient Sydney. Resident recruitment was achieved through a combination of residential letterbox leafleting and targeted advertising via social media (in relevant Facebook groups) and through posters and leaflets in community organisations such as government buildings, public libraries, and other local service facilities. Residents self-identified as impacted by the target event and were screened by Resilient Sydney staff in a brief phone call prior to event registration and confirmation. Residents were provided with a $50 grocery voucher for their time. Service provider recruitment was achieved by accessing existing networks run or contributed to by Resilient Sydney, local emergency management committees, local emergency management officers, and community groups.

Two impacted groups were identified. First, local residents of the area who self-identified as impacted, including residents who had lost their homes, had been evacuated, or who had had to take time off work to care for others. Two resident focus groups were held in Springwood and were attended by participants in the 25–34 to 65+ age range, with annual incomes varying from under $20,799 to over $156,000. All participants were either homeowners or living with home-owning relatives, and the majority had lived in the Blue Mountains LGA for eight or more years. Second, we engaged with professional or volunteer service providers who had been involved in response and recovery efforts associated with the fires. One service provider focus group was held in Lawson and was attended by representatives from an emergency service provider, a community charity, and a community resource organisation.

13.4 The Experience of the Blue Mountains Bushfires

Disruption to place attachment—in terms of both physical place and conception of community—was central to those impacted by the 2013 bushfires. The experience of our interviewees highlights the relationship between physical disruption, ontological disruption, and perceptions of well-being in the way that people experience shock events. We begin with the reflections of residents, followed by service providers. These groups provide different lenses through which to examine shock event disruption as it relates to issues of identity, place, values, and futures associated with places under threat.

It is important to note, as many of our focus group participants did, that bushfire is a longstanding and natural event in the eucalyptus forests of the Blue Mountains. Many spoke of their own previous experiences with fires in their lifetimes, and of the stories of their parents, grandparents, and communities over multiple generations of living in the region. These stories made clear that bushfires are understood as part of the natural place of the Blue Mountains, and part of the identity of both the place and the living within it. They also illustrate that there is a historical role such fires have played in the development of community identity and values—that fire has traditionally brought people closer to both place and each other. And yet, residents and service workers alike talked about a growing fear of the expected increase in disruptive fires. They see larger fires coming in historically unique and threatening ways, and have begun to reflect on new fire regimes due to climate change as potentially threatening the nature of the Blue Mountains, the nature of community, and the traditional rebuilding of lives and houses after such events.

Several participants in our resident focus group lost pets and property as a result of the 2013 fires; some lost their homes after fleeing an oncoming fire, or becoming trapped and narrowly escaping. These were highly traumatic events. Participants communicated long-term emotional impacts associated with the physical impacts of the fire, including sadness over the loss of pets, the anxiety as a result of having to permanently house family members who lost their own homes, the stress of dealing with governments and insurance companies in the rebuilding process, and the ongoing post-traumatic stress disorder of a child. There was evidence of significant feelings of loss, particularly around the loss of animals, as well as photographs as symbols of childhood memories. Many had continuing emotional associations of guilt and regret around the event. Many lost material objects, from homes to photo albums, represented crucial socially constructed meanings and values in their daily lives. One focus group participant whose house was burnt down highlights how this event disrupted feelings of belonging, continuity, and familial meaning, as they could not transfer their emotional associations about place from the old family home to a newly rebuilt one.

> *Long term—... my eldest daughter...hardly ever comes out to the house. She feels really weird.... We've rebuilt in the same place. But she just doesn't feel like it's home.*
> *[Other participant:] She didn't grow up there.*
> *That's not the house she grew up in, yeah. Yeah.*

Focus group commentary on the destruction of material objects suggests that experiences of loss are most strongly felt for items with irreplaceable value, a value that is emotionally constructed through memory, identity, and belonging in place and through time. Objects such as photo albums and memorabilia and indeed houses themselves may enable these particular emotional connections through time, and 'replacements' are never fully embraced. However, the material destruction of the fires was also discussed in comparatively light-hearted ways, highlighting a diversity of emotional impacts between and within participant experiences. This was evident in humorous comments on the benefit of fire in 'decluttering' efforts, as well as reflections on the potentially positive opportunity to rebuild a more suitable home with insurance money. This contrast highlights the limitation of using physical destruction and damage as a blanket measure for loss, as for some the physical destruction of the bushfires redefined or destroyed valued associations with place, while for others it did not.

Beyond the individual or familial experience, the fire has also led to a range of complex reactions to the deeply embedded associations in the Blue Mountains community. Many residents stated that the sense of community cohesion they experienced as a direct result of the fires increased or reinforced their connection to place, with the event emphasising the strong civic life of the community, even as the physical place faced severe damage and change.

> *But I guess the good things were how helpful everyone was afterwards. I don't think I've said thank you so many times in my life.*
>
> *What made it easier was the fact that we had neighbours who had been through all this before, including someone who had been there in '67, there was a major fire, basically went through the entire suburb. And he was great because he was like, "Don't worry, just do this, do this, wait for this, if this happens"—so that was invaluable, having neighbours who were experienced.*
>
> *So in that sense, it's made me very, very happy about where I live because yeah, I have a community life or civic life here that I don't think is available—well, I think it may be available in other places, but I'd never experienced it. I think the fires had a particular role in that in the sense that they made that incredibly apparent. I think the fires didn't create it, but it makes it more visible, like when you drove up and down Hawkesbury Road, it would just be "Thank you RFS, thank you RFS" on every second tree.*

This consistency of participant reflections on the strength of community cohesion during and after the fires as a positive aspect of the experience suggest that this bushfire was not perceived simply as disruptive to place, but also part of an ongoing characteristic of Blue Mountains life and community.

Other comments began to get at the reality that fire was an accepted aspect of living in the Blue Mountains, though something was different and particularly threatening about this one.

> [I]t was just another fire, really. It was just that it was such a big fire and I've never seen a fire event—probably not since 1977—stretch the resources of the fire brigade and everything so much.
>
> [I]t was definitely unprecedented and so a lot of people that have been through many fires behaved or reacted completely different to probably everything they've ever thought they knew how to do.

These comments reflect an understanding of bushfires not as entirely unexpected shocks, but rather as historically present and regular events in the local area. They illustrate that relationship between the physical characteristics of place, experiences of fire, and a sense of historical continuity. Yet, at the same time they demonstrate a sense of a difference and a break from that history.

Many participants expressed this sense of change—that fire is normal, but this event was something different, and it could be seen as an example of a new and more intense fire regime with the potential to upend a long history of fire, recovery, and rebuilding. The ontological significance of place attachment and the historical norm of fire influenced community patterns of response around safety, control, and responsibility for the family home. 'Stay and defend' behaviours were a by-product of historical experiences, as was 'return and rebuild'. But the experience of this particular fire may have changed these historical attachments. The attachments and associations with place, and the emotional responses they trigger, are now articulated in ways that differentiate residents from the historical norm, as the response of one participant conveys:

> I've always stayed and fought to defend my house every single time, but as of this year, if a fire comes to my house this year, I'm leaving. I'm not doing it. I won't do it again because yeah, I don't have the emotional stuff to get through it any more.

This reconstitution of the relationship between emotional attachment and fire and future bushfire events is significant as it reinforces the fluidity of place attachment while also providing insight as to how and why highly entrenched place-based behaviours are being altered and changed in the face of growing climate change-enhanced events. This work mirrors that of Morrice (2013) on the relationship between emotional impact of events and decisions to return home among victims of Hurricane Katrina.

For service workers, the relationship between fire and history, and behavioural norms and values in the Blue Mountains, can also be seen in this exchange between two participants suggesting that changes to the socio-political contexts of fire response have contributed to unexpected feelings of loss of the nature of the community.

Service Provider: *Because a lot of the older members have left because now you've got to do a 13 week course to get your chainsaw certificate and you've got to have a road certificate and - - -*

Service Provider 2:   *You've got to have a licence to order the sandwiches as well.*

P1:   *Yes, it's become so over-regulated. You do. It's like you've got to have food handling to work in the kitchen, you've got to have chainsaw certificates, backhoe certificates, you've got to have - - -*

P2:   *And that used to be the women's auxiliary. That's what the women's auxiliary used to do and none of them had food handling certificates.*

Moderator: *Red tape.*

P1:   *And I don't think any firefighters died because of a bad sandwich.*

This comment is indicative of wider participant discussions around fire response back when everyone pitched in and volunteered their time for communal benefit. Such practices are understood to form the basis of positive community cohesion, valued as a characteristic of the community as a result of historical bushfire threats. The reflection is about loss of community of a different type, and the lack of government attention to the impacts of regulatory structure on community cohesion and responsibility, which has always been a part of the meaning of, and attachment to, place.

Comments such as this from service providers highlight the complexity of place attachment meaning and impact, as well as the intersection of both values and experiences, and of policy structures, that contribute to changes in personal and community identity. To extend this reflection, we draw briefly on the experiences of one service provider and the clear and growing tension between public and private roles after the fire.

Service Provider 3:   *It's just that—yeah, I think for a long time, I lost my identity as a resident and you become the organisation*

	and, you know, you're not a parent anymore and you're not a person, so you get...
Service Provider 4:	*Sucked into a role.*
P3:	*Yeah, you get a lot of—it's hard to just do anything normal in your community and it's hard to just be a parent at the school or do anything normal and like my record was I took three hours to get bread and milk once—I just wanted bread and milk and it—just those things. It just takes a long time to do anything.*

This participant, who assisted in providing both basic needs and counselling, talked about a professional identity that was constantly 'on' in public space, consuming their private 'off-time', even when simply shopping for milk (see also Chap. 12). The reflection illustrates key changes as a result of the event—a growth in the demand for such service providers, a shift in the needs of community members, and a need for more extensive services necessary to address the mental health and social well-being impacts of the event and ongoing threats. But this provider continued about how difficult it was to remain in their own community when trauma became woven into their everyday life—not just their professional life. It was another example of a resident's existential concern about the strain of continuing a life in a threatened place.

Overall, both resident and service worker experiences of the Blue Mountains fires reveal that, for many people, bushfires were in the past understood as physically disruptive but ontologically valued and part of the identity of living in place. After the 2013 events, while there was evidence of significant trauma from the physical damage itself, less obvious sources of existential uncertainty about the meaning of place and the future of the community became part of the conversation. Reflections on the event demonstrated a changing understanding of what fire means to the Blue Mountains, both to the place itself and to people's lives with that place and each other. Fear of loss of the physical place and community connections in future is clear.

13.5 Discussion and Conclusions

We began this investigation into the relationship between resilience and loss through the metaphor of metamorphosis, the term used by Beck (2016) to emphasise that the element of ontological consistency that enables adaptation or transformation may no longer be available in the processes of change shaping modern cities. This is critical, as our case study findings demonstrate that participant experiences of loss and reduced well-being were tied to ontological disruption arising from changed relationships to place.

For residents in the Blue Mountains, the historical familiarity with bushfire had led to strong response practices that benefited community cohesion, but changes in the nature of both fire events and emergency response have changed socio-political expectations. This has led to disruption and detachment, and an ongoing fear of the inevitable future fires. In the aftermath of the event, the shocks themselves caused significant and varied physical damage and emotional trauma. However, participant discussions reveal that the disruption of lived values (Graham et al., 2013), in a way that triggered feelings of loss, was a common result of these social responses and changed meanings.

These events were associated with a wide variety of experiences about the relationship between place, meaning, and well-being. Participants noted the value they put on attachments to place—and the impacts of these shock events on their changed relationship to those places. Primarily, we see that a loss of meaning—or a creation of new meaning—leads to a loss of connection and harm to well-being. The key accelerator of these impacts, in the Blue Mountains, was the fear of the change in and the growth of fires, which are due to increase given the impacts of climate change on Australia.

The changing temporal frame also plays a significant role. For Blue Mountains residents, the temporal loss is of the historical role that bushfires have played in the development of community identity and connectivity in place, and the growing and future existential threat of catastrophic bushfire events. Traditional views of resilience often fail to emphasise these key longer-term temporal dynamics.

We started with the premise that emotional and intangible place-based attachments form one element of a conceptual stability that enables individuals and communities to maintain a cohesive and desired life trajectory in the face of change. This concern with the relationship between place and well-being, place and the provision of capabilities, we argued, is a growing concern in the environmental justice literature. Our results contribute to a growing body of evidence that suggests that when it comes to shock events and disasters, such qualities are not reducible to physical impacts, and require attention to the people- and place-specific meaning of such threats in order to address minimise experiences of trauma and loss. The essential human experience of shock events is about more than property or infrastructure impacts. The key to the distinction between understanding such events in this traditional way and more fully engaging change-as-loss lies in these personal intersections of identity, emotion, and place. For our participants, resilience is a measure of their quality of life, well-being, and connection to place before versus after a shock event. It is this engagement with what people value, and why, in their local environments that inform the critical difference between a traditional material approach to shock events, versus one that addresses the crucial nature of change and loss. Both the justice and resilience question thus become one of connection—how do people remake and redefine the connections to place that support well-being through protection of their places, their identity, their agency, and their idea of their place and role in the world. We need to ask how place attachment is the connective tissue that binds a range of individual and collective goods, and ties together justice and well-being in resilience responses and policies.

While we have examined only a single event here, it is clear that the threat of the impacts of climate change is on the rise. The future is already one of uncertainty and precarity, and these examples and findings illustrate further threats to place, community, attachment, and well-being. The grief that comes with undermining connection to place is real, and it extends far beyond the loss of iconic places like the Great Barrier Reef (Marshall et al., 2019), to every place threatened by new, different, shock events.

In response to these findings, just resilience policies and initiatives must be capable of supporting and encouraging positive place attachment while addressing problematic 'invisibilities', capabilities, and issues of exclusion. Head (2016) discusses numerous examples of how "hope" can be engaged, rather than grief, in response to the impacts of climate change, colonialism, and the Anthropocene in general. The focus is on efforts that can foster attachment and empower individuals and communities to actively shape their own associations in order to minimise feelings and experiences of loss—a longstanding demand of environmental justice movements, and now a core desire in the face of resilience to shock events. As Ojala and Lidskog (2017) highlight, participatory processes that genuinely engage with communities can help uncover place-specific values associated with local understandings and responses to environmental changes. Such processes counter homogenous and infrastructure-focused resilience policy by engaging heterogeneous experiences of complex pre-existing and shock-related vulnerabilities related to the experience of place. These inclusive and attentive policy responses can increase both the efficacy and equity of the governance of the challenges of shock events—and the reality of a just resilience or a just adaptation. Engaging people in threatened places about the meaning of change, and designing new ways to support social well-being in changing places, is key to how we might incorporate place, attachment, identity, and justice into resilience policy in the face of shock events.

REFERENCES

Agyeman, J., Devine-Wright, P., & Prange, J. (2009). Close to the Edge, Down by the River? Joining Up Managed Retreat and Place Attachment in a Climate Changed World. *Environment and Planning A, 41*(3), 509–513.

Agyeman, J., Schlosberg, D., Craven, L., & Matthews, C. (2016). Trends and Directions in Environmental Justice: From Inequity to Everyday Life, Community, and Just Sustainabilities. *Annual Review of Environment and Resources, 41,* 321–340.

Albrecht, G. (2005). 'Solastalgia': A New Concept in Health and Identity. *PAN: Philosophy Activism Nature, 3,* 41–55.

Almedom, A. M., & Tumwine, J. K. (2008). Resilience to Disasters: A Paradigm Shift from Vulnerability to Strength. *African Health Sciences, 8*(Suppl 1), S1–S4.

Amundsen, H. (2015). Place Attachment as a Driver of Adaptation in Coastal Communities in Northern Norway. *Local Environment, 20*(3), 257–276.

Askland, H. H., & Bunn, M. (2018). Lived Experiences of Environmental Change: Solastalgia, Power and Place. *Emotion, Space and Society, 27,* 16–22.

Barnett, J., Tschakert, P., Head, L., & Adger, W. N. (2016). A Science of Loss. *Nature Climate Change, 6*(11), 976.

Barr, S., & Devine-Wright, P. (2012). Resilient Communities: Sustainabilities in Transition. *Local Environment, 17*(5), 525–532.

Beck, U. (2016). *The Metamorphosis of the World: How Climate Change Is Transforming Our Concept of the World.* Chicester: John Wiley & Sons.

Carpenter, S., Folke, C., Scheffer, M., & Westley, F. (2009). Resilience: Accounting for the Noncomputable. *Ecology and Society, 14*(1), 13.

Carrus, G., Scopelliti, M., Fornara, F., Bonnes, M., & Bonaiuto, M. (2014). Place Attachment, Community Identification, and Pro-environmental Engagement. In *Place Attachment: Advances in Theory, Methods and Applications* (pp. 154–164). Abingdon: Routledge.

Craven, L. K. (2017). System Effects: A Hybrid Methodology for Exploring the Determinants of Food In/Security. *Annals of the American Association of Geographers, 107*(5), 1011–1027.

Cunsolo, A., & Ellis, N. R. (2018). Ecological Grief as a Mental Health Response to Climate Change-Related Loss. *Nature Climate Change, 8*(4), 275.

Delaney, D. (2016). Legal Geography II: Discerning Injustice. *Progress in Human Geography, 40*(2), 267–274.

Della Bosca, H., & Gillespie, J. (2018). The Coal Story: Generational Coal Mining Communities and Strategies of Energy Transition in Australia. *Energy Policy, 120*, 734–740.

Devine-Wright, P. (2015). Local Attachments and Identities: A Theoretical and Empirical Project Across Disciplinary Boundaries. *Progress in Human Geography, 39*(4), 527–530.

Friend, R., & Moench, M. (2013). What Is the Purpose of Urban Climate Resilience? Implications for Addressing Poverty and Vulnerability. *Urban Climate, 6*, 98–113.

Fullilove, M. T. (1996). Psychiatric Implications of Displacement: Contributions from the Psychology of Place. *The American Journal of Psychiatry, 153*(12), 1516–1523.

Gee, K. (2010). Offshore Wind Power Development as Affected by Seascape Values on the German North Sea Coast. *Land Use Policy, 27*(2), 185–194.

Glandon, D. (2015). Measuring Resilience Is Not Enough; We Must Apply the Research. Researchers and Practitioners Need a Common Language to Make This Happen. *Ecology and Society, 20*(2), 27.

Graham, S., Barnett, J., Fincher, R., Hurlimann, A., Mortreux, C., & Waters, E. (2013). The Social Values at Risk from Sea-Level Rise. *Environmental Impact Assessment Review, 41*, 45–52.

Groves, C. (2015). The Bomb in My Backyard, the Serpent in My House: Environmental Justice, Risk, and the Colonisation of Attachment. *Environmental Politics, 24*(6), 853–873.

Hassler, U., & Kohler, N. (2014). Resilience in the Built Environment. *Building Research & Information, 42*(2), 119–129.

Head, L. (2016). *Hope and Grief in the Anthropocene: Re-conceptualising Human–Nature Relations.* London: Routledge.

Macfarlane, R. (2016, April 1). Generation Anthropocene: How Humans Have Altered the Planet for Ever. *The Guardian.* Retrieved from https://www.theguardian.com/books/2016/apr/01/generation-anthropocene-altered-planet-for-ever

Magee, L., Handmer, J., Neale, T., & Ladds, M. (2016). Locating the Intangible: Integrating a Sense of Place into Cost Estimations of Natural Disasters. *Geoforum, 77*, 61–72.

Manzo, L. C. (2005). For Better or Worse: Exploring Multiple Dimensions of Place Meaning. *Journal of Environmental Psychology, 25*(1), 67–86.

Manzo, L. C., & Devine-Wright, P. (2013). *Place Attachment: Advances in Theory, Methods and Applications.* London: Routledge.

Marshall, N., Adger, W. N., Benham, C., Brown, K., Curnock, M. I., Gurney, G. G., … Thiault, L. (2019). Reef Grief: Investigating the Relationship Between Place Meanings and Place Change on the Great Barrier Reef, Australia. *Sustainability Science, 14*(3), 579–587.

Milligan, M. J. (1998). Interactional Past and Potential: The Social Construction of Place Attachment. *Symbolic Interaction, 21*(1), 1–33.

Milman, O. (2013, October 22). Bushfires Take Heavy Toll on Wildlife, Including Possums, Koalas and Gliders. *The Guardian*. Retrieved from https://www.theguardian.com/world/2013/oct/22/bushfires-take-heavy-toll-wildlife

Morrice, S. (2013). Heartache and Hurricane Katrina: Recognising the Influence of Emotion in Post-disaster Return Decisions. *Area, 45*(1), 33–39.

Nicolosi, E., & Corbett, J. B. (2018). Engagement with Climate Change and the Environment: A Review of the Role of Relationships to Place. *Local Environment, 23*(1), 77–99.

Ojala, M., & Lidskog, R. (2017). Mosquitoes as a Threat to Humans and the Community: The Role of Place Identity, Social Norms, Environmental Concern and Ecocentric Values in Public Risk Perception. *Local Environment, 22*(2), 172–184.

Pelling, M., & Manuel-Navarrete, D. (2011). From Resilience to Transformation: The Adaptive Cycle in Two Mexican Urban Centers. *Ecology and Society, 16*(2), 11.

Rawluk, A., et al. (2017). Public Values for Integration in Natural Disaster Management and Planning: A Case Study from Victoria, Australia. *Journal Of Environmental Management, 185*, 11–20.

Resilient Sydney. (2016). Preliminary Resilience Assessment 2016. Retrieved from https://www.cityofsydney.nsw.gov.au/__data/assets/pdf_file/0005/263975/2016-503932-Report-Resilient-Sydney-PRA-FINAL-ISSUED.pdf

Schlosberg, D., Craven, L., Della Bosca, H., Dawson, B., & Gabriel, K. (2018). Insights into Community Urban Resilience Experiences. Sydney Environment Institute. Retrieved from http://sydney.edu.au/environment-institute/publications/insights-community-urban-resilience-experiences/

Schlosberg, D., Rickards, L., & Byrne, J. (2018). Environmental Justice and Attachment to Place: Australian Cases. In R. Holifield, J. Chakraborty, & G. Walker (Eds.), *The Routledge Handbook of Environmental Justice* (pp. 591–602). London: Routledge.

Stanley, A. (2009). Just Space or Spatial Justice? Difference, Discourse, and Environmental Justice. *Local Environment, 14*(10), 999–1014.

Tschakert, P., Barnett, J., Ellis, N., Lawrence, C., Tuana, N., New, M., … Pannell, D. (2017). Climate Change and Loss, as If People Mattered: Values, Places, and Experiences. *Wiley Interdisciplinary Reviews: Climate Change, 8*(5), e476.

Tschakert, P., Ellis, N. R., Anderson, C., Kelly, A., & Obeng, J. (2019). One Thousand Ways to Experience Loss: A Systematic Analysis of Climate-Related Intangible Harm from Around the World. *Global Environmental Change, 55*, 58–72.

Tweed, F., & Walker, G. (2011). Some Lessons for Resilience from the 2011 Multi-disaster in Japan. *Local Environment, 16*(9), 937–942.

Wolff, J., & De-Shalit, A. (2007). *Disadvantage*. Oxford: Oxford University Press.

CHAPTER 14

Legal Identity Documenting in Disasters: Perpetuating Systems of Injustice

Kathryn Allan and James Mortensen

14.1 Introduction

This chapter explores the use of a legal identity in formulating disaster response, and argues that response frameworks that rely on legal identity have the capacity to further entrench existing social fragility and inequality within affected communities, and increase social vulnerabilities that further marginalise persons without access to legal identity documents. This marginalisation is concurrent across each phase of disaster response; through disaster risk reduction (DRR) planning, response and recovery.

To understand the use of legal identity throughout a wide variety of disaster settings, this chapter draws upon multiple academic case studies, NGO, UN and government disaster-related reports, post-disaster evaluations, and disaster policy and programming documentation. This wide

K. Allan (✉)
School of Anthropology and Archaeology, ANU College of Arts & Social Sciences, Australian National University, Canberra, ACT, Australia
e-mail: Kathryn.Allan@anu.edu.au

J. Mortensen
National Security College, Australian National University, Canberra, ACT, Australia
e-mail: james.mortensen@anu.edu.au

gamut of sources is brought together to answer the following: How is a legal identity used in each phase of a disaster? What is the impact of the identity frameworks on the disaster response? And how does this affect persons without access to legal identity documentation?

To address these questions, this chapter will first assess how legal identity is used in disaster contexts. It will then assess the points where disaster response frameworks have led to a failure of social protection for those without access to legal identity documentation; as a result it finds that reliance on legal identity documentation has the capacity to not only fail to respond to the needs of some affected by the disaster, but also hold the possibility of further entrenching the affected area's existing vulnerabilities and inequalities. Some of these areas of failure expose those with a lack of legal identity documentation to key high-risk factors that not only prevent them from accessing disaster relief, but also exaggerate existing social vulnerabilities such as poverty, homelessness, access to humanitarian aid, trafficking and exploitation. Finally, it is argued that the marginalisation of those lacking legal identity documentation in disasters does not simply entrench existing issues within affected communities, but imposed notions of justice vindicate that entrenchment and justify the exclusion of already vulnerable individuals.

14.2 Background

The fact that those who are stateless—and especially stateless women and children—are more vulnerable and disproportionately affected by disasters has been well established within the literature (Akerkar, 2007; Bradshaw & Fordham, 2013; Chew & Ramdas, 2005; Fisher, Swithern, & Walmsley, 2018; Van der Gaag, 2013). Children's vulnerability specifically is increased when they lack citizenship, are orphaned, displaced, or disabled (Delaney, 2006; Drumm & Stretch, 2008; Inter-Agency Standing Committee, 2007; M'jid, 2015; Peek, 2008; Penrose & Takaki, 2006; Save the Children, 2007; United Nations Development Programme, 2004). Persons who are rendered 'stateless', lack citizenship or do not have legal identity documentation are exposed to increased risks and vulnerabilities in the disaster context. This is no small issue, according to World Bank statistics there are currently one billion people without legal identity documentation worldwide (World Bank, 2019). A lack of formal identity documentation is intertwined with disasters and their impacts. Disasters increase the likelihood of persons becoming displaced or stateless; in 2016 alone,

24.2 million persons were displaced by climate-induced disasters. Further, those already stateless are of a higher risk due to the lack of inclusion in preparedness and planning (Fisher et al., 2018; UN Environment, 2019). Finally, of those displaced by disasters, those who were already displaced, recognised as refugees, informally displaced, or stateless have an increased chance of being excluded in disaster risk reduction and response frameworks (Akerkar, 2007; Fisher et al., 2018).

Without access to legal identity documentation, individuals within a disaster zone often cannot address basic human needs. If such individuals are reliant on humanitarian assistance, for example, a lack of such documentation can prevent them from accessing that aid. Further, incorrectly recorded, lost, destroyed or simply out of date identifying documents can have drastic effects on an individual's capacity to access relief or redress a disaster's effect on their lives; those living in irregular or makeshift housing, for example, are unlikely to be able to demonstrate what they have lost, or even that they lived in the disaster area at all (Akerkar, 2007; International Commission of Jurists, 2016; Sanderson, Knox Clarke, & Campbell, 2012).

This notion is strongly linked to Verchick's (2012) "disaster justice" concept where natural disasters are not a 'natural' phenomenon but rather occur within a social construct where the most devastating disasters in terms of loss of life are not necessarily those that have the most abstract of natural occurrences; for example, the higher on the Richter scale an earthquake is does not necessarily correlate to a more devastating impact on the society that suffers from it. Instead, the impact of a disaster is intertwined with the demographic, social, economic and political realities of the affected society prior to the disaster (Verchick, 2012). As a result, how an individual's legal identify—or lack thereof—is handled within a disaster response framework is crucial to mitigating these issues.

14.3 How Is Legal Identity Currently Used in the Disaster Context?

14.3.1 Pre-disaster Planning and Risk Reduction

In order to effectively plan for a disaster, governments and NGOs rely on data to assess the potential needs of the population. Where persons are undocumented or lack legal identity, they are excluded from this centralised data meaning that there are significant gaps in the data and therefore needs assessments. These gaps led to a lack of accurate data in regards to planning for aid provision, evacuation and conducting needs assessments (Boyer, 2011; Brown et al., 2014; Egal, 2011; Fisher et al., 2018; Knox Clarke & Ramalingam, 2012). As such, even in the preliminary stages, current disaster risk reduction (DRR) and response frameworks have the capacity to privilege those with a legal identity inasmuch as they cannot properly account for persons without a legal identity. Therefore, under-identified and 'stateless' people suffer from increased vulnerability and further marginalisation in many DRR frameworks (Fisher et al., 2018; Gaillard, 2010; Knox Clarke & Ramalingam, 2012; Macauslan & Phelps, 2012; Sanderson et al., 2012).

This shortfall in planning manifests in different ways across geographical and social lines. In urban areas, especially those with sporadic population fluctuation, DRR programmes must contend with populations in which there exists a large amount of undocumented migration. For example, Oxfam Great Britain struggled to know how to plan its programming to include informal settlements, especially undocumented households and the homeless population in its programs in Nairobi (Macauslan & Phelps, 2012, p. 11). Such issues can be confused or exacerbated by imperfect (and even conflicting) data; in Kibera, Narobi, population data had been collected by the government, UN organisations and NGOs, all which have different estimates, problematising the delivery of services to the population (Macauslan & Phelps, 2012, p. 11).

This gap in data further marginalises those who lack legal identity documentation and registration where they are unable to be included in policy and planning. A study on undocumented street children undertaken in Dhaka, Kathmandu, Manila and Jakarta revealed that a lack of government data on street children meant that NGOs were expected to fill the gap in data collection and aid delivery not only in a disaster context, but in everyday life as well (Brown et al., 2014, p. 29; West, 2003). Moreover, in this case study, when children were asked about their experiences with disasters, concerns were focused on everyday hazards and relief, rather than DRR planning. Where everyday lives are consistent with struggle and their own hazards, DRR is often not given precedence so when a disaster does occur children are not only excluded from government and NGO planning, but are underprepared themselves (Brown et al., 2014). However this state of affairs makes it even more important for NGOs and governments to be inclusive of undocumented persons—especially children—in DRR planning, as the absence of such planning compounds existing issues with basic aid delivery (Walters & Gaillard, 2014, p. 216).

A further example of extenuated vulnerabilities for those lacking legal identity is the exclusion of the homeless population from DRR planning. Across various Indian cities, the government conducted a 'mega-mock earthquake drill' as part of its DRR planning. However, in this drill the homeless population where excluded, which is indicative of the marginalisation and overlooking that this section of society is exposed to in daily life (Walters & Gaillard, 2014, p. 216). Wisner et al. also note the particular lack of consideration for refugees, homeless and 'socially invisible' populations in DRR planning and policy (Wisner, Gaillard, & Kelman, 2012, p. 175). This exclusion in the DRR planning phase of a disaster is directly connected to the resourcing and support supplied in the disaster itself (Walters & Gaillard, 2014, p. 216).

14.3.2 Response Phase: Humanitarian Aid Distribution, Healthcare, Social Benefits, Freedom of Movement

The immediate 48 hours following a disaster are the most critical in minimising the effects of a disaster on the population (M'jid, 2011). It is during this period that in which rapid response assessments are conducted, and immediate humanitarian assistance is delivered. During this phase of the disaster, identity frameworks are used as a means for states to collect quantitative data on the persons affected by a natural disaster, including those in evacuation centres (Seneviratne, Amaratunga, Haigh, & Pathirage, 2012). Identity verification is an integral part of the system and refers to the process of confirming identity details with systemic data or other resources (Harper, 2009, pp. 43–45).

Current methods and policy in dispensing aid and relief in disaster responses are reliant on government information and identity systems to inform the distribution of aid to areas affected by disaster. This includes the need to provide identity documents when receiving humanitarian aid in the post-disaster context. Whilst these identity frameworks in a disaster aim to be ad-hoc in nature, they are intrinsically linked to a State-based fixed identity system used in non-disaster contexts (Fisher et al., 2018; Harper, 2009; International Commission of Jurists, 2016). As such, due to this reliance on government data—which often excludes those without a legal identity—the disaster response during this phase is often ineffective in serving the needs of the entire population.

In the 2015 earthquake that struck Nepal a reoccurring theme noted by disaster victims was a lack of ability to access humanitarian aid or basic services throughout the response (International Commission of Jurists, 2016). The issue of legal documentation is entrenched in Nepalese society where certain minorities and marginalised groups do not have equal access to obtaining legal identity documentation in daily life, an issue which is exasperated in the disaster context. Even when there was specific dedicated 'relief' efforts to issue new legal documentation, those who had never had access to legal documents were unable to obtain 'relief identification' or the 'Earthquake Victim Identity Card' (EVIC). The fact these identity cards were tied to government-based identity frameworks was decried by NGOs and the 'victims' themselves, as those without citizenship certificates, proof of land ownership or legal residence were ineligible for an EVIC and therefore blocked from receiving humanitarian assistance (International Commission of Jurists, 2016: 8–9). Further, despite the fact that the head of a Nepalese household is traditionally male, EVICs were registered by household rather than individual, so even if women did have identity documentation, if there was no male present, or they were one of multiple wives in marriages not recognised by the State, then both they and their children were denied humanitarian relief (International Commission of Jurists, 2016, p. 10).

The Pakistan earthquake was similarly reliant on government-based identity frameworks in its disaster response identity framework. During the earthquake, victims had to show a National Identity Card (CNIC) to access any government-sponsored relief or recovery, this included "*compensation for death and personal injury, livelihood cash grants, housing reconstruction compensation and distribution of land to the landless*" (Harper, 2009, p. 27). Such top-down systems-based approach to humanitarian provisioning in a disaster response not only means that the most vulnerable are not able to access basic necessities but also encourages family units to separate in order to increase their chances of receiving basic assistance. For example, in 2010 in Haiti, families were maximising their chances of receiving aid by having individual family members disperse across different 'Internal Displacement Camps' (IDCs) in order to increase their likelihood of getting food (Schuller, 2010).

14.3.3 Post-disaster Response Phase

In most circumstances post-disaster recovery requires personal documentation to access relief assistance to rebuild housing, banking and financial assistance, re-entering into the education system, continued healthcare, and resume to life as it was in the 'pre-disaster' context. The requirements for legal identity in the post-disaster response framework further marginalise the population without legal identity documentation.

In this phase of the disaster, single women and women-headed households are disproportionately affected by a lack of legal identity documentation to access disaster recovery assistance including monetary and housing. Post-tsunami in Thailand in order to receive assistance entire households had to register, however men are traditionally the 'heads' of households which meant that they were able to gain access to housing documentation before women. In Akerkar's 2007 study one Thai women explained how her son registered himself as the head of the household and was granted both property rights and financial assistance, even though she had previously been the homeowner. This essentially left the mother homeless and heightened the potential for eviction as she no longer owned the rights to a home. The assumption from the institutionalised identity system that men should be granted property documentation over women discriminated against and negatively affected pre-existing rights of women, leaving women significantly worse-off and with fewer rights than they had pre-tsunami (Akerkar, 2007, p. 370).

Women were further marginalised in the Indian and Sri Lankan 2004 disaster context where families are unable to access compensation for the death of a family member without a death certificate. The identification process and framework for this is unclear in the post-disaster context and so means that the granting of a death certificate doesn't necessarily happen, this creates increased vulnerabilities in households that rely on the income of the often male, deceased family member to survive and are unable to access compensation (Akerkar, 2007, p. 371). The intersection of gender and lacking legal identity documentation further exasperates social discrimination in a disaster. Furthermore, in the Sri Lankan context legal identity documentation was required in order for orphaned and displaced children to be reunited with family members, specifically elderly family members who were more likely to not have access to such documentation (Mulqueeny, Fernando, & Bonifacio, 2010). This further marginalised not only orphaned children but their elderly family members in the reunification process.

14.3.4 Conclusion to the Review of Identity in Disasters

Overwhelmingly in the disaster context legal identity and personal documentation is required at every phase of a disaster, including the 2004 Tsunami that affected Thailand, Sri Lanka, and Indonesia; the 2015 Nepal Earthquake; and 2010 Delhi Earthquakes (Akerkar, 2007; Fisher et al., 2018; International Commission of Jurists, 2016; Walters & Gaillard, 2014). Legal identity is a tool that is used to protect against false claims of need by unaffected individuals, ensure a fair distribution of resources amongst the affected community, and streamline aid processes to ensure a swift response and promote resilience amongst the community in DRR planning, however persons who do not have access to a legal identity or legal identification documentation are exposed to increased vulnerabilities. Current frameworks for identity management in disaster response are inadequate because they limit the capacity of persons lacking legal identity documentation to access emergency relief, humanitarian aid, reunite with family members, gain access to healthcare or social benefits, and limit freedom of movement (Fisher et al., 2018; Inter-Agency Standing Committee, 2011; International Committee of the Red Cross, 2004).

Where one does not legally exist, one is not considered in disaster risk reduction frameworks, or at least not given equal standing nor priority. Stateless persons are often rendered 'invisible' as they are not functioning members of the identity management system of a nation. In terms of the International Federation of Red Cross and Red Crescent Societies 2018 they would be defined as "hidden people" which is defined as "*people without the necessary documents to qualify as eligible for assistance, for example basic proof of identification, school certificates or proof of tenure*" (Fisher et al., 2018, p. 29).

14.4 SYSTEMIC ISSUES

Ultimately, many of the wider systemic issues raised by the use of legal identity in disaster response are extensions of existing issues found within volatile and vulnerable communities. Issues such as entrenched inequality and marginalisation, the protection of women, children and other vulnerable individuals, and risks of criminal behaviours such as identity theft and compromise all have the capacity to be exacerbated in a disaster zone. However, reliance on identity in DRR can inflame these already exacerbated issues.

14.4.1 Entrenching Social Inequality; Women, Social Mobility, Political Vulnerability

The experiences of individual disaster victims in Nepal, Haiti, Ecuador, Indonesia and the US (Alfirdaus, 2014) highlight the fact that traditional systems of identity and aid management can be prejudicial to delivering outcomes for those in need in disaster areas. This risk is compounded for those lacking legal identity, and especially those in urban areas, which tend to have higher proportion of unidentified persons, whether through unregistered migration or a simple lack of registration (Cross & Johnston, 2011). These people are not only often the most vulnerable, they are also often the hardest to find, and may in some cases actively be avoiding identification by the government—an issue that further exacerbates the importance of support to these people, as well as the capacity of identity structures to further marginalise them (Macauslan & Phelps, 2012, p. 16; Sanderson et al., 2012, p. 9).

Similarly, the Thai government's response to the 2004 tsunami was such that no support was given to individuals such as Burmese migrants or communities like the Moken sea gypsies who lacked legal identity (Akerkar, 2007, p. 366). Whether by design, intention or accident, this selective response induced a situation in which these already marginalised communities were pushed further away. Left poorer and even more vulnerable than before the event, by basing the response on notions of identity, the DRR in this scenario became as much a function of political identity as of disaster response (Akerkar, 2007, p. 372).

However marginalisation through legal identity in disaster response can be used within citizen groups as well. As discussed above, those that are homeless (Macauslan & Phelps, 2012, p. 11), women in patriarchal societies (International Commission of Jurists, 2016) and unaccompanied children (Brown et al., 2014, p. 29) are all at risk of further marginalisation even within communities to which they nominally belong. In addition to this, such marginalisation also has the capacity to unfold on socio-economic lines, with the more affluent within a society more likely to have access to identification and information that allows them to gain more out of a DRR program. As a respondent to a study by Sharma et al. makes plain, in some cases, *"[o]nly higher class people have the information when the relief is coming and they can manipulate the resources"* (Sharma, Kc, Subedi, & Pokharel, 2018, p. 784).

14.4.2 Vulnerabilities to Exploitation and Trafficking

In the current disaster framework, a lack of legal identity documentation can inhibit the protection and registration process of children which heightens their risk to exploitation and trafficking in the post-disaster environment (M'jid, 2011).

After the 2010 earthquake in Haiti, thousands of children were exposed to a multitude of risks, where a lack of legal identity documentation led to children being trafficked across international borders (Binford, 2011; Reyes & Charles, 2010; Selman, 2011). Prior to the earthquake approximately 380,000 children were living in orphanages (UNICEF, 2010), and 2000 children were reported to be trafficked annually (Binford, 2011). UN organisations, NGOs and international humanitarian aid organisations prioritised unaccompanied children, including through registration (UNICEF, 2011). However, human rights experts Najat M'Jid Maalla, Joy Ngozi Ezeilo and Marta Santo Pais stressed that children who were vulnerable pre-disaster, especially children living in orphanages, were exposed to increased risks of trafficking and exploitation in the disaster context, especially unregistered children (UN News Centre, 2010; UNICEF, 2011).

In the 2004 Tsunami, Burmese migrants in Thailand were exposed to and exploited due to their lack of identity documentation. According to Akerkar's 2007 study this disproportionately affected women whose work permits went missing and were exposed to the risk of deportation (Akerkar, 2007, p. 369). Whilst these women were legally allowed to be in Thailand, the loss of their legal documents in the Tsunami meant that they could no longer work within the regulated system and had to seek work "illegally" and as such were exploited for cheap labour (Akerkar, 2007, p. 369). Further exposing them to exploitation was the necessity and desperate-nature of work that is linked to a lack of citizenship that denied them access to any relief or support from the Thai government (Akerkar, 2007, p. 369).

14.4.3 Risks to Identity Compromise

The chaotic nature of the disaster in disrupting normal social protections mean that there is increased risk for criminal behaviour; current identity frameworks in a disaster fail to adequately address the risk of identity compromise in relation to victims of the disaster (Lacey, 2015).

The preferencing of legal identities can not only disrupt the delivery of aid to individuals lacking such documentation, but also has the capacity to undermine the provision of a legal identity to those who need it, both in the short and long term. By granting benefits to only those with a recognised identity, risks of theft, fraud, coercion or even kidnapping for vulnerable members of a community could increase as unidentified individuals are forced to choose between anti-social behaviour and going without basic necessities. These risks are all the more pressing given the drive by these same organisations to provide those most at risk with identities. The UNHCR, UNICEF and others have made it clear that women and children, especially those who are unaccompanied, should be fast-tracked on the path to legal identification when attempting to gain access to humanitarian services. While great efforts are made to protect such individuals from exploitation, disaster zones are high-stake environments—a situation that may raise the risk of theft, abuse or coercion and control.

While less drastic, a far more likely problem is the theft or fraudulent acquisition of legal identity. Again, unidentified persons must choose between the improper and ultimately counterproductive avenues of identity theft or fraud, and their access to basic aid and services. In a disaster setting in which legal identity is necessary for the acquisition of aid, individuals could find practises of identity theft and fraud a more attractive endeavour, thus undermining the value of a legal identity and further hampering future efforts in identifying the individual correctly. In some circumstances, the misappropriation of legal identities of deceased persons have fed ad-hoc black markets in which victims of disaster trade what little they have for a legal identity that might secure more certainty in the immediate future.

Concerns surrounding identity theft and fraud are compounded in the context of sudden disasters given the widespread capacity of individuals co-opting or seizing control of the identities of the recently deceased. Such behaviour is more problematic in countries that lack consistent or centrally processed identity management, and is particularly acute in disasters that result in a large and immediate loss of life. While responders are attempting to quantify the loss of life and identify the bodies of the deceased, the removal and misappropriation of legal documents hampers these efforts. Thus any stresses placed on the survivors of a disaster that might induce them to steal or fraudulently obtain an identity from a deceased citizen not only negatively impact the correct identification of the survivor, but also run the risk of problematising the identification of the deceased.

14.5 Recommendations

Given the issues and problems raised in this chapter, making meaningful policy recommendations is difficult to say the least; the most impactful changes that could be made in mitigating the failure of legal identity in disaster justice would be to require a complete overhaul of existing social inequality, the imposition of NGOs on sovereign governments, or both. A possible short-term mitigation of the issues raised here would be a careful implementation of alternatives to legal identity, likely by NGOs. Properly distributed 'disaster identities' might assist in the dispensation of basic aid in the early stages of a disaster response, and might even be able to be used as the basis of an ongoing effort to identify stateless or other at risk individuals. One immediate roadblock to such efforts, however, would be the relationship between these 'disaster identities' and the legal identities of governments; NGOs would require permission from local governments to disregard their own identity efforts, something that might problematise the relationship between governments and NGOs.

Even if the use of such short-term mitigations proved to be effective, the unfortunate reality is that as long as compounding factors such as inequality, statelessness, poor identity management and systemic social bias or indifference remain, there is little capacity for legal identities to deliver holistic disaster justice outcomes. Improvements to the provision of identity documentation and systems can certainly go some way to mitigating these risks in many circumstances, however such improvements sit firmly outside the scope of DRR planning. Similarly, the addressing of issues such as social inequality and statelessness must also be addressed outside the context of disaster relief—in short, if we wish to deliver the best relief possible to many disaster affected areas, at-risk communities need to be assisted before disasters strike, not after.

14.6 Conclusion

As Verchick notes, disasters are not simply the initial event, but are instead the full gamut of societies, communities, responses and complications that are affected and realised through the disaster event, and as such efforts towards delivering disaster justice cannot simply be limited to a response to the effects of the event. While legal identities may be seen as the best way to manage injustice in aid provision (by preventing fraudulent behaviour, stockpiling or black markets), they can also serve to reinforce existing injustice by preventing already marginalised individuals from receiving aid. Further, while the use of legal identity seeks to ensure justice in provision of aid by cleanly demarcating those directly affected from those who are not, by concentrating on only those who have been directly affected by the physical disaster institutions fail to address the social vulnerabilities that put communities at risk of the very issues such institutions are trying to alleviate.

While the use of legal identity in providing aid can deliver 'just' outcomes in the sense of ensuring that aid is only granted to those who 'deserve' it, in terms of offering holistic relief from the suffering disasters can cause, such measures are insufficient. The immediate effects of the event might be mitigated for those who can prove they were directly affected, but the issues that undermined the resilience of that community and the compounding effects those issues present run the risk of being ignored. Put simply, delivering aid in this way does justice to the disaster, not necessarily to the individuals affected by it.

Taken together, the above issues reinforce the fact that in many cases legal identities do not serve or empower the individuals who possess them so much as allow individuals to be recognised by holders of power—institutions that often come from different countries and bring with them different conceptions of justice, and enjoy a privileged position in comparison to those whom they seek to assist. There is little doubt that legal identities are an important part of robust bureaucracies, and no doubt at all that the institutions that might rely on legal identities have the best of intentions regarding the members of the community they seek to help, however the dynamics of power and the way justice may manifest in light of that dynamic cannot be overlooked.

The unfortunate reality is that the issues faced by those lacking legal identity in disaster zones is not confined to disaster events. Even without the chaos of an unfolding disaster, threats such as trafficking, aid provision, political vulnerability and entrenched poverty are an everyday reality for many people lacking legal identification. On one hand, this demonstrates the immensity of the problem; if tackling issues of identity management is problematic under regular circumstances, it is unfair to presume that disaster respondents have the capacity to address the problem.

However on the other hand, we must acknowledge that current disaster response frameworks adhere to institutions that we know to be not simply problematic, but exclusionary and dangerous even under regular circumstances. The fact that these already marginalised and vulnerable people must endure the perils of being 'hidden' as part of everyday life in no way justifies the magnification of those perils in times of desperate need. Indeed, given the problematic relationship those lacking legal identity have with the community which they must navigate to survive, it is precisely when that community is under threat that the 'hidden' people need help the most. As we have seen, even legally recognised individuals can be further marginalised through the selective provision of disaster support, and it would be naïve to assume that the capacity for similar marginalisation or exploitation was not increased for those lacking legal recognition.

While the use of legal identity might seek to ensure that the needs of victims are genuine, such an expectation relies on the privilege of just governments, robust legislatures and effective bureaucracies. Individuals and communities that were operating with citizens lacking legal identities, abodes lacking titles and markets lacking transaction records cannot be delivered a 'just' response by using expectations borne from more reliably documented communities. Relying on legal identities in delivering aid to disaster-affected communities is a decisive failure to deliver 'disaster justice', especially in the case of developing nations. Developing nations often face a double-edged sword when it comes to disaster response, lacking the domestic resources for pre-disaster planning or prevention, for short-term relief of a disaster's immediate effects, or for long-term issues stemming from the event. Not only does it stand to exclude the most vulnerable people affected by the disaster, but it also stands as another dynamic through which the more vulnerable and fragile areas of the world—namely developing countries—remain vulnerable.

These are also often environments in which unregistered migration is common, where legal documentation is patchy and inconsistent, and where centralised records are not kept or cannot be quickly accessed. For external aid agencies to then use legal identities in its distribution of aid, they are essentially reinforcing a status quo—a status quo that in the case of developing countries includes the reinforcement of their existing societal issues. This is especially true in the case of communities with large populations of undocumented people, in which the restoration of living standards to legally identified individuals would do little to restore the community as a whole.

As such, humanitarian responders should consider what message a reliance on legal identity sends to victims in disaster areas. While aid-givers might see a necessary correlation between legal identity and the proper provision of humanitarian assistance, a reliance on legal identity runs the risk of calling into question how the 'humanitarian' aspect of aid and assistance is defined. For an undocumented individual struggling to find food, strict reliance on legal identity could mean that they are essentially excluded from 'humanitarian' assistance based on their failure to demonstrate their personhood—or at least demonstrate it in the manner that is acceptable to an agency or government. Beyond the obvious immediate and physical ramifications, such exclusion also serves to reify and reinforce the primacy of institutions, bureaucracy and legality over and above the needs of the living human person.

References

Akerkar, S. (2007). Disaster Mitigation and Furthering Women's Rights: Learning from the Tsunami. *Gender, Technology and Development, 11*(3), 357–388. https://doi.org/10.1177/097185240701100304

Alfirdaus, L. (2014). Disaster and Discrimination: The Ethnic Chinese Minority in Padang in the Aftermath of the September 2009 Earthquake. *Journal of Social Issues in Southeast Asia, 29*(1), 159–183. https://doi.org/10.1355/sj29-1f

Binford, W. (2011). Saving Haiti's Children from Hell. *6 Intercultural Human Rights Law Review 11*. Retrieved from SSRN https://ssrn.com/abstract=1829722 or https://doi.org/10.2139/ssrn.1829722

Boyer, B. (2011). *Kabul/Port-au-Prince, Reflections on Post Crisis Aid Operations in Urban Environments*. Humanitarian Aid on the Move, Newsletter #8. Plaisians: Groupe URD.

Bradshaw, S., & Fordham, M. (2013). *Women, Girls, and Disasters. A Review for DFID*. Middlesex University, pp. 1–54.

Brown, D., Dodman, D., & International Institute for Environment and Development. (2014). *Understanding Children's Risk and Agency in Urban Areas and Their Implications for Child-Centred Urban Disaster Risk Reduction in Asia: Insights from Dhaka, Kathmandu, Manila and Jakarta*. London: International Institute for Environment and Development.

Chew, L., & Ramdas, K. (2005). *Caught in the Storm: The Impact of Natural Disasters on Women*. San Francisco: The Global Fund for Women.

Cross, T., & Johnston, A. (2011). *Urban Cash Transfer Programmes: Desk Review of Existing Program*. Cash Learning Partnership.

Delaney, S. (2006). *Protecting Children from Sexual Exploitation & Sexual Violence in Disaster & Emergency Situations*. Bangkok: ECPAT International.

Drumm, L., & Stretch, J. (2008). Identifying and Helping Long Term Child and Adolescent Disaster Victims. *Journal of Social Service Research, 30*(2), 93–108.

Egal, F. (2011). *Managing Crises in Urban Areas: Food and Nutritional Security and Rural – Urban Links.* Humanitarian Aid on the Move, Newsletter #8. Plaisians: Groupe URD.

Fisher, D., Swithern, S., & Walmsley, L. (2018). *World Disasters Report 2018: Leaving No One Behind.* International Federation of Red Cross and Red Crescent Societies.

Gaillard, J. C. (2010). Vulnerability, Capacity, and Resilience: Perspectives for Climate and Development Policy. *Journal of International Development,* 22, 218–232.

Harper, E. (2009). *International Law and Standards Applicable in Natural Disaster Situations.* Rome, Italy: International Development Law Organization.

Inter-Agency Standing Committee (IASC). (2007). *IASC Guidelines on Mental Health and Psychosocial Support in Emergency Settings.* Retrieved from http://www.humanitarianinfo.org/iasc/

Inter-Agency Standing Committee. (2011). *Human Rights and Natural Disasters.* Operational Guidelines and Field Manual on Human Rights Protection in Situations of Natural Disaster. The Brookings – Bern Project on Internal Displacement.

International Commission of Jurists. (2016). *Nepal: Human Rights Impact of the Post-Earthquake Disaster Response.* Retrieved from https://doi.org/10.1163/2210-7975_HRD-0088-2016008

International Committee of the Red Cross. (2004). *Inter-Agency Guiding Principles on Unaccompanied and Separated Children.* Geneva: Red Cross.

Knox Clarke, P., & Ramalingam, B. (2012). *Meeting the Urban Challenge: Adapting Humanitarian Efforts to an Urban World.* ALNAP Meeting Paper. London: ALNAP/ODI.

Lacey, D. (2015). Beyond Virtual and Physical Borders: Extending the Conceptualization of the Identity Ecosystem. *ID360 2015: The Identity Economy Conference,* University of Austin, Texas.

M'jid, N. (2011). *Report of the Special Rapporteur on the Sale of Children, Child Prostitution and Child Pornography, Najat Maalla M'jid.* Human Rights Council Nineteenth Session Agenda Item 3 Promotion and Protection of All Human Rights, Civil, Political, Economic, Social and Cultural Rights, Including the Right to Development. Geneva: UN Human Rights Council, pp. 1–24. Retrieved from https://reliefweb.int/report/world/report-special-rapporteur-sale-children-child-prostitution-andchild-pornography-najat

M'jid, N. (2015). *Report of the Special Rapporteur on the Sale of Children, Child Prostitution and Child Pornography,* Najat Maalla M'jid. Human Rights Council Nineteenth Session Agenda Item 3 Promotion and Protection of All Human Rights, Civil, Political, Economic, Social and Cultural Rights, Including the Right to Development. Geneva: UN Human Rights Council, pp. 1–24.

Macauslan, I., & Phelps, L. (2012). *Oxfam GB Emergency Food Security and Livelihoods Urban Programme Final Report* (No. Final). UK: Oxfam Policy Management.

Mulqueeny, K., Fernando, H., & Bonifacio, S. (2010). *Empowering People After Natural Disasters: Lessons from the Post-Tsunami Legal Assistance, Governance, and Anticorruption Project* in Sri Lanka (No. Brief 2). Asian Development Bank.

Peek, L. (2008). Children and Disasters: Understanding Vulnerability, Developing Capacities, and Promoting Resilience-an Introduction. *Children Youth and Environments, 18*(1), 1–29.

Penrose, A., & Takaki, M. (2006). Children's Rights in Emergencies and Disasters. *The Lancet, 367*(9511), 698–699.

Reyes, G. & Charles, J. (2010). Trafficking, Sexual Exploitation of Haitian Children in the Dominican Republic on the Rise. *Miami Herald*.

Sanderson, D. & Knox Clarke, P. with Campbell, L. (2012). *Responding to Urban Disasters: Learning from Previous Relief and Recovery Operations*. ALNAP Lessons Paper. London: ALNAP/ODI.

Save the Children. (2007). *Legacy of Disasters: The Impact of Climate Change on Children* (pp. 1–16). London: Save the Children.

Schuller, M. (2010). Haiti's Disaster after the Disaster: The IDP Camps and Cholera. *The Journal of Humanitarian Assistance*. Retrieved from http://jha.ac/2010/12/13/haiti% E2, 80.

Selman, P. (2011). Intercountry Adoption After the Haiti Earthquake: Rescue or Robbery? *Adoption & Fostering, 35*(4), 41–49.

Seneviratne, K., Amaratunga, D., Haigh, R., & Pathirage, C. (2012). Managing Disaster Knowledge: Identification of Knowledge Factors and Challenges. *International Journal of Disaster Resilience in the Built Environment, 3*(3), 237–252.

Sharma, K., Kc, A., Subedi, M., & Pokharel, B. (2018). Challenges for Reconstruction After Mw 7.8 Gorkha Earthquake: A Study on a Devastated Area of Nepal. *Geomatics, Natural Hazards and Risk, 9*(1), 760–790. https://doi.org/10.1080/19475705.2018.1480535

UN Environment. (2019). *Global Environment Outlook 0 GEO-6: Healthy Planet, Healthy People*, Nairobi. https://doi.org/10.1017/9781108627146.

UN News Centre. (2010). *Haiti's Children at Increased Risk of Abduction, Slavery, and Trafficking*, UN Experts Warn, February 2, 2010. Retrieved from http://www.un.org/apps/news/story.asp?NewsID=33650&Cr=haiti&Cr1#

UNICEF. (2010). *State of the World's Children Report 2010 (Statistical Annex)*. New York: UNICEF.

UNICEF. (2011). *Children in Haiti: One Year After – The Long Road from Relief to Recovery*. Retrieved from http://www.unicef.org/republicadominicana/english/Children_in_Haiti_One_Year_After_-_The_Long_Road_from_Relief_to_Recovery(3).pdf [Hereinafter, UNICEF Children in Haiti: One Year After].

United Nations Development Programme. (2004). *Reducing Disaster Risk: A Challenge for Development*. Bureau for Crisis Prevention and Recovery, New York, NY: United Nations Development Programme.

Van der Gaag, N. (2013). *Because I Am a Girl: The State of the World's Girls 2013. In Double Jeopardy: Adolescent Girls and Disasters*. Plan International (pp. 1–224).

Verchick, R. R. (2012). Disaster Justice: The Geography of Human Capability. *Duke Environmental Law Policy Forum, 23*(1), 23–72.

Walters, V., & Gaillard, J. C. (2014). Disaster Risk at the Margins: Homelessness, Vulnerability and Hazards. *Habitat International, 44*, 211–219. https://doi.org/10.1016/j.habitatint.2014.06.006

West, Andrew. (2003). *At the Margins: Street Children in Asia and the Pacific*. © Asian Development Bank. Retrieved from http://hdl.handle.net/11540/2287. License: CC BY 3.0 IGO.

Wisner, B., Gaillard, J. C., & Kelman, I. (2012). *Handbook of Hazards and Disaster Risk Reduction*. Abingdon: Routledge.

World Bank. (2019). Identification for Development (ID4D) Global Data Set (Database: Last Accessed June 1, 2019).

CHAPTER 15

Justice, Resilience and Participatory Processes

Claudia Baldwin

15.1 Introduction

In spite of priorities for action in the Hyogo Framework for Action 2005 to ensure that disaster risk reduction is both a national and local priority with a strong institutional basis for implementation, *"disasters have continued to exact a heavy toll"* and *"continue to undermine efforts to achieve sustainable development"* (UNISDR, 2015, pp. 10–11). The United Nations Office of Disaster Risk Reduction (UNISDR) responded to ongoing disturbing trends with the Sendai Framework for Disaster Risk Reduction 2015–2030 (UNISDR, 2015). Among its four priorities for action, the Framework identified the need to: improve understanding of disaster risk; strengthen disaster risk governance; and enhance disaster preparedness for effective recovery, rehabilitation and reconstruction. The framework identified a *"shared responsibility between governments and relevant stakeholders"* including non-State stakeholders (UNISDR, 2015, p. 23). Its principles and actions reinforce the themes of inclusion and empowerment at a local level:

C. Baldwin (✉)
Urban Design and Town Planning, and Sustainability Research Centre,
University of the Sunshine Coast, Maroochydore, QLD, Australia
e-mail: cbaldwin@usc.edu.au

> It is necessary to empower local authorities and local communities to reduce disaster risk, including through resources, incentives and decision-making responsibilities, as appropriate. (s19(f), p. 13)

It required an "all-of-society engagement and partnership", requiring empowerment and inclusive participation, paying special attention to people disproportionately affected by disasters (s19(d), ibid., p. 13). This raises the challenge of how to build shared responsibility and empowerment, especially among the most vulnerable, within the context of, and in contrast to, centralised disaster governance. Indeed, a trend since Agenda 21's "Think Global and Act Local" has been towards more locally situated responses towards resilience in communities, and through bottom-up democratic processes focused on legitimation and consensus (Barr & Devine-Wright, 2012). In fact, with continued climate trends, *"there is likely to be a diverse geography of resilience within and between communities that maps onto existing social, economic and political systems and inequalities"* (Barr & Devine-Wright, 2012, p. 531).

Disaster governance literature provides some insight into these challenges. Ahrens and Rudolph (2006) identify institutional failure to commit to sustainable development practices as intensifying disaster risk. They suggest that accountability, participation, predictability and transparency are key features of a governance structure that supports risk reduction. Shared responsibility and empowerment referred to in the Sendai Framework rely on effective community access to bonding and bridging social capital, social coordination capacity of community organisations and stability of social networks within the community (Grube & Storr, 2014). The concept of strong interagency networks for both formal and informal collaboration is also highlighted as one of several strategies to address disaster risk (Howes et al., 2015). One might suggest that if government agencies have difficulty collaborating amongst themselves, there is little hope for collaboration with the community.

This sets the context for further exploration of the concepts of community resilience, vulnerability, and fairness, responding to Brown (2014, p. 107), who found that there has been *"little analysis of social difference and resilience, and there are continuing tensions between normative and analytical stances on resilience"*. As part of this exploration, I also interrogate the traditional understanding of vulnerability.

15.2 The Concepts: Resilience, Vulnerability and Fairness

15.2.1 Resilience

Resilience is a term commonly referred to in academic literature as well as policies and strategies in relation to global change, whether as a result of disasters, slow onset impacts from climate change, or generally in terms of goals for a community. Academics have drawn "social" into the original ecological concept of resilience since Adger (2000), with continuing discussion and debate about meaning and components of social and community resilience (Davidson, 2010; Maclean, Cuthill, & Ross, 2014; Magis, 2010). The concept of resilience has broadly evolved over time through independent bodies of literature including:

- ecological multi-level systems that are able to cope with disturbances, while remaining essentially the same system; and
- socio-psychological which focus on coping, with an emphasis on individual and community strengths—it is not about 'bouncing back', or returning to some 'normal' but coping with and adapting to change;
- towards an integrated view (Berkes & Ross, 2013, 2016; Brown & Westaway, 2011) about interdependency and resilience of people and environment (social-ecological systems), including collective capabilities and processes that enable a community to adapt in self-determined ways (Davidson et al., 2016).

Davidson et al. (2016) also differentiate among three types of resilience: a basic resilience involving persistence and bouncing back; to an adaptive form which includes self-organisation and adaptability; and thirdly to transformative resilience including preparedness and innovation, capitalising on new opportunities—that is, a transition, either purposeful or unintended.

To illustrate the subtle yet varied discussion about resilience, three resilience perspectives are described here: social-ecological resilience; disaster resilience; and community resilience (Table 15.1). These three concepts do not directly represent the evolution of the concept (see Davidson et al., 2016 for in-depth discussion of the evolution). Baird et al. (2016) suggest that resilience types might be overlapping and not mutually exclusive; in addition they may have different meanings to academics and stakeholders.

Table 15.1 Three resilience concepts

Concepts	Description
Social-ecological resilience	"*the capacity to adapt or transform in the face of change in social-ecological systems, particularly unexpected change, in ways that continue to support human well-being*" (Chapin et al., 2010; Biggs, Schluter, & Schoon, 2015 from Folke, Biggs, Norström, Reyers, & Rockström, 2016, p. 41).
Disaster resilience	How well people and societies deal with disruptive change, to proactively prevent, adapt to and recover from hazards and risks through capacities for anticipation, adaptation and innovation (adapted from Davidson et al., 2016).
Community resilience	A resilient community is "*one that takes intentional action to enhance the personal and collective capacity of its citizens and institutions to respond to and influence the course of social and economic change*" (Colussi, 2000, pp. 1–5). "*the existence, development, and engagement of community resources by community members to thrive in an environment characterized by change, uncertainty, unpredictability, and surprise. Members of resilient communities intentionally develop personal and collective capacity that they engage to respond to and influence change, to sustain and renew the community, and to develop new trajectories for the communities' future*" (Magis, 2010, p. 102).

They all though involve system characteristics referring to enabling change (Davidson et al., 2016) and/or the capacity to adapt to change (Magis, 2010, p. 408) but the emphasis and limitations are different in each type.

Social-ecological resilience expanded beyond ecological resilience to acknowledge the inextricable linkage among social and ecological systems, with social including "*the human dimension in its diverse facets, including the economic, political, technological, and cultural*" (Folke et al., 2016, p. 41). Folke's (2006) earlier view was that social-ecological resilience was about a system remaining within the same state, capable of self-organisation, and had capacity for learning and adaptation (Folke, 2006, p. 259).

Living with complexity and uncertainty requires management and governance principles for building resilience in social-ecological systems in the face of change. Among Biggs et al.'s (2015 from Folke et al., 2016) seven principles, three are particularly relevant here: to encourage learning; broaden participation; and promote polycentric governance. There are no panaceas for building resilience, and before applying such principles, careful consideration needs to be given to who may benefit or lose, to avoid entrenching or exacerbating existing inequalities. Without intending to diminish the intention to study humans and nature as an integrated whole, not as separated parts, one limitation of the social-ecological type is that, in relation to environmental change, "social" parameters are conservative (MacKinnon & Derickson, 2013), quite broad and still poorly developed conceptually and in application.

Disaster resilience recognises that willingness to change can be prompted by preparing in advance to the threat of a disaster and/or for recovery after a disaster. It can include large scale multi-scalar compound events that have long lasting cascading effects. The emphasis on crisis and disaster tends to, though, downplay change that arises from slow onset processes.

Community resilience is an aspirational but often ill-defined goal of governments/institutions at all levels, for example, World Bank, International Panel of Climate Change (IPCC), Australian, State and local governments. Resilience aims to ensure that communities can prepare for and respond effectively to multiple challenges: whether rapid onset disasters, slow incremental changes due to climate change, migration or other issues. They may involve continuous and cumulative effects and require ongoing adaptation. Magis (2010) suggests that the most appropriate response to system disruption will vary from maintenance, to adaptation, to occasional transformation. Tran et al. (2016, p. 2) refers to community resilience as the resources a community has and the ability to access and mobilise them to address all development needs, not just to respond to climate hazards. That is, it involves the resources a community uses to cope in times of crisis (access to money and disaster centres) and how they are accessed (social networks). The proposed outcome is that when a disaster or challenge occurs, a community can cope and take action to reduce the impact in the future. Thus a key difference between community resilience and other resilience types is the *"unique capacity and important role of communities in developing their own resilience"* by active agency building (Magis, 2010, p. 405).

An objective in supporting community resilience is to build adaptive capacity, collective efficacy and agency, reflecting Bandura's (2000) view of the importance of personal efficacy and a shared belief in the power to bring about change with collective action. Various authors suggest that community resilience and adaptive capacity can be built through participatory research and joint fact-finding that involves reflexive and social learning (Amundsen, 2012; Christensen and Krogman, 2012; Goldstein, 2009; Olsson, Folke, & Berkes, 2004; Robinson & Berkes, 2011; Ross & Berkes, 2014; Wilson, 2012). It can be supported by building community capitals, including social capital through bonding, bridging and linking networks.

There is a need to repack the cross disciplinary nature of community resilience, in the context of environmental change. Indicators of resilience, discussed later in this chapter, are needed in order to better target action, to monitor and track system change (Davidson et al., 2016) and to understand better how to support it in an adaptive way. Factors for achieving resilience need to be better understood in order to achieve fair outcomes. Responses to disasters can be inequitable. There may be uneven access to resources for recovery, unequal representation of community members in recovery programs and unfair modes of compensation and aid distribution, including insurance and legal services that assist compensation.

Thus improved understanding of resilience can lead to better targeted resilience-building measures and activities that are fair and lead to just outcomes as a result of disaster. Before examining some of these measures, let's look briefly at the concept of vulnerability and a case study that demonstrates challenges in addressing vulnerability to enable just solutions to a disaster.

15.2.2 Vulnerability and Fairness

Geographic locational disadvantage can lead to increased vulnerability to risk. Examples include those living in a coastal area potentially subject to tsunamis or longer-term sea level rise; in a seismically active area; on a flood plain; or in an area prone to wild fire. People live or work in these areas for a number of reasons which may include historical factors; lack of awareness, underestimation or intentional downplaying of the real risk; or little choice due to economic, social-cultural or even political reasons.

The IPCC (2013) also refers to social vulnerability, drawing specific attention to social equity. Those who are considered to be the most vulnerable to climate change for example, are the elderly, infants, people with disability and those with compromised health (IPCC, 2013). In addition, people with poor education, low income, unemployed and from culturally and linguistically diverse backgrounds can be at a disadvantage in preparing for and recovering from disaster. Vulnerability may also be gendered in some areas/regions. Social vulnerability can also be affected by locational disadvantage. Including the vulnerable in participatory processes can not only contribute local knowledge to solutions, but strengthen stakeholder networks. For example if vulnerable residents of the US Gulf Coast had been consulted about hazard planning in New Orleans, they would have quickly identified the lack of transportation for evacuation as a critical issue (Horney, Simon, Grabich, & Berke, 2015).

It may, however, be difficult to predict who in a community is vulnerable and who might be motivated to participate in community processes. It is important to gain better understanding of the variability in social vulnerability and associated justice issues, because inadequate or inappropriate responses by institutions will only serve to further alienate or disadvantage those affected.

15.3 Case Study: Illustration of Vulnerability to Flooding in Rocklea Industrial Area

The case study of flooding of Rocklea industrial area during the Brisbane floods in January 2011 is used to illustrate less identifiable forms of vulnerability and prompt discussion of ways that building community resilience can address the needs of everyone in a community.

The Rocklea industrial area is located on the Oxley Creek floodplain, a geographically vulnerable location that has become increasingly flood prone due to intensive development with large paved surfaces and roof areas, leading to concentrated stormwater runoff. The heavy rains preceding the January 12, 2011 flood resulted in a saturated Brisbane River catchment (van den Honert & McAneney, 2011). The swollen Brisbane River prevented Oxley Creek, a tributary, from draining. Although flooding in general was not as high as in the previous 1974 floods, in Rocklea some areas were inundated higher than levels predicted by Brisbane City Council. The completion of Wivenhoe dam in the upper Brisbane River in 1985 was widely expected to mitigate future downstream flooding, giving a false sense of security.

However more than 12,000 residences and 2500 commercial properties were flood-affected within Brisbane itself. Among several studies of flood impacts (Bird et al., 2013; George, 2013), little research focused on the severely affected industrial businesses in Rocklea. My study included in-depth interviews of eight business owners in the Rocklea area and documentation of impacts through their photos which they gave permission to use in publications (Fig. 15.1).

In general, there were huge environmental, economic, social and health impacts. Elevated concentrations of industrial contamination were recorded with high levels of herbicides, pesticides, polycyclic aromatic hydrocarbons (PAHs) and hydrocarbons, heavy metals, and enterococci in Oxley Creek; almost 2000 hazardous containers moved by floodwaters were retrieved from the flooded area by HAZMAT (hazardous materials) teams within a month (Baldwin, 2013). Few owners and volunteers were

Fig. 15.1 Inundated light industrial building recently renovated (P1)

adequately prepared for the clean-up resulting in respiratory disease and skin rashes from contaminated waste. Economic costs included loss of expensive industrial assets and computer equipment, damage to buildings and property, loss of financial records, temporary and permanent business closures, lost business income and jobs and delayed retirement. This had a serious effect on the mental health of those affected and the ability to claim compensation and on insurance (Fig. 15.2).

Key features of the Brisbane floods were release of Wivenhoe Dam water and the late public advice of the seriousness of the flooding by the responsible authorities. Managers became increasingly aware of the extent of the flood threat over the day on Wednesday—12 January. By noon on 12 January, most workers had already been sent home due to a notification that an access road would likely be inundated. Floodwaters peaked at 5 pm on 12 January with a slightly higher peak at 3am on Thursday, 13 January. So by the time business owners were informed that the entire area would be severely impacted, no staff were available and it was impossible to move relocatable equipment such as forklifts, containers and utility trucks, let alone large expensive industrial machinery and materials.

An unexpected finding of the research was about individual characteristics of some of the business owners that increased their vulnerability. Traditional forms of communication (television, radio and social media) about the flood did not necessarily work for them. Interviewees had hearing impairments from industrial noise but did not consider that they had a disability and so did not have adapted telecommunication and could not

Fig. 15.2 Recently rebuilt to meet Council flood standards but hazard chemical containers were tossed around by floodwaters (P2)

read sign language which was available via television broadcasts. Others had dyslexia and could not read maps showing risk; yet others did not know how to use a computer to link to a website, or did not otherwise use social media which was widely used during the emergency.

This situation is in stark contrast with a simultaneous but quite different response anecdotally reported by a couple with young children living further downstream, not far from the Brisbane River, but at the edge of the area identified by flood mapping as vulnerable to extreme flood events. Their story illustrates the difference made by education, financial and physical means. Their expertise in remote sensing and map-reading, river management and computer literacy backed by educated local parental support, led them to identify a high risk of staying in their home even before it was widely understood or communicated that the Brisbane River might peak much higher than expected. They were in a financial position and physically able to hire a moving truck to remove their possessions out of the home by noon on 12 January. The first floor of the house was flooded that evening and they would have lost most of their possessions. As it was, they had the inconvenience and additional cost of replacing much of the interior including the kitchen. Insurance covered much of the cost, as the policy covered overbank river flooding. In contrast, because flooding in Rocklea was from stormwater back-up, many business owners found themselves not covered by insurance, reinforcing both physical and social vulnerability of business owners in Rocklea.

This case study illustrates that the capacity to manage risk and adapt to change is unevenly distributed across communities and often hard to predict. This has *"implications for assigning responsibility for managing change to individuals, communities, governments, or other sets of social actors"* (Brown & Westaway, 2011, p. 335). It prompts questions about how to build adaptive capacity and individual and collective agency amongst vulnerable individuals and communities.

15.4 Disaster Justice

Disaster justice is concerned with anthropogenic interventions in nature that incubate environmental crises and magnify their socially and spatially uneven impacts (Douglass & Miller, 2018, p. 271). Calls for justice are more visible during disasters which affect societies socially, spatially and economically unevenly (Douglass & Miller, 2018, p. 272). Disasters occur in the context of political spaces and power, and governance approaches influence preparedness, response and redress to risk. While illustrating contrasting vulnerabilities, the case study above also illustrates the role of responsible institutions as enablers or inhibitors of effectiveness of community response, simplistically illustrated in Fig. 15.3. Importantly it means that communities cannot become resilient in isolation; they need an enabling environment, reinforced by Biggs et al.'s (2012) principles such as managing connectivity, encouraging learning, broadening participation and promoting polycentric governance. Community resilience is inseparable from governance and power.

Particularly troubling is that communities are expected to be resilient in a time of austerity and reinforced neoliberalism (MacKinnon & Derickson, 2013, p. 262). Neoliberal governance places responsibility on social actors for their own outcomes. Combined with government downsizing and streamlined processes which reduce public involvement, transparency and trust in governments are seen as inadequate to safeguard environmental, cultural and social interests (Baldwin et al., 2019). Likelihood for transformative outcomes is increased when capacity building is joined with genuine participation, yet neoliberalism has resulted in less advantaged regions and people experiencing challenges in building stakeholder capacity (Davidson & Lockwood, 2009).

The collective agency needed to effect transformative change focuses on three dimensions: underlying social and spatial processes leading to uneven patterns of vulnerability; participatory forms of disaster governance; and

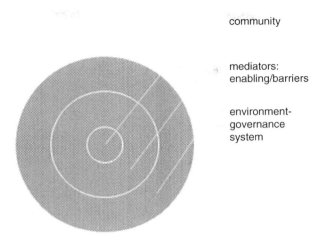

Fig. 15.3 Governance systems as mediators of community resilience

just distribution of resources to support recovery and social resilience (Douglass & Miller, 2018, p. 271). According to Brown (2014), resilience studies to date have not adequately considered whose needs are being met or the politics of their distribution and management. A lack of social science input has inhibited understanding of how to apply fairness principles to disaster resilience, including the root causes of vulnerability (Brown, 2014). Bolitho and Miller (2017, p. 685) for example, suggests that:

> *Addressing the underlying causes of vulnerability to extreme heat would likely reduce its impacts as well as reduce social inequalities associated with health, housing, incomes, mobility and access to essential services and infrastructure.*

This contribution suggests, as per Brown (2014, p. 113), the need to "*negotiate a vision for resilience which puts concerns for social justice and social capital at the centre*". To achieve this, we look to procedural, distributional and interactive justice parameters from Lukasiewicz and Baldwin (2017) for guidance and refer to the case study to illustrate application (Table 15.2).

Following Table 15.2 is a discussion of how resilience can be fostered, not just in relation to preparation for disasters, such as evacuation plans, but how social and participatory processes contribute to 'just' resilience to disasters.

Table 15.2 Social justice

Justice types (Lukasiewicz and Baldwin 2017)	Key features	Application to Rocklea case study
Procedural—How decision-making is structured: who participates, level of participation, power to influence and rules (transparency, impartiality and timeliness)	• Inclusive • Recognition of those who deserve representation, including those affected (Adger, 2016). • Discourse to reveal values (Dryzek, 2013); local knowledge. • Meaningful dialogue promotes cross-sector understanding (Cross, McCarthy, Garfin, Gori, & Enquist, 2013), knowledge sharing and social learning (Pahl-Wostl et al., 2007) • Raises awareness of climate change impacts (Burton & Mustelin, 2013; Coffee, Parzen, Wagstaff, & Lewis, 2010) • Interchange horizontally and vertically (up and down and across).	Transparency and accuracy of information—predictions of flood impact too late and imprecise for appropriate response; business owner communication difficulties made it difficult to access information and understand risk. Few networks for sharing information and social learning.
Distributional—outcomes meet needs, equitable, efficient, fair [self/public-interest considerations]	• Greater equity in knowledge and understanding • Enhanced adaptive capacity—critical to building resilience, solutions and capacity to take action (Cross et al., 2013; Sheppard, 2012) and address unknowns (Seidl & Lexer, 2013). • Builds commitment, ownership and responsibility (Bardsley & Rogers, 2010) • Effective, feasible, practical and holistic outcomes (Anguelovski & Carmin, 2011; Rosenzweig et al., 2011) • Burden of impacts are recognised and shared more equitably.	Inequity in knowledge and understanding of risks. Distribution of costs and benefits affected by governments, large corporations including SEQWater and insurance agencies. Traumatised victims need greater support in recovery.
Interactive justice—relationship—how participants are treated (trust, respect, recognition of social standing and trustfulness)	• Transparency and accountability of government decision-making (Anguelovski & Carmin, 2011; Aylett, 2010; Kithiia & Dowling, 2010; Rosenzweig et al., 2011). • Builds trust between stakeholders and with institutions (Pahl-Wostl et al., 2007).	Inadequate flood warnings affected trust in dam operator. The Flood Commission of Inquiry made useful recommendations relevant to Rocklea, including regarding transparency of risk information.

15.5 Role of Research in Building Community Resilience

Building community resilience can benefit from action research and monitoring. In spite of growing research on resilience, application and testing of the concept is limited by little agreement on a comprehensive measurement framework (Asadzadeh, Kötter, Saleh, & Birkmann, 2017) and still developing indicators. For example, Davidson et al. (2013) identified indicators, such as public perceptions of risk, social capital networks, trust, social-ecological learning, openness to change and innovation, education and experience. However, they suggested that these were not intended to be adequate for data collection. They did illustrate the complexity and diversity of indicators and the need to vary indicators depending on the situation.

Other authors have also explored components of community resilience and how different disciplines can contribute to our understanding. Cavaye and Ross (2019), for example, suggest that community development approaches which already contribute to capacity building, empowerment and building networks could be expanded to deal with dynamically changing systems. Such approaches include building adaptive capacity, managing complexity, enhancing community values and identity, managing multiple level systems and supporting community agency. Berkes and Ross (2013) suggest that concepts in the psychology of development and mental health areas would contribute a more integrated approach to understanding community resilience. Likewise, Imperiale and Vanclay (2016) argue that social impact assessment can play a role in strengthening community resilience. Saja, Teo, Goonetilleke, and Ziyath's (2019) critical review of social resilience studies identified key dimensions that, among other measures, include social cohesion and social networks (social capital), social trust, community engagement, community values (sense of community) and community competence (understanding of risk). Similar attributes were identified in a New Zealand study (Hayward, 2013): collective efficacy, ability to adapt to change, community inclusiveness, social networks and network connectedness, leadership, social support, sense of community and shared values, trust and knowledge of risks and consequences. Many of these same features of justice align with community resilience. Magis (2010) too, in the USA, identified eight "dimensions" of community resilience and associated metrics: community resources, active agents, collective action, strategic action, equity, impact, resource engagement and resource development. In contrast, a UK study examined place attachment, leadership, community cohesion and efficacy, community networks and knowledge and learning (Faulkner, Brown, & Quinn, 2018). Berkes and Ross (2016) suggest that these characteristics are not just additive, we need to know about the processes of agency and self-organising that draw communities into a resilience process. Thus a high priority is the need for application, further testing and refinement of monitoring criteria and more detailed indicators.

Measuring and monitoring community resilience are important for a number of reasons. These procedures provide a baseline and determine direction of change in relation to condition. Monitoring can trigger remedial action and adaptive management. It might stimulate stakeholder change by identifying windows of opportunity, informing capacity building to cope with uncertainty. It can indicate preparedness for change, existence of multi-level networks and stakeholder inclusiveness (Davidson et al., 2013, p. 4). It can be used by communities to track and strengthen their resilience. It can assist community development organisations and policymakers to test the efficacy of various interventions on improving community resilience and identify how best to invest funds (Magis, 2010). Evaluation can provide information to decision makers on complex issues and consumer preferences. Social learning can be enhanced through involving stakeholders in an indicator development process.

Here I suggest that insight can be drawn from research on community engagement and participatory research for which procedural, distributional and interactive justice are core, as indicated in Table 15.2.

Access to scientific expertise may provide a degree of legitimacy. On the other hand, though, smaller local organisations may contribute valuable place-based local knowledge. There is a need for both. For processes to be considered fair, there must be an opportunity for the perceptions of diverse stakeholders to influence decisions about design and implementation of policy (Hamilton, 2018).

15.6 Lessons Learned: Applying Justice Principles to Build Community Resilience

In summary, justice in relation to disasters is affected by the level of community resilience and capacity of vulnerable sectors. Certain sectors such as the aged, people with disabilities and those in low socio-economic conditions are more vulnerable pre-, during and post-disaster. The case study showed though, that it is not possible to reliably predict who are the most vulnerable in a community.

The Rocklea case study provides lessons about procedural, distributional and interactive justice. It provides insights about the relationship between vulnerability, adaptive capacity, community resilience and governance, including power and leadership.

In terms of procedural justice in relation to disasters, it is essential to understand the community. Multiple methods of communication need to be used and tailored to the purpose and for the range of abilities in the community. Culturally appropriate ways of engaging are needed, recognising diversity and vulnerabilities. Governments have a responsibility to communicate transparently about risks to environment and health from hazards and should aim to build trust through information accuracy. Complex issues need time and expertise to enable dialogue and discussion. Insurers need to use 'plainspeak' about what is covered and in their reason for denial of a claim.

Participatory approaches can draw on specific attributes identified in resilience studies (Maclean et al., 2014; Ross, Cuthill, Maclean, Jansen, & Witt, 2010; Walton, McCrea, Leonard, & Williams, 2013) such as:

- building and making use of local knowledge and skills and fostering social learning as a strengths-based approach;
- building meaningful relationships and networks within communities as well as with decision-makers (social capital; engaged governance); and
- supporting and drawing on people-place connections.

Distributional justice requires sincere governance that is focused on all responsible parties taking a share of costs, as well as benefits. Traumatised communities need additional support to be involved in decisions that affect their lives, re-building agency in times of uncertainty, and being offered new opportunities to mobilise transformations (Brown, 2014). Change proposals need to be anchored in culture and place, recognising identity and attachment; while also working across scales and recognising politics and power.

To serve interactional justice, the system of governance should not just be concerned with displaying technical information and undertaking disaster or adaptive planning, but maintain better connection with the individuals, households and businesses that form a community, enabling participation in strategic decisions and so they can respond better.

Partnerships are important: a diversity of groups can act as bridging organisations, operating at different scales and in different areas. Benefits of a third party knowledge broker (such as universities and other education and research institutions) are that they can:

- facilitate provision of information with neutrality. Researchers must meet ethical standards and their work is subject to peer review.

- experiment with different participatory techniques, for example, visual, interactive and assist in understanding stakeholders' interests, needs and values.
- provide independent evaluation to improve processes and show community and decision-makers the benefits of engagement and how stakeholder views were taken into account in decision-making. This relies on further development of monitoring criteria and indicators.
- share learning for improving future processes.

15.7 Conclusion

The original entry point for this chapter was that policies at multiple levels promote shared responsibility by governments and the community to support building of resilient communities to minimise the impact of disasters and recover quickly post-disaster. This includes intentionally developing personal and collective capacity (and agency) to sustain and renew the community, to influence change and to develop new trajectories for the communities' futures.

Resilient communities will be better prepared, but to become resilient, communities need an enabling environment of institutions that support procedural, distributional and interactive justice.

Each disaster is different. Building and maintaining resilience that supports fairness is an ongoing process. It takes time. New circumstances require new directions and solutions that institutions and communities will be better able to address if they embrace procedurally just participatory processes, capacity and network-building accompanied by fair distribution and sharing of information and knowledge.

References

Adger, W. (2000). Social and Ecological Resilience: Are They Related? *Progress in Human Geography*, *24*(3), 347–364. https://doi.org/10.1191/030913200701540465

Adger, N. (2016). Editorial: Place, Well-Being, and Fairness Shape Priorities for Adaptation to Climate Change. *Global Environmental Change*, *38*, A1–A3.

Ahrens, J., & Rudolph, P. (2006). The Importance of Governance in Risk Reduction and Disaster Management. *Journal of Contingencies and Crisis Management*, *14*(4), 207–220.

Amundsen, H. (2012). Illusions of Resilience? An Analysis of Community Responses to Change in Northern Norway. *Ecology and Society*, *17*(4), 46. https://doi.org/10.5751/ES-05142-170446

Anguelovski, I., & Carmin, J. (2011). Something Borrowed, Everything New: Innovation and Institutionalization in Urban Climate Governance. *Current Opinion in Environmental Sustainability*, *3*(3), 169–175.

Asadzadeh, A., Kötter, T., Saleh, P., & Birkmann, J. (2017). Operationalizing a Concept: The Systematic Review of Composite Indicator Building for Measuring Community Disaster Resilience. *International Journal of Disaster Risk Reduction, 25*, 147–162.

Aylett, A. (2010). Participatory Planning, Justice, and Climate Change in Durban, South Africa. *Environment and Planning A: Economy and Space, 42*(1), 99–115.

Baldwin, C. (2013). Impact of 2011 Flood on Brisbane Industrial Area Presented at the National Climate Change Adaptation Research Facility Conference, Sydney, 25–27 June 2013.

Baldwin, C., Marshall, G., Ross, H., Cavaye, J., Stephenson, J., Carter, L., ... Syme, G. (2019). Hybrid Neoliberalism: Implications for Sustainable Development. *Society and Natural Resources, 32*(5), 566–587. https://doi.org/10.1080/08941920.2018.1556758

Bandura, A. (2000). Exercise of Human Agency Through Collective Efficacy. *Perspectives on Psychological Science, 9*(3), 75–78.

Bardsley, D., & Rogers, G. (2010). Prioritizing Engagement for Sustainable Adaptation to Climate Change: An Example from Natural Resource Management in South Australia. *Society & Natural Resources: An International Journal, 24*(1), 1–17.

Barr, S., & Devine-Wright, P. (2012). Resilient Communities: Sustainabilities in Transition. *Local Environment, 17*(5), 525–532. https://doi.org/10.1080/13549839.2012.676637

Berkes, F., & Ross, H. (2013). Community Resilience: Toward an Integrated Approach. *Society and Natural Resources, 26*(1), 5–20.

Berkes, F., & Ross, H. (2016). Panarchy and Community Resilience: Sustainability Science and Policy Implications. *Environmental Science & Policy, 61*, 185–193.

Biggs, R., Schluter, M., Biggs, D., Bohensky, E., BurnSilver, S., Cundill, G., ... West, P. (2012). Toward Principles for Enhancing the Resilience of Ecosystem Services. *Annual Review of Environment and Resources, 37*, 421–448.

Biggs, R., Schluter, M., & Schoon, M. (Eds.). (2015). *Principles for Building Resilience: Sustaining Ecosystem Services in Social-Ecological Systems*. Cambridge: Cambridge University Press. https://doi.org/10.1017/CBO9781316014240

Bird, D., King, D., Haynes, K., Box, P., Okada, T., & Nairn, K. (2013). Impact of the 2010–11 Floods and the Factors That Inhibit and Enable Household Adaptation Strategies. NCCARF, Gold Coast, February 2013.

Bolitho, A., & Miller, F. (2017). Heat as Emergency, Heat as Chronic Stress: Policy and Institutional Responses to Vulnerability to Extreme Heat. *Local Environment, 22*(6), 682–698. https://doi.org/10.1080/13549839.2016.1254169

Brown, K. (2014). Global Environmental Change: A Social Turn for Resilience? *Progress in Human Geography, 38*(1), 107–117.

Brown, K., & Westaway, E. (2011). Agency, Capacity, and Resilience to Environmental Change: Lessons from Human Development, Well-Being and Disasters. *Annual Review of Environment and Resources, 36*, 321–342.

Burton, P., & Mustelin, J. (2013). Planning for Climate Change: Is Greater Public Participation the Key to Success? *Urban Policy and Research, 31*(4), 399–415.

Cavaye, J., & Ross, H. (2019). Community Resilience and Community Development: What Mutual Opportunities Arise from Interactions Between the Two Concepts? *Community Development.* https://doi.org/10.1080/15575330.2019.1572634

Chapin, F., III, Carpenter, S., Kofinas, G., Folke, C., Abel, N., Clark, W., ... Swanson, F. (2010). Ecosystem Stewardship: Sustainability Strategies for a Rapidly Changing Planet. *Trends in Ecology and Evolution, 25,* 241–249.

Christensen, L., & Krogman, N. (2012). Social Thresholds and Their Translation into Social-Ecological Management Practices. *Ecology and Society, 17*(1), 1–9.

Coffee, J., Parzen, J., Wagstaff, M., & Lewis, R. (2010). Preparing for a Changing Climate: The Chicago Climate Action Plan's Adaptation Strategy. *Journal of Great Lakes Research, 36*(Suppl. 2), 115–117.

Colussi, M. (2000). *The Community Resilience Manual.* Canadian Centre for Community Renewal, B.C., Canada.

Cross, M., McCarthy, P., Garfin, G., Gori, D., & Enquist, C. (2013). Accelerating Adaptation of Natural Resource Management to Address Climate Change. *Conservation Biology, 27*(1), 4–13.

Davidson, D. (2010). The Applicability of the Concept of Resilience to Social Systems: Some Sources of Optimism and Nagging Doubts. *Social and Natural Resources, 23*(12), 1135–1149. https://doi.org/10.1080/08941921003652940

Davidson, J., Jacobson, C., Lyth, A., Dedekorkut-Howes, A., Baldwin, C., Ellison, J., ... Smith, T. (2016). Interrogating Resilience: Toward a Typology to Improve Its Operationalization. *Ecology and Society, 21*(2), 27. https://doi.org/10.5751/ES-08450-210227

Davidson, J., & Lockwood, M. (2009). Interrogating Devolved Natural Resource Management: Challenges for Good Governance. In M. B. Lane, C. J. Robinson, & B. Taylor (Eds.), *Contested Country: Local and Regional Environmental Management in Australia* (pp. 75–89). Melbourne: CSIRO Publishing.

Davidson, J., van Putten, I., Leith, P., Nursey-Bray, M., Madin, E., & Holbrook, N. (2013). Toward Operationalizing Resilience Concepts in Australian Marine Sectors Coping with Climate Change. *Ecology and Society, 18*(3), 4. https://doi.org/10.5751/ES-05607-180304

Douglass, M., & Miller, M. (2018). Disaster Justice in Asia's Urbanising Anthropocene. *Environment and Planning E: Nature and Space, 1*(3), 271–287.

Dryzek, J. (2013). *The Politics of the Earth: Environmental Discourses.* Oxford: Oxford University Press.

Faulkner, L., Brown, K., & Quinn, T. (2018). Analyzing Community Resilience as an Emergent Property of Dynamic Social-Ecological Systems. *Ecology and Society, 23*(1), 24. https://doi.org/10.5751/ES-09784-230124

Folke, C. (2006). Resilience: The Emergence of a Perspective for Social-Ecological Systems Analyses. *Global Environmental Change, 16,* 253–267.

Folke, C., Biggs, R., Norström, A. V., Reyers, B., & Rockström, J. (2016). Social-Ecological Resilience and Biosphere-Based Sustainability Science. *Ecology and Society, 21*(3), 41. https://doi.org/10.5751/ES-08748-210341

George, N. (2013). It Was a Town of Friendship and Mud. *Australian Journal of Communication, 40*(1), 41–56.

Goldstein, B. (2009). Resilience to Surprises through Communicative Planning. *Ecology and Society, 14*(2), 33. Retrieved from http://www.ecologyandsociety.org/vol14/iss2/art33/

Grube, L., & Storr, V. (2014). The Capacity for Self-Governance and Post-Disaster Resiliency. *Review of Austrian Economics, 27*, 301–324.

Hamilton, M. (2018). Understanding What Shapes Varying Perceptions of the Procedural Fairness of Transboundary Environmental Decision-Making Processes. *Ecology and Society, 23*(4), 48.

Hayward, B. (2013). Rethinking Resilience: Reflections on the Earthquakes in Christchurch, New Zealand, 2010 and 2011. *Ecology and Society, 18*(4), 37.

Horney, J., Simon, M., Grabich, S., & Berke, P. (2015). Measuring Participation by Socially Vulnerable Groups in Hazard Mitigation Planning, Bertie County, North Carolina. *Journal of Environmental Planning and Management, 58*(5), 802–818.

Howes, M., Tangney, P., Reis, K., Grant-Smith, D., Heazle, M., Bosomworth, K., & Burton, P. (2015). Towards Networked Governance: Improving Interagency Communication and Collaboration for Disaster Risk Management and Climate Change Adaptation in Australia. *Journal of Environmental Planning and Management, 58*(5), 757–776.

Imperiale, A., & Vanclay, F. (2016). Using Social Impact Assessment to Strengthen Community Resilience in Sustainable Rural Development in Mountain Areas. *Mountain Research and Development, 36*(4), 431–442.

IPCC. (2013). Summary for Policymakers. In T. Stocker, D. Qin, G. Plattner, M. Tignor, S. K. Allen, J. Boschung, A. Nauels, Y. Xia, V. Bex, & P. Midgley (Eds.), *Climate Change 2013: The Physical Science Basis. Contribution of Working Group I to the Fifth Assessment Report of the Intergovernmental Panel on Climate Change*. Cambridge, UK; New York, NY: Cambridge University Press.

Kithiia, J., & Dowling, R. (2010). An Integrated City-Level Planning Process to Address the Impacts of Climate Change in Kenya: The Case of Mombasa. *Cities, 27*(6), 466–475.

Lukasiewicz, A., & Baldwin, C. (2017). Voice, Power, and History: Ensuring Social Justice for All Stakeholders in Water Decision-Making. *Local Environment, 22*(9), 1042–1060.

MacKinnon, D., & Derickson, K. (2013). From Resilience to Resourcefulness: A Critique of Resilience Policy and Activism. *Progress in Human Geography, 37*(2), 253–270. https://doi.org/10.1177/0309132512454775

Maclean, K., Cuthill, M., & Ross, H. (2014). Six Attributes of Social Resilience. *Journal of Environmental Planning and Management, 57*(1), 144–156.

Magis, K. (2010). Community Resilience: An Indicator of Social Sustainability. *Society and Natural Resources, 23*, 401–416.

Olsson, P., Folke, C., & Berkes, F. (2004). Adaptive Co-management for Building Resilience in Social–Ecological Systems. *Environmental Management*, *34*(1), 75–90.

Pahl-Wostl, C., Craps, M., Dewulf, A., Mostert, E., Tabara, D., & Taillieu, T. (2007). Social Learning and Water Resources Management. *Ecology and Society*, *12*(2), 5. Retrieved from http://www.ecologyandsociety.org/vol12/iss2/art5/

Robinson, L., & Berkes, F. (2011). Multi-level Participation for Building Adaptive Capacity: Formal Agency-Community Interactions in Northern Kenya. *Global Environmental Change*, *21*(4), 1185–1194.

Rosenzweig, C., Solecki, W., Blake, R., Bowman, M., Faris, C., Gornitz, V., … Zimmerman, R. (2011). Developing Coastal Adaptation to Climate Change in the New York City Infrastructure-Shed: Process, Approach, Tools, and Strategies. *Climatic Change*, *106*(1), 93–127.

Ross, H., & Berkes, F. (2014). Research Approaches for Understanding, Enhancing and Monitoring Community Resilience. *Society and Natural Resources*, *27*(8), 787–804.

Ross, H., Cuthill, M., Maclean, K., Jansen, D., & Witt B. (2010). *Understanding, Enhancing and Managing for Social Resilience at the Regional Scale: Opportunities in North Queensland*. Report to the Marine and Tropical Sciences Research Facility. Reef and Rainforest Research Centre Ltd, Cairns. Retrieved from http://www.rrrc.org.au/publications/social_resilience_northqueensland.html

Saja, A., Teo, M., Goonetilleke, A., & Ziyath, A. (2019). An Inclusive and Adaptive Framework for Measuring Social Resilience to Disasters. *International Journal of Disaster Risk Reduction*, *28*, 862–873.

Seidl, R., & Lexer, M. (2013). Forest Management under Climatic and Social Uncertainty: Trade-Offs between Reducing Climate Change Impacts and Fostering Adaptive Capacity. *Journal of Environmental Management*, *114*, 461–469.

Sheppard, S. (2012). *Visualizing Climate Change – A Guide to Visual Communication of Climate Change and Developing Local Solutions*. Abingdon, Oxon: Routledge.

Tran, T., Tran, P., Tran, T. A., & Jacobson, C. (2016). *Community Resilience Assessment and Climate Change Adaptation Planning: A Vietnamese Guidebook*. Maroochydore, OLD: University of the Sunshine Coast.

UNISDR. (2015). *Sendai Framework for Disaster Risk Reduction 2015–2030*. Geneva: UNISDR.

van den Honert, R., & McAneney, J. (2011). The 2011 Brisbane Floods: Causes, Impacts and Implications. *Water*, *3*, 1149–1173.

Walton, A., McCrea, R., Leonard, R., & Williams, R. (2013). Resilience in a Changing Community Landscape of Coal Seam Gas: Chinchilla in Southern Queensland. *Journal of Economic & Social Policy*, *15*(3), 4.

Wilson, G. (2012). Community Resilience, Globalization, and Transitional Pathways of Decision-Making. *Geoforum*, *43*, 1218–1231.

CHAPTER 16

The Theory/Practice of Disaster Justice: Learning from Indigenous Peoples' Fire Management

Jessica K. Weir, Stephen Sutton, and Gareth Catt

16.1 Introduction

In this chapter we discuss the implications of the fire management experiences of three groups of Aboriginal people that have been shared with us and how this might inform and be informed by the theory/practice of Disaster Justice.[1] Our focus is Australia where we live, but we have also

---

[1] We use the term 'Aboriginal' to refer to Indigenous peoples who burn country in Australia, which is the preference of some of our Aboriginal fire manager colleagues. We note that this is not strictly accurate as there are likely Aboriginal fire managers who also identify as Torres Strait Islander people. We use the term 'First Nations' in a settler-colonial context, and 'Indigenous' in a global context. We also use 'traditional owner' and 'traditional custodians' where appropriate to the case studies.

J. K. Weir (✉)
Institute for Culture and Society, Western Sydney University,
Rydalmere, NSW, Australia

Fenner School of Environment and Society, Australian National University,
Canberra, ACT, Australia
e-mail: J.Weir@westernsydney.edu.au

accepted the opportunity to learn from the experiences of and documentation by other First Nation, Indigenous and settler-colonial contexts, especially the United States, Canada and Aotearoa/New Zealand. In these contexts, the authority and priorities of the First Nations are not necessarily supported by the land management and natural hazard institutions of the nation state (Thomassin, Neale, & Weir, 2018). However, in Australia this is changing with respect to Aboriginal peoples' fire management practices. The 'reigniting' of these fires on diverse land tenures with diverse land holders has been accompanied by a proliferation of fire management activity between government authorities and Aboriginal people (Neale, Carter, Nelson, & Bourke, 2019; Prober, Yuen, O'Connor, & Schultz, 2013; Russell-Smith, Whitehead, & Cooke, 2009). This chapter considers what these collaborations mean for Disaster Justice, through a conceptual/material study that understands that knowing/doing is embedded together (Bawaka Country et al., 2015). We engage with both thinking about thinking, and what we have learnt from our research and work experiences with Aboriginal people in the Western Desert, Central Arnhem Land and the Australian Capital Territory. Critically, this deliberately continental expanse challenges discriminatory standpoints that place Indigenous peoples and their interests as bounded to the local, discrete and marginal to the 'mainstream', to instead demand serious consideration in nation state fora dominated by others (Watson, 2017; Weir, 2016). This continental approach also draws out the commonalities of the matters raised by Aboriginal peoples across 'settled' and 'remote' Australia.

We write from our non-Indigenous standpoint, seeking to act as allies and perhaps translators, whilst acknowledging the colonial privilege that has influenced the lives of all involved. We affirm that engaging with what Indigenous people are saying is required of non-Indigenous people as part of addressing their responsibilities to support decolonisation (Land &

S. Sutton
College of Health and Human Sciences, Charles Darwin University, Darwin, NT, Australia
e-mail: stephen.sutton@cdu.edu.au

G. Catt
10 Deserts Project, Desert Support Services, East Perth, WA, Australia

Vincent, 2005). We appreciate the sentiment expressed in the Aboriginal resistance meme from Australia:

> *Dear White People, No one is asking you to apologise for your ancestors. We are asking you to dismantle the systems of oppression they built, that you maintain and benefit from.*[2]

Apologies are important, but so too is ensuring that justice research and work has material outcomes for Indigenous people (Tuck & Yang, 2012). This requires non-Indigenous people who are concerned about injustice to both identify and address it, including through reflexivity about what is considered normal and appropriate from different standpoints.

In this chapter, we argue that non-Indigenous peoples and institutions need to do more than acknowledging Indigenous people as 'stakeholders', 'participants' or (ostensibly) 'partners' within and in relation to the dominant status-quo. The fundamental conceptual/material terms of engagement need to be considered deeply and then addressed, in order to ensure such engagements are respectful, meaningful and just. Significantly, we have learnt from many Indigenous peoples that ecological life and political-legal self-determination are inseparable justice matters (e.g. Graham, 2008; Watson, 2017; Whyte, 2013). That is, social and ecological justice issues need to be considered together. They are not separate elements that occasionally mix but co-constituted socio-natural worlds within which we all live (Nightingale, 2018). For non-Indigenous people to not appreciate this is to misunderstand what many Indigenous people are saying, why they are saying it and what is needed in response.

Helpfully, Australia's emerging fire management engagements are important opportunities for positive learning and collaboration, with potential to grow recognition of Indigenous peoples' priorities, as well as to help decolonise relationships between peoples and with nature. We identify three reasons for this. First, natural hazards tangibly express the power of natural forces and make obvious the inextricable fates of nature and society. This is clearly expressed, for example, in the natural hazard sector's focus on natural hazard risk, with risk being the consequential combination of hazards, communities and the environment (COAG, 2011). Second, the life and death context prioritises social equity matters that might be intractable or neglected in other policy contexts. Third, the collegial, material and placed-based context of fire management work

---

[2] Unknown, no date. Sighted on Facebook in 2018.

establishes a community of practice which supports learning and exchange through knowing/doing together. These possibilities are aided by broad shifts in Australian society, where there has been a cultural and political resurgence of Indigenous peoples, their rights and responsibilities. For example, whilst marginalised in democratic governance, Indigenous people are now Australia's largest land holders, both as individual land holding polities and in collective terms. There has also been a change and growth in values within the wider society that Indigenous peoples' culture, language and understandings of the landscape is something to be proud of and respected. These shifts are accompanied by an agenda of redressing historic and contemporary wrongs.

The expertise that supports this chapter is supported by specific research and work relationships. Jessica Weir and Steve Sutton have research partnerships with Aboriginal people and the natural hazard sector in southern and northern Australia respectively, with funding from the Bushfire and Natural Hazards Cooperative Research Centre (BNHCRC). Gareth Catt writes with the support of the Martu people in the Western Desert, with whom he worked with as part of the Indigenous Desert Alliance.

16.2　Justice Considered and Reconsidered

At a workshop on Disaster Justice organised by the editors of this book, participants agreed on a broad definition featuring the interaction of justice and the natural hazard event:

> *Disaster Justice recognizes that disasters can expose, magnify and deepen existing injustices in society, which can then lead to further injustices. This perspective purposefully situates the disaster event in relation to past and present social choices, and also acknowledges that the disaster itself is a dynamic opportunity to investigate perceived injustice and vulnerability using different dimensions of justice. As humans live with and within nature, these matters of justice include nature and consideration of our shared futures.*[3]

This broad definition provides a platform for the diverse concerns arising out of disasters, as well as the plurality of meanings around 'justice', 'disaster' and 'nature'. Here we focus on different understandings of justice and nature and their relations. We only touch upon the conceptual/

---

[3] Lukasiewicz, Anna, Steve Dovers and Claudia Baldwin, 'Disaster Justice Workshop', 19–20 November 2018, Australian National University, Canberra.

material framing of 'natural hazard' in this chapter, but appreciate natural hazards are generally understood to be spatio-temporal events such as wildfires, floods, tsunamis, cyclones and earthquakes.

Justice is a normative term used to call out wrongs and argue for what is right. Indigenous people in settler-colonial countries have leveraged different understandings of justice in different forums, including in relation to their self-determination rights, and the effects of environmental change on their homelands and ecologies. For example, Kyle Whyte argues for a forward-looking justice framework centred around 'collective continuance', which he defines as: "*a community's capacity to be adaptive in ways sufficient for the livelihoods of its members to flourish into the future*" (Whyte, 2013, p. 518).

Collective continuance embeds natural and social justice together. It requires that Indigenous peoples contest colonial hardships and pursue robust living through relational responsibilities that belong to larger systems of responsibilities, ranging from 'interspecies relationships to government-to-government partnerships' (Whyte, 2013, p. 518). This justice work is constrained by 'obstructive political orders', including policy frameworks that are against the existence of Indigenous peoples' polities (Moran, 2005; Whyte, 2013, p. 522). This justice work is also intensified by the persistence of political systems originally designed to weaken First Nation resistance to colonial expansion (Whyte, 2013, p. 522). Further, the immense consequences of climate change compel a broad adoption of this justice agenda. Critically, Whyte notes that collective continuance is important for all communities to flourish. This is a common argument made by Indigenous people: That what they are saying is not just about their own lives and territories but is important for all people and all life on earth.

Justice has a different focus in the Keynesian-Westphalian frame promoted by democratic nation states and the political economy of capitalism.[4] In Western liberal democracies, notions of 'justice' became important after World War II and Holocaust discussions about the relationships between citizens and nation states, and as part of supporting peace amongst nations (Robin, 2017). From the Keynesian-Westphalian perspective, Nancy Fraser defines justice as parity of participation amongst peers, with justice traditionally concerned with two interrelated dimensions: Socio-economic

---

[4] We draw on Fraser's use of this term to signal the political imaginary of mutually recognising sovereign territorial states—an international system whereby these states are presumed to have sovereignty over 'domestic' affairs (Fraser, 2007, 30, footnote 2).

'distributive justice' (to address maldistribution) and cultural 'recognition justice' (to address status inequality) (2007, p. 19). Subsequently, Fraser critiqued the problematic conflation of justice with the nation state and added a third dimension which she terms 'representation justice'. This facilitates a deeper political consideration of the 'who' and 'how' of the setting of the parameters of justice, including reconsidering these matters with respect to both distributive and recognition justice. As Fraser writes, "*Far from being of marginal importance, frame setting is among the most consequential of political decisions*", it is "*the process by which first-order political space is constituted*" (2007, pp. 22, 27). For example, she identifies that

> *Meta-political misrepresentation arises when states and transnational elites monopolize the activity of frame setting, denying voice to those who may be harmed in the process, and blocking creation of democratic fora where the latter's claims can be vetted and redressed.*

Disrupting assumptions that the nation state is the justice framework is a critical step for being able to hear what many First Nations peoples say about their political-legal standing in relation to settler-colonial states. For example, Audra Simpson has analysed how the Mohawk Nation sits both within and apart from Canada and the United States (2014). She argues that these are 'nested sovereignties' that do not negate each other but stand in 'terrific tension', posing "*serious jurisdictional and normative challenges*" (2014, p. 10). In the Age of Imperialism, the British common law required that settler-colonial authorities confirm their own jurisdictional authority through negotiations with First Nations, to establish the terms for territory and power sharing. In Aotearoa/New Zealand, Maori customary titles were considered to be inferior to 'matured law' and could not be asserted without the beneficent imposition of such superior law (Frame, 1999). In Australia, the colonial and imperial authorities held the view that the First Nations had no concept of land ownership, and thus no treaties were made (Moreton-Robinson, 2015; Watson, 2017).[5] The arrival of the British common law and the 1901 founding of the Federation of Australia did not just dismiss the First Nations, it depended on their absence. The idea of Australia as terra nullius, land belonging to no one, was overturned in 1992 when Indigenous peoples' pre-existing and ongoing polities and territories were partially recognised by the High Court as 'native title'.[6]

---

[5] There is one long-disregarded exception. In 1835 a treaty was made between a farmer and the First Nation for the land that is now known as Melbourne.

[6] *Mabo v Queensland (No 2)* [1992] HCA 23.

Scholarly work to reset the political-legal justice framework is, however, merely one dimension of a broader conceptualisation of the interactions between First Nations, settler-colonial societies and the places where we all live together. Support for the concept of collective continuance requires a deeper reflexivity, to also be reflexive about nature and our human relationships with it. For example, consider the effect of recognising political-legal rights over lands and waters that have been diminished, polluted and/or destroyed (Grinde & Johansen, 1995). How can interspecies kinship, intergenerational learning, livelihoods, ceremony and more be practised and enjoyed without the ecological life that supports it? Profoundly, the substance of what is being recognised is missing. Yet this co-constitution of nature and society routinely goes ignored, as Irene Watson writes:

> *Current human-rights frameworks ignore our core role as carers for the land for future generations; the fullness of Indigenous epistemologies is misunderstood and also ignored. Our inherent connectedness to the natural world is ignored and remains largely unfathomable to the non-Indigenous world.* (2017, p. 217)

This epistemological standpoint is very different to Western standpoints that insist on the separation of nature from society and have become predominant as a defining achievement of the modern world (Mitchell, 2000). The reflexivity needed to reconnect society with nature, however, also needs to reconnect nature with society. We illustrate what we mean in relation to the North American Environmental Justice movement.

Environmental Justice has a particular political-social meaning and form in the United States, where most of the literature comes from, although it can be generally understood as both an organisational movement and a body of scholarship (Schlosberg & Collins, 2014). An important issue for this Environmental Justice movement is how environmental 'bads' (e.g. pollution and toxic waste) are disproportionally distributed amongst minority and low-income communities, including on Native American lands (e.g. Grijalva, 2011)—a linking of environmental concerns with equity (i.e. distributive justice) and racial (i.e. recognition justice) concerns. This Environmental Justice has also always been allied with Indigenous peoples' justice concerns, such as through joint position statements.[7] Whilst Indigenous peoples' justice concerns have long been

---

[7] For example, the First National People of Color Environmental Leadership Summit, The Washington Court on Capitol Hill, United States, October 24–27, 1991.

part of the organisational movement, the US scholarship has neglected theorising a more intimate relationship between justice (amongst humans) and nature. As Schlosberg and Collins write, this theoretical work grew substantively after Hurricane Katrina when the connection was made between the toxic industries located between New Orleans and Baton Rouge, the warming of the Gulf from climate change, and the strength of Katrina (2014, p. 363). The Environmental Justice literature made the conceptual step to understand that *"the environment and climate system are not simply symptoms of existing injustice, but instead the necessary conditions for the achievement of social justice"* (2014, p. 363).

This is a significant shift in where justice sits in relation to the environment. Humanity, our actions and our potentialities are situated within the environments we live. However, more than the utilitarian pragmatism of placing humans within nature's energy flows, effects and food webs, many Indigenous people speak of nature as kin, law, polity, knowledge and more (Bawaka Country et al., 2015). This is an understanding of nature as part of cultural and ethical domains, which provides more connected relationships for consideration (Plumwood, 2002). As Mary Graham (2008, p. 181) has written:

> *The land is a sacred entity, not property or real estate; it is the great mother of all humanity. The Dreaming is a combination of meaning (about life and all reality), and an action guide to living. The two most important kinds of relationship in life are, firstly, those between land and people and, secondly, those amongst people themselves, the second being always contingent upon the first. The land, and how we treat it, is what determines our human-ness. Because land is sacred and must be looked after, the relation between people and land becomes the template for society and social relations. Therefore all meaning comes from land.*

This co-constituted understanding of society and nature, with nature as all-encompassing, is also evident in how Indigenous people seek to change the national/global conversations about the environment. For example, the recognition of the legal status of personhood over rivers in Aotearoa/New Zealand (Ruru, 2018) and the Universal Declaration of the Rights of Mother Earth which Indigenous people and others are lobbying for the United Nations to adopt. It is also evidenced in their fire management practices in Australia.

16.3 Aboriginal Peoples' Fire Management in Australia

In Australia, Aboriginal peoples' lighting of landscape fires has a long and diverse history of being practised as part of sacred and secular responsibilities that First Nations have towards each other and both with and within 'Country', a common term used to describe Indigenous peoples' homelands. Country involves systems of responsibilities as described by Graham above, which give meaning and purpose to fire management activities such as burning to reduce wildfire risk, claim territory, access areas, support plant growth, hunt, protect important sites, and so on (e.g. Prober et al., 2013; Garde et al., 2009). As with any practice, Indigenous or otherwise, Aboriginal peoples' fire management is not fixed in the past but always forming and re-forming to meet contemporary circumstances (Russell-Smith et al., 1997). These fires are co-constituted with the activities and interests of all, including the newer presences of peri-urban settlements, fire weeds and carbon markets.

Aboriginal peoples' burning is often raised in public debates in relation to prescribed burning for biodiversity values and/or wildfire risk mitigation, including being recruited to competing scientific positions and ideological politics (Buizer & Kurz, 2016; Neale, 2018). Many of these debates pivot around different ideas of what nature should be and are often inflected by speculations and research about Aboriginal peoples' pre-colonial fire management practices. In the 1950s, prescribed burning replaced fire suppression and exclusion as the public sector practice for mitigating wildfire risk, a result of learning from, inter alia, Aboriginal peoples' fire practices. At times, fire agencies and foresters have made arguments for prescribed burning based on it being the way Aboriginal people have always managed the land, although not necessarily empowering Aboriginal people to do so themselves, or even engaging with contemporary Aboriginal land managers (Neale, 2018, pp. 82, 87). Historian Bill Gammage's celebrated environmental history posits Aboriginal peoples' burning and their role as land managers as fundamental to how we understand Australia, albeit glossing over these diverse practices as undifferentiated and omnipotent, as well as a panacea for catastrophic wildfires (Gammage, 2011; Neale, 2018). Whilst there is much interest in specific cool burning techniques and their ecologies, our focus is with how Aboriginal peoples' contemporary fire management might be understood as part of a Disaster Justice that supports the socio-natural forward-looking justice of collective continuance.

Significantly, public debates about prescribed burning rarely grasp the agenda that Aboriginal people bring. This agenda arises out of their diverse cultures, histories and geographies, as well as their unique standing and shared experiences with:

- settler-colonial violence to exclude their polities and their access to Country, including through warfare, massacre, disease, poisoning and the violence of legal arguments and other knowledge practices, including *terra nullius*;
- the diminishment of their Country through the effects of land use change, introduced species, pollution and climate change and the curtailment of their roles in taking care of Country; and,
- the labour and time needed to re-group and address the trauma of being co-located with a settler-colonial nation which is predicated on the absence of their political-legal rights and their eventual 'disappearance'. This includes claiming the emotional, physical, spatial and temporal space to enjoy Country and be Aboriginal.

To further illustrate, we have prepared summaries of three contexts where we have learnt from working and researching with Aboriginal fire managers. Two of these are in northern and central Australia, where Aboriginal people have substantial territorial land holdings and usually live as a majority on their homelands. The third is in temperate southeast Australia, which is where most Aboriginal people live but as a minority with many other people and with more limited access to and ownership of land.

16.3.1 Martu Country: Gareth Catt

Western Australia's Western Desert lands are where some of the planet's largest wildfires occur, with fires burning in excess of two million hectares every few years; this is also an area strongly held by Aboriginal people, including the Martu. In addition to climate change, this wildfire context is likely driven by fuel build up due to the changed nature of Indigenous peoples' burning in response to colonisation pressures; and, in some occurrences, the influence of introduced grasses such as buffel (*Cenchrus ciliaris*) (Bird, Codding, Kauhanen, & Bird, 2012).

Fire management in the Western Desert occurs within a complex set of parameters that align and conflict. There are traditional rights and obligations, such as the need to hunt and interact with Country, including as part of Jukurrpa. This is desert law passed down from ancestors and ancestral beings, which profoundly informs ways of knowing and being with Country. There are the imposed requirements of Australian governments across different types of desert land tenures. For example, permits from shires, prescriptions for burning from state agencies and training for specialised burning techniques such as aerial incendiary. Where conflict arises, it can have a detrimental effect on Aboriginal peoples' fire use, disempowering the traditional land holders. For Martu, this is a particularly extraordinary situation given the very recent arrival of the colonial frontier in their Country, with the last Martu people coming into contact with 'white Australia' in the 1960s. The widening application of fire across the landscape recommenced in the early 1980s and was facilitated through the establishment of 'communities' by Aboriginal people. These periods of fire management have been called '*pujiman (bushman) times*' and '*community times*'. From the mid-2000s, Martu have accessed both public and private environmental monies to set up ranger teams for a range of activities.

While Martu have lived on their land for millennia, their property rights have only begun to be recognised with the successful determination of several native title applications in the twenty-first century. Native title is a new regulatory layer, which requires navigating the interaction of Martu's laws and customs, their recognised native title rights, the common law, Commonwealth and State/Territory statutory law, as well as government policy. In Western Australia, the recognition of native title has ambiguous consequences for the attribution of the fire management responsibilities of land holders (Weir & Duff, 2017).

16.3.2 Central Arnhem Land: Steve Sutton

Central Arnhem Land is in the monsoonal north-west of the Northern Territory, with the majority of residents being Aboriginal people living on or near their ancestral territories. Wildfire risk is fuelled by Wet season grass growth and subsequent curing, and is mitigated through cool burns conducted early in the Dry season. This entire region was declared Aboriginal land under the *Aboriginal Land Rights (Northern Territory) Act 1976* (Cth) and is owned by Land Trusts comprising the families of all the 'traditional owners'—a term used to describe the people of Country. Consequently, native title recognition has not been a priority.

While Aboriginal ownership has been maintained for generations, large areas of Arnhem Land were violently 'de-populated' in the late nineteenth century and again in the twentieth century (Dewar, 1992). After World War II, demographic and economic pressures combined with government policies to draw Aboriginal people into a series of townships on the peripheries of the Land Trust areas, and these forces are still in play today (Altman & Kerins, 2012; Cooke, 2009). The diaspora led to a breakdown in fire management by the traditional owners, and this was not arrested until the beginning of the twenty-first century. During this interruption, massive wildfires were common across the entire region (Russell-Smith et al., 2009). Through the revival of traditional approaches to burning, Aboriginal people have found a sound working relationship with government fire management agencies. This has established a node of trust centred on this community of practice and the shared objectives of landscape scale fire regime planning and execution. Nevertheless, when asked about what they thought their natural hazards were, one Elder pointedly said that it is not nature that places them at risk, but the government that is the biggest hazard of all (Sithole et al., 2018, p. 273).

Despite ongoing pressures of colonialism, Aboriginal people have continually found ways to connect and reconnect with both the land and its traditions. Senior community members emphasise the importance of the performance of ceremony while acknowledging the value of aspects of western education. The expressed objectives of the traditional owners have remained a clear choice to continue their ancient traditions of connection between people and place, while accepting useful elements of modern western society and technology. Collaborative efforts to manage fire are entirely consistent with this resurgent cultural assertion.

16.3.3 The Australian Capital Territory: Jessica Weir

The temperate lands of the Australian Capital Territory (ACT) have long been a meeting place for people travelling between the sea, the tablelands and alpine mountains, and the Ngunnawal people identify as the traditional custodians of Country and continue to assert their authority with both ACT and Federal governments. Built as a 'Bush Capital', and bordered to one side by steep forested mountains, Canberra's wildfire risk is high and detailed government regulation has been developed in response.

Carved out of New South Wales in 1911 as the capital for the new nation, the ACT is the only mainland jurisdiction with neither land rights legislation nor the recognition of native title. It is also the smallest.[8] The first Europeans arrived in the area in the 1820s as pastoralists, and from the late 1800s onwards Aboriginal people came under ever more organised forms of centralised control by colonial and then nation state authorities. They were moved onto missions and forbidden to speak their language and practise their traditional culture. In neighbouring New South Wales, a few of the mission lands were re-vested as land rights lands. The ACT is now home to 400,000 people, many of whom have come here for work, including Aboriginal people from other parts of Australia.

An ACT government 'cultural burning' program was developed after being suggested by then Senior Aboriginal Ranger, Ngunnawal man Adrian Brown, and supported by managers for its risk mitigation potential as well as being 'the right thing to do'. These cultural burns are listed within the ACT's annual bushfire operational plan alongside 'hazard reduction' and 'ecological' burns. The cultural burns are undertaken on public lands by staff from the ACT Parks and Conservation Service, who are predominantly Aboriginal people from other parts of Australia. Ideally, the traditional custodians identify the priority burn areas and a traditional Ween Bidja (fire boss) lights the fire. As imperfect and limited as this arrangement is, it is nonetheless changing government practice as to why land is burned, by whom, how, where and when. It remains to be seen what a closer engagement with the traditional custodians might bring.

16.4 Framing Disaster Justice

These brief summaries illustrate how Aboriginal peoples' fire management engagements with the natural hazard sector are constrained by misframings—the meta-injustices of meta-political misrepresentation (Fraser, 2007, p. 22). For example, this is evident in the framings of the key terms' 'hazard' and 'risk':

What is a hazard? The naming of the government as the hazard by an Elder from Central Arnhem Land highlights how Aboriginal people experience government actions that undermine and/or ignore their governance structures and priorities as threats to their collective

---

[8] The longest and widest points are 88 kilometres and 30 kilometres respectively.

continuance. This feedback clearly situates fire management engagements within the context of settler-colonial violence and ongoing jurisdictional and normative tensions.

What is at risk? Public sector risk mitigation prioritises lives, property and 'the environment', which is different to the diverse relationships Aboriginal people have with Country. For example, there continues to be a strong perception in southern Australia that a fire in 'remote' Australia doesn't affect anything, which is why there is no public outcry about continental scale wildfires in the desert. This is a failure to appreciate the value of Country.

The term meta-political misrepresentation is appropriate to use because these justice issues are not those of another stakeholder group within the nation state, but the polities of the First Nations—the first people to form nations on this continent, as organised through their laws and customs in relation to their lands and waters. To address the meta-political misrepresentation of Keynesian-Westphalian frames, settler-colonial governments need to acknowledge that they have monopolised the influential activity of frame setting, the injustices that have arisen out of this, including the dismissal of the voices of those who have been harmed, and respond with appropriate fora and material redress (after Fraser, 2007, p. 27). In Australia, there is much 'unfinished business' to attend to.[9] This agenda is not just driven by Indigenous peoples and their allies, but also by those governments and publics who recognise the authority of the First Nations and wish to meet and work together on more just terms. Indeed, there is an expectation that the traditional owners are available to be traditional owners in relation to a suite of public interests and concerns.

Further, this work is set around the importance of collective continuance. With the understanding that nature/society is co-constituted, relationships of importance can be identified and responded to, including the increased frequency and intensity of wildfires as part of rapid climate change. For example, compare how public sector agencies perceive natural hazard events as probabilistic phenomena about which we have no control, but for which we can prepare and respond; however, in places such as Arnhem Land, many Aboriginal people feel both personal and societal responsibility for natural hazards. That is, there is a strong belief that

---

[9] Unfinished business is a term used in Australia to describe the outstanding work needed to prepare the foundations for good relationships between Indigenous and non-Indigenous people.

correct practice can obviate the onset of a disaster, and this belief is strongly borne out in the case of wildfire. The traditional owners understand that there exists a suite of behaviours that effectively prepare for disasters and mitigate their effects including performing ceremony, visiting and maintaining sacred sites and traveling through and observing the state of Country. These actions build on existing cultural and personal strengths and retain the deep understanding of the landscape and the effects of hazard events consistent with Pyles' (2016) recommendations to decolonise disaster management. Whilst the public sector viewpoint understands that there is a consequential nature, these Aboriginal people have embraced this understanding within systems of responsibility that permeate their way of life. Such learnings cannot be heard within unreflexive frames, such as those prescribed burning debates that limit Aboriginal peoples' involvement in fire management to the utility of their 'traditional ecological knowledge'; that is, their usefulness on the terms of others (Neale et al., 2019; Thomassin et al., 2018).

These justice matters play out in terms of the 'who' and 'how' of managing landscape scale fires. The natural hazard sector is resourced to do this work, and by and large the First Nations are not. For example, in Central Arnhem Land, the traditional owners have always had clear structures for decision-making (Sutton, 2018), but have never been formally supported for the many services this leadership provides. What is extraordinary in the ACT is that the government is re-framing its fire management program in response to justice principles, a pathway that is facilitated by assumed synergies with ecological and risk goals; however, a much more comprehensive response is needed to support the authority and governance of the traditional custodians. More than partnerships and contracts, First Nations need to be supported to be First Nations through the sharing of resources and jurisdictional power. This requires the nation state to more comprehensively address the matter of nested sovereignties.

As stated in our Introduction, we see natural hazard collaborations as an important and dynamic place for addressing these issues, with the different politics set around the life and death consequences of overwhelming natural hazard events. Helpfully, regardless of the cultural background of natural hazard practitioners, from our experiences this sector has a bona fide interest in achievement on the ground, including through its volunteer capacity, which facilitates engagement in ways not immediately recognisable in other public sector agencies. The new field of Disaster Justice can help document, analyse and inform this important work. Critically, this academic field brings to the foreground the normative matter of what is right. To minimise the inscription and re-inscription of colonial privilege, it is important that this scholarship itself embraces a deep reflexivity about justice.

16.5 Conclusion

Disasters can expose, magnify and deepen inequalities, and we have shown how this is a justice agenda that is not limited to humans, but embedded within our socio-natural worlds. The case studies demonstrate the amount of work that is involved for Aboriginal fire managers to navigate and negotiate fire regulation regimes that do not necessarily align with their own governance priorities or their territories. This energy could be better placed. The normative focus of Disaster Justice, the spatial-temporal forces of natural hazards and the community of practice that is fire management are all important opportunities for reframing and redressing along more just lines between natures and peoples.

Acknowledgements We thank all the people who have supported our research and work, especially our Aboriginal friends and colleagues. We hope that we have been some small aid to an improvement in Disaster Justice for Aboriginal people by representing their shared information, but the responsibility for any errors or omissions is our own.

References

Altman, J. C., & Kerins, S. P. (2012). *People on Country: Vital Landscapes Indigenous Futures*. Annandale: Federation Press.

Bawaka Country, B., Wright, S., Suchet-Pearson, S., Lloyd, K., Burarrwanga, L., Ganambarr, R., ... Sweeney, J. I. (2015). Co-becoming Bawaka. *Progress in Human Geography, 40*(4), 455–475.

Bird, R. B., Codding, B. F., Kauhanen, P. G., & Bird, D. W. (2012). Aboriginal Hunting Buffers Climate-Driven Fire-Size Variability in Australia's Spinifex Grasslands. *Proceedings of the National Academy of Sciences, 109*(26), 10287–10292.

Buizer, M., & Kurz, T. (2016). Too Hot to Handle: Depoliticisation and the Discourse of Ecological Modernisation in Fire Management Debates. *Geoforum, 68*, 48–56.

COAG. Commonwealth of Australian Governments. (2011). *National Strategy for Disaster Resilience*. Policy Document. Canberra: Council of Australian Governments.

Cooke, P. M. (2009). Buffalo and Tin, Baki and Jesus: The Creation of a Modern Wilderness. In J. Russell-Smith, P. Whitehead, & P. Cooke (Eds.), *Culture, Ecology and Economy of Fire Management in North Australian Savannas* (pp. 69–84). Collingwood: CSIRO Publishing.

Dewar, M. (1992). *The 'Black War' in Arnhem Land; Missionaries and the Yolŋu 1908–1940*. Australian National University North Australian Research Unit.

Frame, A. (1999). Property and the Treaty of Waitangi: A Tragedy of the Commodities? In J. McLean (Ed.), *Property and the Constitution: A Practical Guide to Company Investigations* (pp. 224–234). London: Hart Publishing.

Fraser, N. (2007). Reframing Justice in a Globalising World. In J. Connolly, M. Leach, & L. Walsh (Eds.), *Recognition in Politics: Theory, Policy and Practice* (pp. 16–35). Newcastle-upon-Tyne: Cambridge Scholars Press.

Gammage, B. (2011). *The Biggest Estate on Earth: How Aborigines Made Australia*. Crows Nest: Allen & Unwin.

Garde, M., Nadjamerrek, B. L., Kolkkiwarra, M., Kalarriya, J., Djandomerr, J., Birriyabirriya, B., … Biless, P. (2009). The Language of Fire: Seasonality, Resources and Landscape Burning on the Arnhem Land Plateau. In J. Russell-Smith, P. Whitehead, & P. Cooke (Eds.), *Culture, Ecology and Economy of Fire Management in North Australian Savannas: Rekindling the Wurrk Tradition* (pp. 85–164). Collingwood, VIC: CSIRO Publishing.

Graham, M. (2008). Some Thoughts on the Philosophical Underpinnings of Aboriginal Worldviews. *Australian Humanities Review, 45*, 181–194.

Grijalva, J. M. (2011). Self-Determining Environmental Justice for Native America. *Environmental Justice, 4*(4), 187–192.

Grinde, D., & Johansen, B. (1995). *Ecocide of Native America: Environmental Destruction of Indian Lands and Peoples*. Santa Fe: Clear Light Publishers.

Land, C., & Vincent, E. (2005). Thinking for Ourselves, *New Matilda*, June 29.

Mitchell, T. (2000). The Stage of Modernity. In T. Mitchell (Ed.), *Questions of Modernity* (pp. 1–34). Minneapolis: University of Minnesota Press.

Moran, A. (2005). White Australia, Settler Nationalism and Aboriginal Assimilation. *Australian Journal of Politics and History, 51*(2), 168–193.

Moreton-Robinson, A. (2015). *The White Possessive: Property, Power, and Indigenous Sovereignty*. Minneapolis; London: University of Minnesota Press.

Neale, T. (2018). Digging for Fire: Finding Control on the Australian Continent. *Journal of Contemporary Archaeology, 5*(1), 79–90. https://doi.org/10.1558/jca.33208

Neale, T., Carter, R., Nelson, T., & Bourke, M. (2019). Walking Together: A Decolonising Experiment in Bushfire Management on Dja Dja Wurrung Country. *Cultural Geographies*. https://doi.org/10.1177/1474474018821419

Nightingale, A. J. (2018). The Socioenvironmental State: Political Authority, Subjects, and Transformative Socionatural Change in an Uncertain World. *Environment and Planning E: Nature and Space, 1*(4), 688–711.

Plumwood, V. (2002). Decolonising Relationships with Nature. *PAN: Philosophy Activism Nature, 2,* 7–30.

Prober, S., Yuen, E., O'Connor, M., & Schultz, L. (2013). *Ngadju Kala: Ngadju Fire Knowledge and Contemporary Fire Management in the Great Western Woodlands.* Western Australia: CSIRO in Partnership with the Ngadju Nation, Goldfields Land and Sea Council and Department of Parks and Wildlife WA.

Pyles, L. (2016). Decolonising Disaster Social Work: Environmental Justice and Community Participation. *British Journal of Social Work, 47*(3), 630–647.

Robin, L. (2017). A History of Global Ideas About Environmental Justice. In A. Lukasiewicz, S. Dovers, L. Robin, S. Schilizzi, & S. Graham (Eds.), *Natural Resources and Environmental Justice: Australian Perspectives* (pp. 13–25). Canberra: CSIRO.

Ruru, J. (2018). Listening to Papatūānuku: A Call to Reform Water Law. *Journal of the Royal Society of New Zealand, 48*(2–3), 215–224.

Russell-Smith, J., Lucas, D., Gapindi, M., Gunbunuka, B., Kapirigi, N., Namingum, G., … Chaloupka, G. (1997). Aboriginal Resource Utilization and Fire Management Practice in Western Arnhem Land, Monsoonal Northern Australia: Notes for Prehistory, Lessons for the Future. *Human Ecology, 25*(2), 159–195.

Russell-Smith, J., Whitehead, P., & Cooke, P. (2009). *Culture, Ecology and Economy of Fire Management in North Australian Savannas Rekindling the Wurrk Tradition.* Collingwood, VIC: CSIRO Publishing.

Schlosberg, D., & Collins, L. B. (2014). From Environmental to Climate Justice: Climate Change and the Discourse of Environmental Justice. *Wiley Interdisciplinary Reviews: Climate Change, 5*(3), 359–374.

Simpson, A. (2014). *Mohawk Interruptus: Political Life across the Borders of Settler States.* Durham: Duke University Press.

Sithole, B. with Hunter-Xenie, H. H., & Daniels, H. C., Daniels, G., Daniels, K., Daniels, A., Daniels, G., Daniels, D., Turner, H., Daniels, C. A., Daniels, T., Thomas, P., Thomas, D. (Ngukurr Community Based Research Team), & Yibarbuk, D., Campion, O. B., Namarnyilk, S., Narorroga, E., Dann, O., Dirdi, K., Nayilibibj, G., Brown, C. (Gunbalanya Community Based Research Team). (2018). Living with Widdijith – Protocols for Building Community Resilience in Remote Communities in Northern Australia. In D. Paton & D. Johnston (Eds.), *Disaster Resilience: An Integrated Approach* (2nd ed.). Springfield, IL: Charles C Thomas.

Sutton, S. A. (2018). North Australian Bushfire and Natural Hazard (BNH) Training: A Case Study in Building Remote Community Resilience. In J. Russell-Smith, G. James, H. Pedersen, & K. K. Sangha (Eds.), *Sustainable Land Sector Development in Northern Australia: Indigenous Rights, Aspirations, and Cultural Responsibilities.* Boca Raton: CRC Press.

Thomassin, A., Neale, T., & Weir, J. K. (2018). The Natural Hazard Sector's Engagement with Indigenous Peoples: A Critical Review of CANZUS Countries: Critical Review of CANZUS Countries. *Geographical Research.* https://doi.org/10.1111/1745-5871.12314

Tuck, E., & Yang, K. W. (2012). Decolonization Is Not a Metaphor. Decolonization: Indigeneity. *Education & Society, 1*(1), 1–40.

Watson, I. (2017). What Is the Mainstream? The Laws of First Nations Peoples. In R. Levy, M. O'Brien, & S. Rice (Eds.), *New Directions for Law in Australia: Essays in Contemporary Law Reform* (pp. 213–220). Canberra: Australian National University Press.

Weir, J. K. (2016). Hope and Farce: Indigenous Peoples' Water Reforms During the Millennium Drought. In E. Vincent & T. Neale (Eds.), *Unstable Relations: Indigenous People and Environmentalism in Contemporary Australia.* Crawley: UWA Publishing.

Weir, J. K., & Duff, N. (2017). Who Is Looking After Country? Interpreting and Attributing Land Management Responsibilities on Native Title Lands. *Australian Journal of Public Administration, 76*(4), 426–442.

Whyte, K. P. (2013). Justice Forward: Tribes, Climate Adaptation and Responsibility. *Climatic Change, 120*(3), 117–130.

CHAPTER 17

Inclusion: Moving Beyond Resilience in the Pursuit of Transformative and Just DRR Practices for Persons with Disabilities

Emma Calgaro, Michelle Villeneuve, and Genevieve Roberts

17.1 Introduction

As argued by Lukasiewicz (in Chap. 1), disasters act as trigger points that expose existing (and often deep-rooted) procedural and distributive inequalities and injustices that influence every aspect of daily life. Those who are disproportionately impacted by disasters triggered by hydro-

---

E. Calgaro (✉)
University of Sydney, School of Geosciences and Sydney South East Asia Centre (SSEAC), Sydney, NSW, Australia
e-mail: emma.calgaro@sydney.edu.au

M. Villeneuve
The University of Sydney, Centre for Disability Research and Policy, Sydney, NSW, Australia
e-mail: michelle.villeneuve@sydney.edu.au

G. Roberts
The Deaf Society, Parramatta, NSW, Australia
e-mail: groberts@DeafSociety.com

meteorological and climatological events are also those who are socially, economically, culturally, politically and institutionally marginalised (IFRC, 2018; IPCC, 2014). This includes people with disabilities[1] who account for 15% of the world's population (approximately one billion people), making them the world's largest minority group (IFRC, 2007b; World Health Organization & The World Bank, 2011). The high impact of disasters on people with disabilities is reflected in disaster mortality rates (Chou et al., 2004; Osaki & Minowa, 2001). People with disabilities are four times more likely to die when a disaster strikes than those without disabilities (UNESCAP, 2014b, 2017). Disasters also worsen pre-existing disabilities and create new disabilities (Kelman & Stough, 2015; Sheppard & Landry, 2016; UNESCAP, 2014b). Six percent of all disaster-affected people acquire physical, cognitive or psychological disabilities from the event (UNESCAP, 2014b).

International frameworks and legally binding conventions (e.g. the United Nations (UN) Convention on the Rights of People with Disabilities, Sendai Framework for Disaster Risk Reduction 2015–2030 and the Incheon Strategy 2013–2022) recognise that inclusion in disaster risk reduction (DRR) processes is a human right and must be factored into DRR policy and practice in order to significantly reduce risk (UNESCAP, 2012; UNESCAP & UNISDR, 2012; UNISDR, 2015a). People with disabilities must be afforded (1) the same rights to risk reduction and management measures and (2) the right of equal participation in the design and implementation of DRR policies and practices as any other citizen. However, pathways to achieving inclusion and greater disaster justice for people with disabilities are unclear and fragmented and there are few documented cases of success (Weibgen, 2015). Using disability and disability-inclusive DRR (DiDRR) as a lens, this chapter begins with an overview of common barriers to inclusion for people with disabilities and uses three projects to present a series of

---

[1] The term 'disability' is a contested term and there is no one agreed definition. Here, we define disability as "...*those who have long-term physical, mental, intellectual or sensory impairments which in interaction with various [attitudinal, environmental and institutional] barriers [that] may hinder their full and effective participation in society on an equal basis with others*" (UNCRPD, 2006, p. 4). This definition aligns most closely with interactional models of disability that see disability as an outcome of their impairments and the socially constructed environment (see World Health Organisation & World Bank, 2011, p. 4).

grounded solutions used in different contexts and scales (e.g. local community; state; and international) to demonstrate how to overcome some of these barriers to achieve greater inclusion and disaster justice at all stages of DRR. The chapter concludes with a reflection on our collective experiences of doing DiDRR and highlights common attributes for success.

17.2 Disability, Disasters and Barriers to Inclusion and Greater Justice

The vulnerability of people with disabilities to hazard-related risk and low levels of preparedness is due to multiple and reinforcing systemic sociocultural, economic and political barriers that they face in their daily lives (Gaskin et al., 2017; IFRC, 2007b; Sheppard & Landry, 2016; World Health Organization & The World Bank, 2011). People with impairments are dis-abled throughout their daily lives by normalised structural and attitudinal barriers and by their bodies and minds (Shakespeare, 2013). This also affects their participation in DRR (King, Edwards, Watling, & Hair, 2019). The types of barriers that hinder participation in daily life (generally) and DRR practices and processes (specifically) are summarised in Table 17.1.

People with disabilities are largely unseen, unheard and unaccounted for in all levels of disaster management due to normalised exclusionary policies and practices of communities, governments and disaster assistance organisations (IFRC, 2007b; King et al., 2019). Reasons for this include: a lack of disaggregated data on people with disabilities and their risk-related needs (Kelman & Stough, 2015; Paudel, Dariang, Keeling, & Mehata, 2016; World Health Organization & The World Bank, 2011); stigma, discrimination and marginalisation rooted in cultural and religious beliefs perpetuate negative views about the capability of people with disabilities and hamper their participation in everyday life and DRR processes (Brown, Haun, & Peterson, 2014; Ha, 2016; Kelman & Stough, 2015; King et al., 2019); an ongoing disconnect between DRR policies and practices and disability rights-based laws (ASB & Handicap International, 2011; Twigg, Kett, & Lovell, 2018); DRR policy, planning and consultation processes are largely top-down, rigid and lack adequate mechanisms to ensure the inclusion and voice of

Table 17.1 Barriers to the inclusion of people with disabilities in DRR processes

Contextual factors operating in 'place'		• Stigma, discrimination and marginalisation perpetuate negative views about the capability of people with disabilities and hamper their participation in everyday life and DRR processes (Brown et al., 2014; Ha, 2016; Kelman & Stough, 2015; King et al., 2019).
		• In many cultures, people with disabilities are robbed of their human agency and a public 'voice' by normalised medical and charity models, the latter of which is shaped by cultural and religious beliefs. This undermines rights-based approaches that see people with disabilities as capable agents (Belser, 2015; Gartrell & Hoban, 2013).
		• Women and girls with disabilities are at greater risk of violence, physical abuse and sexual exploitation after disasters due to displacement and unsafe shelters and public spaces (Anam, Khan, Bari, & Alam, 2002; National Council on Disability Affairs, 2009; Smith et al., 2012).
		• People with disabilities also face compounding problems of intersectionality that is, multiple threats of discrimination when an individual's identities overlap with a number of minority classes such as race, class, sexual orientation, gender and religion (ASB & Handicap International, 2011; Gaskin et al., 2017; Sonpal, 2017; Twigg et al., 2018; World Health Organization & The World Bank, 2011).
Economic factors		• People with disabilities are generally poorer than the general populace due to social discrimination and exclusion from opportunities to escape poverty; laws, customs, practices and attitudes and inaccessible public and work spaces exclude them from educational and livelihood opportunities (IFRC, 2007b; Stough & Kang, 2015; UNESCAP, 2017; UNISDR, 2015b).
		• Due to lower incomes, they typically live in areas that are highly exposed to hazard events where their low-quality housing is more likely to be damaged, and their access to the resources needed to support a timely evacuation and to rebuild is limited (Peek & Stough, 2010; Wisner et al., 2004).
		• People with disabilities are overlooked and marginalised further by humanitarian aid distribution processes and practices, leading to unequal access to aid provisions following a disaster event (IFRC, 2007b; King et al., 2019).
Human and social barriers	People with disabilities	• People with disabilities often have little knowledge of hazard risk and how to reduce their risk (Engelman & Deardorff, 2016; Sheppard & Landry, 2016) due to:
		– Disproportionately low literacy levels due to limited or interrupted access to education (IFRC, 2007b; UNESCAP, 2017);
		– Access to information in accessible formats (e.g. sign language, simplified language and braille) is limited and mediums used to deliver emergency messages (e.g. TV and radio alerts, door-to-door messages and social media) are inaccessible (Murray, 2011; Neuhauser et al., 2013; Takayama, 2017); and
		– Deaf people find it difficult to communicate effectively with emergency responders due to differences in languages used (sign language and/or home signs versus dominant spoken language of the 'hearing world') and a shortage of sign language interpreters (Engelman et al., 2013; Tannenbaum-Baruchi et al., 2014).
	Disability support services	• Disability support organisations are often ill-equipped to engage fully with DRR processes that is, they often lack adequate information on risks, appropriate responses and DRR policy frameworks, they are largely unaware of existing DRR programmes and lack the funding needed to advocate for greater support (Engelman et al., 2013; Gaskin et al., 2017; Priestley & Hemingway, 2007).
	DRR Stakeholders	• Governments lack reliable disaggregated data on people with disabilities that is, how many there are, where they are located and what types of disabilities they have (Paudel et al., 2016; Ronoh, 2017; World Health Organization & The World Bank, 2011).
		• DRR actors, emergency responders and humanitarian aid organisations have little knowledge and training on what people with different disabilities need before, during and after events, do not know how to support DiDRR and lack the skills and capacity needed to mainstream DiDRR (ASB & Handicap International, 2011; Hunt, Chung, Durocher, & Henrys, 2015; Kendall-Tackett & Mona, 2005; King et al., 2019, National Council on Disability, 2005; Twigg et al., 2011; World Health Organization, 2011).

Physical & environmental barriers	• DRR infrastructure—evacuation routes, shelters, transportation and early warning systems—is often inaccessible to people with different disabilities (Good, Phibbs, & Williamson, 2016; Kelman & Stough, 2015; Sheppard & Landry, 2016; Twigg et al., 2011). Consequently, people with disabilities (particularly those with mobility disabilities) have been left behind in emergency situations because they were not assisted or accommodated within building and emergency plans (National Council on Disability, 2006; Rooney & White, 2007).
• People with disabilities may be outright barred from entering shelter facilities due to the inaccessibility of shelter facilities and overcrowding (IFRC, 2007b; Twigg et al., 2011).
• When people with disabilities do gain access to shelters, they often face discrimination, harassment and harm at the hands of other survivors (Anam et al., 2002; Busapathumrong, 2013; IFRC, 2007b; Takahashi et al., 1997).
Political & governance-related (procedural) barriers	• Current disaster paradigms and response plans are biased towards helping the already privileged or physically abled; disaster plan provisions for those with disabilities are often included as an add-on to core plans that are often seen by DRR actors as a costly burden that requires specialist services and expertise (Hunt et al., 2015; King et al., 2019; Lunga, Pathias Bongo, van Niekerk, & Musarurwa, 2019; Wisner, 2003).
• There is an ongoing disconnect between DRR policies and practices and disability rights-based laws and a lack of robust guidelines and examples on how to mainstream DiDRR (ASB & Handicap International, 2011; EU Humanitarian Aid and Civil Protection et al., 2014; King et al., 2019; Nick et al., 2009; Ronoh et al., 2017; Twigg et al., 2011; UNESCAP, 2014a; UNISDR, 2015a).
• A lack of baseline data and clear legal and practical DiDRR directives undermine enforcement, accountability and prohibit the measurement of achievement (World Health Organization & The World Bank, 2011).
• DRR policy, planning and consultation processes are largely top-down, rigid and lack adequate mechanisms (such as cross-sector policies and structures) to ensure the inclusion, active participation and voice of people with disabilities in design, implementation and monitoring (Gaskin et al., 2017; Tannenbaum-Baruchi et al., 2014; Villeneuve, 2018).
• In government, there is often limited (or no) application of universal design principles (i.e., designing environments, products and communications that are accessible to everyone) (Boon, Pagliano, Brown, & Tsey, 2012; Gaskin et al., 2017; White et al., 2007).
• Disability and rehabilitation support providers have the capacity to connect people with disabilities and emergency services and support DiDRR but have not traditionally been included in DRR processes (Subramaniam & Villeneuve, 2019).
• There are few platforms to link DRR stakeholders with those who have the skills and knowledge they lack—specifically, disability support organisations and people with disabilities themselves (ASB & Handicap International, 2011; IFRC, 2007a; National Council on Disability, 2005; World Health Organization, 2011). Consequently, DRR actors and disability support organisations often act in isolation from one another, which impedes coordination and implementation (National Council on Disability, 2005).
• There is often no single overarching authority tasked with ensuring adherence to inclusive policies (where they exist) at the country and local levels, leading to low policy implementation accountability (Gartrell, Calgaro, Goddard, & Saorath, 2017).

people with disabilities in design, implementation and monitoring (Gaskin et al., 2017; Tannenbaum-Baruchi, Feder-Bubis, Adini, & Aharonson-Daniel, 2014; Villeneuve, 2018); and a lack of robust guidelines on how to mainstream DiDRR (ASB & Handicap International, 2011; Calgaro, Allen, Craig, Craig, & Dominey-Howes, 2013; EU Humanitarian Aid and Civil Protection, ActionAid International, Handicap International, & Oxfam, 2014; Ronoh, Gaillard, & Marlowe, 2017; UNESCAP, 2014a; UNISDR, 2015a).

DRR stakeholders and emergency responders have little knowledge and training on what people with different disabilities need before, during and after events, do not know how to support DiDRR and lack the skills and capacity needed to mainstream DiDRR (ASB & Handicap International, 2011; Kendall-Tackett & Mona, 2005; National Council on Disability, 2005; Twigg, Kett, Bottomley, Tan, & Nasreddin, 2011; World Health Organization, 2011). Without data and a solid working relationship with disability support organisations, emergency responders (1) cannot accurately identify, quantify and locate vulnerable populations before or during an emergency, (2) the social, human, economic, physical and political resources needed to respond effectively to risk are not adequately allocated to assist people with different disabilities and their needs and (3) people with disabilities are left out of pre-disaster training and post-disaster response efforts (ASB & Handicap International, 2011; IFRC, 2007b; Nick et al., 2009; Smith, Jolley, & Schmidt, 2012; Twigg et al., 2018).

17.3 A Systems Approach to Achieving Disaster Justice

The complex nature of vulnerability, resilience, procedural inclusion and justice for people with disabilities in the DRR space demands multi-faceted solutions that simultaneously address the multiple issues identified in Table 17.1. Systems approaches do this. Embedded in theories of resilience and sustainability science, systems approaches provide a holistic lens through which to identify and engage with the multiple, contextualised and intertwined socio-cultural, political, economic and environmental factors and processes that collectively influence differential vulnerability and resilience patterns over time and space (Adger,

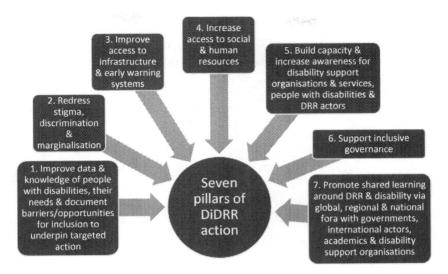

Fig. 17.1 Seven pillars of action to achieve DiDRR

2006; Calgaro, Lloyd, & Dominey-Howes, 2014; Folke, 2006). From this perspective, focusing action on individual components of the coupled human-environment system in isolation is ineffective; doing so creates an incomplete picture of the complexity and dynamism of vulnerability and resilience, which in turn, leads to inappropriate or unsuccessful resilience-building practices (Calgaro, Dominey-Howes, & Lloyd, 2014; Larsen, Calgaro, & Thomalla, 2011). Adopting this holistic, all-system approach, we have identified seven key pillars of action (Fig. 17.1) that help to redress the interlinked socio-cultural, economic, physical and environmental and political and procedural barriers identified in Table 17.1. Yet the proof of solution feasibility and impact in facilitating greater inclusion and justice is in the doing. Therefore, the remaining sections outline our experiences of *doing* DiDRR in three projects that target different pillars of action at different scales, in diverse contexts and populations. From this, we pinpoint the common elements that successfully foster sustainable inclusion and disaster justice.

17.4   *Doing* DiDRR: Demonstrating Change Through Action Using Three Projects as Grounded Examples

17.4.1   Project 1: Get Ready! A Model for Deaf Community Leadership (2015–2016)

Table 17.2   Project 1: Get Ready! A model for Deaf Community Leadership (2015–2016)

Project lead	The Deaf Society of New South Wales (NSW) (Genevieve Roberts)
Partners	NSW State Emergency Service (SES), NSW Rural Fire Service (RFS), Fire and Rescue NSW (FRNSW), Red Cross and the University of Sydney.
Funder	Community Resilience Innovation Program (CRIP), a programme funded through Australia's joint State and Commonwealth Natural Disaster Resilience Program
Targeted pillars of action	Pillars 2, 4, 5, 6, 7
Scale	State-level
Focal population	NSW Deaf Community with an operational focus on metropolitan Sydney, the Blue Mountains, Illawarra, Central Coast, the Hunter Valley, Coffs Harbour, Tweed Heads and Lismore.

17.4.1.1   Project Overview and Aims

The Get Ready! Project was designed to implement key recommendations from a preceding research project entitled "Increasing the resilience of the Deaf Community in NSW to Natural Hazards Project" funded by the National Disaster Resilience Program (2011–2013) (Calgaro & Dominey-Howes, 2013). Multiple recommendations were made but the most pressing needs were to: (1) support Deaf, Deaf-blind and hard of hearing people to increase their preparedness and response to disasters and (2) increase the capacity of NSW emergency services to support and work with Deaf people. To achieve this, the project focused on four key outputs:

1. The recruitment and training of volunteer Deaf Liaison Officers (DLOs) to form a bridge of trust, cross-cultural knowledge and collaborative action between the Deaf Community (a cultural and linguistic minority defined by their use of sign language as their dominant language) and emergency services;

2. The creation of accessible Auslan emergency preparedness videos, thereby providing Deaf Community members with crucial preparedness information in their preferred language;
3. The delivery of Deaf Awareness training to emergency services personnel by trained, culturally aware Deaf experts that is, DLOs; and
4. The delivery of accessible and culturally appropriate workshops to the Deaf Community by the DLOs in partnership with emergency services personnel.

17.4.1.2 *Approach and Method*

Effective risk communication and disaster preparedness strategies are inclusionary, decision-relevant, two-way, and they foster trust, awareness, understanding and motivation to act (Calgaro & Dominey-Howes, 2013). Acknowledging this, the Get Ready! programme was founded on three core principles that are born out of an asset-based community development, capacity-building approach:

- Partnership to facilitate genuine buy-in, accountability and ownership by the emergency services, Deaf support organisations and the Deaf Community, all of which are operating at multiple scales that is, the local, regional (within the state) and state levels;
- Empowerment to acknowledge that Deaf people are the experts in their own lives and must be afforded agency and the opportunity to become agents of change in DiDRR and in their communities; and
- Leadership to ensure that (1) Deaf people have the skills, knowledge and access to legitimate and effective platforms to lead on DiDRR and (2) take ownership of their resilience and that of their community.

The key vehicle through which these principles were implemented was volunteerism.

The six steps taken to achieve greater inclusion of Deaf people of NSW in DiDRR practices were:

1. Convene Advisory Committee comprising of state-level decision-makers from all partner organisations that were in positions to facilitate change. Retaining, where possible, the same representatives that worked on the original research project, preserved institutional memory and collegiality.

2. Engage with the Deaf Community through Deaf clubs: Deaf social clubs—popular forums where Deaf Community members naturally gather—were used to: engage with the community; inform them of the project and its aims; and to identify potential DiDRR champions and leaders i.e. Deaf Liaison Officers.
3. Recruit and train Deaf Liaison Officers (DLOs): The training of DLOs to be skilled leaders and partners in DiDRR was the cornerstone of the project's success. DLOs were recruited via an advert that was circulated widely through Deaf Community networks in both Australian Sign Language (Auslan) and plain English for accessibility. Applicants were also able to submit their application in Auslan by video. The training included two distinct phases: (1) an induction that covered shared values between the individual, the Deaf Society and the project and community development and its inherent link to community resilience-building; and (2) basic training in emergency preparedness and management for different situations and hazards by government and our emergency partners. DLOs were then introduced to the community via the same Deaf clubs and Deaf community networks used in the recruiting process. DLOs explained their role, what to expect from the project's process and began a community conversation around emergency preparedness, current knowledge levels and expertise.
4. Develop Auslan emergency preparedness resources: Seven Auslan videos were produced based on the Australian Red Cross Emergency Preparedness Redi Plan, a national multi-hazard (and multi-agency) plan. The videos can be used as standalone pieces or as a suite and are linked to the full Redi Plan document that is accessible via the Red Cross website. They are used as a teaching tool by DLOs in the Deaf Community Emergency Preparedness Workshops and are a resource for the Deaf Community and emergency services.
5. Undertake Deaf Awareness Training (DAT) for Emergency Services Personnel: This is a vital step in achieving DiDRR; it ensures that emergency services personnel have a culturally sensitive understanding of Deaf people's lived experiences, their culture, the way they view themselves and their place in society, how to communicate respectfully and effectively with them and their support needs.

6. Deaf Community Emergency Preparedness Workshops: the Community workshops were the culmination of all previous steps. Workshops were hosted by the emergency service responsible for the most prevalent hazard for that region. The Redi Plan was delivered by DLOs in Auslan in partnership with local emergency service representatives. The presence of Auslan interpreters enabled full collaborative participation for all—those who are culturally Deaf, signing Deaf, hard of hearing, emergency services personnel and their grassroots volunteers. This provided a forum for cross-cultural shared learning and relationship building. The DLOs, Deaf and hard-of-hearing people benefitted from locally relevant emergency expertise and experience shared by the emergency services representatives whilst emergency services representatives benefited from exposure to Deaf culture. Each workshop ended with a barbeque or sharing of food with roving interpreters to ensure deep community engagement and relationship building.

17.4.1.3 Outcomes

The award-winning Get Ready! project achieved two 'big wins'. First, DLOs are a source of critical capacity. DLOs are culturally embedded DRR specialists and resilience champions that form a bridge between the Deaf Community and the emergency services. Having nine active DLOs also creates a critical mass of experience, collegial support and common purpose. They: (1) design and deliver both Community Preparedness Workshops to Deaf Community members and Deaf Awareness Training to emergency services personnel; (2) advise emergency services staff on accessibility and cultural appropriateness when developing new resources and campaigns and have assisted in the making of SES preparedness videos; and they (3) advise and take part in emergency services staff training, which gives staff first-hand experience of interacting with Deaf in simulated emergency situations. Second, Get Ready!'s demonstrated success has facilitated transformation in DiDRR in the emergency services in NSW. Led by the Deaf Society of NSW, the Get Ready! team are working with the NSW SES to establish a Multi-hazard Multi-agency Deaf Unit—a first of its kind. It is hoped that the Deaf Unit will become a permanent home for the DLOs.

17.4.2 Project 2: DiDRR in NSW: Increasing Access to DRR for People with Disabilities Through Collaboration (2015–2017)

Table 17.3 Project 2: DiDRR in NSW: Increasing access to DRR for people with disabilities through collaboration (2015–2017)

Project lead	Centre for Disability Research and Policy and Natural Hazards Research Group, The University of Sydney (Michelle Villeneuve)
Partners	Stakeholders in three local government areas of NSW including, people with disabilities in NSW and their representatives—Disabled Pope's Organisations (DPOs), disability support providers, community health care providers, local government (community development officers) and local emergency managers.
Funder	Community Resilience Innovation Program (CRIP)
Targeted pillars of action	Pillars 1, 4, 5, 6, 7
Scale	Local/community level
Focal population:	Local emergency managers working in partnership with disability service providers to share knowledge for DiDRR development

17.4.2.1 Project Overview and Aims

DiDRR is the shared responsibility of the emergency management sector and local communities, requiring effective cross-sector collaboration between emergency managers and community services personnel to remove barriers that stop people with disabilities engaging with DRR activities (COAG, 2011; Centre for Disability Research and Policy and Natural Hazards Research Group, 2017). However, little is known about how to engage in a cross-sector collaboration between community services and emergency services. This project, therefore, aimed to: (a) leverage the collective expertise of the disability support sector and emergency managers to ascertain how to increase the participation of people with disabilities, their families and carers in local community DRR activities; and (b) establish best practice in cross-sector engagement required to facilitate local community-level DIDRR though shared learning and collaboration.

17.4.2.2 Approach and Method

The project was guided by four principles of DiDRR (Villeneuve, Robinson, Pertiwi, Kilham, & Llewellyn, 2017):

- Universal accessibility of comprehensive disaster risk reduction information to underpin preparedness for people with different disabilities and policies in emergency management;
- Active participation and engagement of people with disabilities in disaster risk reduction in the local community;
- Collaboration of multiple stakeholders learning and working together to ensure DiDRR; and
- Non-discrimination, requiring fair and unprejudiced treatment of people with disabilities in DRR policy and action.

The process of collaborative cross-sector engagement began with the formulation of an Advisory Panel, comprising of state and national representatives from including decision makers in the disaster and emergency management sector, Disabled People's Organisations (DPOs), disability service organisations (DSOs) and community health care organisations. This Committee supported and guided the project from inception to delivery.

Three participatory workshops were conducted to generate innovative ways to overcome the barriers to DIDRR. This approach enabled stakeholders to generate new, shared understanding and increases their capacity to work together. Drawing upon developmental work research (Engeström, 2008; Patton, 2010) and appreciative inquiry (Cooperrider & Whitney, 2005), the workshops facilitated a cycle of engaging with local knowledge, policy and service provision and the literature to create strategies to increase the participation of people with disabilities in the local community level disaster preparedness. In these discussions, participants reflected on DiDRR principles and deliberated on how those principles could be applied in practice whilst the research team analysed the participants' reflections in the context of the policy, literature and local service contexts. Emergency managers learnt about the function-based support needs of people with disabilities and health whilst disability providers learnt about local natural hazard risks and emergency preparedness. This format enabled a shared focus on community strengths, challenges and resources for DIDRR, created a local knowledge base on DIDRR from which strategies for cross-sector collaboration between local emergency managers and disability support providers was drawn and encouraged networking between stakeholders to sustain collaborative DIDRR actions beyond the life of the project.

17.4.2.3 Outcomes

The culmination of this collaborative cross-sector process was the development of the Local Emergency Management Guidelines for DIDRR in NSW (Centre for Disability Research and Policy and Natural Hazards Research Group, 2017). This is a first for NSW and Australia. The foundational DIDRR framework (Fig. 17.2), included in the guidelines, pairs the four DIDRR principles—accessibility, participation, collaboration and non-discrimination with specific actions that local emergency managers can take to engage effectively with the disability community and their networks in emergency preparedness in all phases of disaster risk management that is, preparedness, response and recovery.

Whilst the DiDRR Guidelines are directed at local emergency managers, these guidelines emphasise collaborative action and partnership between disability providers and emergency managers that is consistent with the Council of Australian Governments (COAG) (2011) principle of shared responsibility for disaster risk reduction. Two key values guide the guidelines' application. First, there is not one distinct starting point for local community engagement in DiDRR; there are tools and practical tips to enable emergency

Fig. 17.2 DiDRR framework. (Centre for Disability Research and Policy and Natural Hazards Research Group, 2017)

managers and DSOs to start at any point in the process. These include tips on hosting disability-inclusive community-based programmes and activities action-oriented strategies to enable participation of people with disabilities in local DRR education and awareness events. Second, local disaster resilience begins when emergency managers, people with disabilities and DSOs learn and work together. Accompanying the guidelines is a video entitled Disability Inclusive Disaster Preparedness in NSW.[2] The video shares the DiDRR principles from the perspective of people with disabilities, emergency managers and disability support providers and offers examples of how they were enacted in this project.

17.4.3 Project 3: Disability and Disasters: Empowering People and Building Resilience to Risk

Table 17.4 Project 3: Disability and disasters: Empowering people and building resilience to risk

Project lead	The University of Sydney (Emma Calgaro)
Partners	Stockholm Environment Institute–Asia (SEI), Craigs Consultants International (CCI), KPC Consultant Co. Ltd, Inclusive Development and Empowerment Agenda (IDEA), Monash University, Cambodian Disabled People's Organization (CDPO)
Funder	Global Resilience Partnership (GRP) through GRP's inaugural Global Resilience Challenge
Targeted pillars of action	Pillars 1, 2, 4, 5, 6, 7
Scale	Local, national and regional (Southeast Asia)
Focal population:	People with disabilities in Thailand, The Philippines and Cambodia

17.4.3.1 Project Overview and Aims

The vision for the Disability and Disasters project is for inclusion to be at the center of DRR and disaster responses for the benefit of all (Calgaro et al., 2015). This requires a shift in DRR strategy and practice. DiDRR demands full integration—disability can no longer be an 'add on' to existing DRR approaches. To achieve this, the international multi-partner team worked with people with disabilities, DPOs and DRR actors at the local, national and regional level to: (1) collaboratively identify needs; (2) provide people with disabilities with institutional platforms and social support to respond

---

[2] Accessed on YouTube via https://www.youtube.com/watch?v=qCWUEiAVii8&t=2s

effectively to disasters; and (3) up-skill and empower people with disabilities to be champions of resilience and change within their communities.

17.4.3.2 Approach and Method

Taking a holistic systems approach, the Disability and Disasters team targeted transformative change on three levels (see Calgaro et al., 2015 for details on individual methods and activities):

1. Inclusivity: People with disabilities were the centrepiece of their approach. Five team members have disabilities which: (1) harnesses cultural relevance and lived expertise; (2) enables people with disabilities to lead on DiDRR action; and (3) builds capacity, enabling people with disabilities to become DiDRR champions.
2. Application of a systems approach to DiDRR: Recognising the complexity of DiDRR and coupled human-environment systems, the team sought change in three areas:

   (a) Increasing knowledge and skills: Research identified the challenges people with disabilities face when responding to risk in each country and enabled DPOs and people with disabilities to co-produce knowledge and solutions. This empirical data directly informed the content of a DiDRR ToolKit that comprised innovative training methods and manuals to (1) up-skill and empower people with disabilities and DPOs to increase their resilience to risk using their solutions, (2) train people with disabilities to lead DiDRR activities (via train-the-trainer programmes and manuals), and (3) educate DRR stakeholders on the needs of people with disabilities and how best to support them in collaboration with DPOs and support services;

   (b) Shifting attitudes and beliefs: Through training workshops, the team: (1) supported people with disabilities and DPOs to become active leaders in DRR-specific change and resilience; and (2) increased the visibility of people with disabilities and their contribution in DRR via direct interactions between people with disabilities, DPOs and DRR actors. This changed the way DRR actors perceived and engaged with people with disabilities (from 'stigmatised' to 'capable actors'); and

   (c) Improving DiDRR governance: By producing two complimentary tools—Inclusive Resilience Scorecard and Roadmap to Inclusive and Resilient Cities—governments and organisations

can assess their inclusivity and follow a step-by-step roadmap on how to be more inclusive.
3. Demonstrating DiDRR best practice: To ensure relevance and impact, the team developed the DiDRR Toolkit in collaboration with end-users and demonstrated how to do DiDRR by (1) testing and refining the toolkit and methods with end-users and (2) providing working examples of DiDRR in action through all phases of disaster risk management.

17.4.3.3 Outcomes

This project increased the resilience of over 22,000 people with disabilities in Thailand, Cambodia and The Philippines. The following key achievements have created the foundations to support effective and durable DiDRR and provide a systems-based model of DiDRR best practice:

1. Twenty-eight DiDRR training/awareness sessions for people with disabilities, provincial and national disaster management officers, non-government organisations and DPOs across three countries were completed. Their capacity-building activities (training workshops and multi-stakeholder advisory committees consisting of people with disabilities, DPOs and key DRR stakeholders) ensured legacy and ongoing impact by: (1) facilitating the co-creation of grounded DiDRR solutions and greater ownership by end-users; and (2) creating DiDRR champions, trainers, advocates and educators;
2. A DiDRR toolkit comprising accessible DiDRR learning and training materials (six manuals, eight videos) was produced in multiple formats (braille and sign language) to (1) increase risk awareness and preparedness for people with disabilities, train people with disabilities to undertake DiDRR research and be DiDRR trainers and (2) teach DRR actors how to do DiDRR in a systematic and holistic way—a first in the target countries;
3. The production of the Inclusive Resilience Scorecard (rating how inclusive governments/organisations are) and the complementary Roadmap to Inclusive and Resilient Cities (step-by-step roadmap showing organisations how to achieve greater inclusivity) supports action planning and budgeting on DiDRR/climate change adaptation (CCA) programmes, projects and activities; and
4. Local and national policies and practice have been changed in Thailand and The Philippines through the adoption of the initiative's model.

The impact of this project extends much further than the three focal countries. This holistic approach and demonstration of DiDRR best practice has been recognised internationally and is being up-scaled in partnership with the United Nations Development Programme (UNDP) and UN Climate Resilience Initiative (A2R) to create a multi-country, multi-phase Disability and Climate Change Programme.

17.5 Discussion: Facilitating Greater Inclusion and Disaster Justice

Transforming systems to be more equal, just and inclusive is complex due to the myriad of interlinked factors and processes that influence outcomes through positive and negative feedbacks (see Calgaro, Dominey-Howes, & Lloyd, 2014; Calgaro, Lloyd, & Dominey-Howes, 2014; Matin, Forrester, & Ensor, 2018; Turner et al., 2003). Some barriers to greater equity, justice and inclusion such as systemic violence against women and poverty due to exclusion from education and work opportunities (see Table 17.1) are more difficult to change than others (such as information dissemination and inclusive engagement). However, our experiences of doing DiDRR have shown that multi-faceted action on DiDRR is possible through consensus, shared ownership and collective will. From these experiences, we have identified five common attributes that were key in ensuring DiDRR success. This final section presents each attribute and explains their importance using illustrative examples taken from our project-based experiences.

17.5.1 Capacity-Building Activities and Tools That Are Rights-Based, Culturally Sensitive and Focus on Empowerment, Self-Determination and Leadership Are Essential for Success

Empowering and supporting people with disabilities to be champions of transformative change in DiDRR (specifically) and in their communities (more broadly) is critical for achieving inclusion and disaster justice. As per Fig. 17.1 (the seven pillars of action), this requires action on multiple levels and scales i.e.: including people with disabilities, DPOs and disability support organisations in DRR activities and processes from inception to execution; enabling self-determinism; increasing knowledge and capacity levels of people with disabilities and their support networks (commonly their families and support organisations and services) and DRR actors; changing how disaster actors and broader society see and engage with people with disabilities; and removing the normalised structure and processes of everyday life that rob them of their human agency and exclude them from DRR practices. Reflecting the complexity of systems approaches and the task at hand, the most successful activities are interlinked and often have dual roles. We achieved this overarching goal through the execution of three main activities.

First, we increased the risk and preparedness knowledge, confidence and capacity levels of people with disabilities and their support networks by including people with disabilities as project team members (in the case of the GRP Disability and Disasters project) and developing a series of training programmes and developing capacity-building tools and materials with end-users. Specifically:

1. The training of DLOs (in Get Ready!) and DiDRR Training-of-Trainers (ToT) (in Disability and Disasters) provided people with disabilities with the skills, confidence and self-belief to be leaders in and advocates for DiDRR. Other skills obtained included strengthening identity and community membership, presenting, facilitation, mentoring and translating;
2. The Deaf Community Emergency Preparedness Workshops (Get Ready!), DiDRR, First Aid training and DiDRR Mainstreaming training (Disability and Disasters) and the DiDRR Guidelines (DiDRR in NSW) focussed on increasing risk awareness, preparedness and capacity to effectively respond to future hazards; and
3. The DiDRR Guidelines (DiDRR in NSW) provided local emergency managers with a process for working in collaboration disability support providers to increase the participation of people with disability in local community-level DRR. It also raised awareness of the knowledge, tools and resources needed by community health and disability support providers to engage in DiDRR practices.

Second, the capacity of DRR actors to better understand the needs and capabilities of people with disability and support them in appropriate ways was increased through (1) Disability/Deaf Awareness Training for Emergency Services personnel led by people with disabilities (Get Ready! and Disability and Disasters) and (2) the production of training manuals and guidelines (DiDRR in NSW and Disability and Disasters). Third, The Disability Awareness trainings and Deaf Community Emergency Preparedness Workshops undertaken by newly trained DLOs and emerging leaders within the disability sector (Get Ready! and Disability and Disasters) doubled as platforms for empowerment, leadership, shared learning and cultural exchange; where direct interactions between people with disabilities, DPOs and DRR stakeholders facilitated a change in the way DRR stakeholders perceive and engage with people with disabilities.

17.5.2 Provide Comprehensive Information in Accessible Formats That Is Delivered by Trusted Sources That Are Embedded in Their Communities

People with disabilities and disability support providers must have access to risk and disaster preparedness information in accessible and readily available forms in order to understand and evaluate their risk levels and effectively prepare for hazards. Resources should be: focused on the support needs and capabilities of people with disabilities; in formats that are accessible to them (e.g. diverse languages, easy read format, braille, and screen reader enabled); and be delivered in manageable amounts. This task requires understanding of local community needs and individual capabilities. Trust in the mode and source of information delivery is also paramount. For example, systemic societal marginalisation and enduring stigmas coupled with negative past experiences with emergency service personnel and first responders has left Deaf Community members in NSW highly distrustful of the abilities of emergency services to understand them—their culture and their needs—and support them in emergency and disaster situations (Calgaro & Dominey-Howes, 2013). The DLOs provide Deaf Community members with that trusted and culturally embedded link between emergency services and themselves.

17.5.3 Develop Capacity in the Disaster Risk Management Sector for DiDRR

Emergency management policy and practices must reflect DiDRR principles but *"people cannot find answers to problems that they don't fully understand or acknowledge"* (Calgaro et al., 2013, p. 61). Consequently, there is a need to build capacity and increase disability awareness in the emergency services sector and create tools and platforms that encourage inclusive governance practices in DRR. The Get Ready! and Disability and Disasters projects increased capacity in doing DiDRR and disability awareness for

DRR actors and emergency service personnel via a series of Deaf Awareness Training and DiDRR Training sessions respectively. Crucially, DLOs and people with disabilities working with the GRP team (who were trained as part of the Train-the Trainer GRP activity for this purpose) led these sessions. This enabled DRR actors and responders to see people with disabilities in action; as capable and active partners in DRR processes via the process of shared learning, which helped break down socio-cultural barriers that hinder inclusion i.e. stereotypes and discomfort in working or communicating with people with disabilities on one side and mistrust in mainstream officials on the other (see Craig, Craig, Calgaro, Dominey-Howes, & Johnson, 2019). It is also imperative that DRR actors and responders from all levels take part—from grassroots volunteers to high-level decision-makers.

17.5.4 Create Supportive Platforms and Tools to Facilitate Inclusive Governance and Practice

Our collective work has demonstrated that there are numerous ways to achieve more inclusive governance to ensure that the diverse perspectives of people with disabilities are heard and acted upon. The first is the creation of formal and informal platforms that support partnerships and sustained engagement between disability support organisations, people with disabilities and DRR actors and emergency responders. Networking across sectors (e.g. disability and emergency management) enhances knowledge, support and capacity for DiDRR at all levels of DRR engagement and embeds this at the local level where support is delivered. Networking between disability organisations is also important; it can reduce duplication and leverage the knowledge, capacity and preparedness of early DiDRR adopters within the disability sector. The Get Ready! Deaf Community Emergency Preparedness Workshops and Disability and Disasters DiDRR Training sessions created spaces for shared learning, networking and socio-cultural understanding between disability support organisations, people with disabilities and DRR actors and responders. The DiDRR in NSW project achieved this through the collaborative process (via workshops and focus group discussions) of creating the DiDRR Guidelines for NSW. However, it is important to formalise these processes by including people with disabilities on local, state, national and (where possible) regional disaster and emergency management committees. Representation fosters respect of people with disabilities and their capabilities. Learning directly from people with disabilities enables emergency managers to identify and address the barriers that prevent DiDRR.

The second is to provide disability support organisations and DRR institutional actors with the tools needed to guide and measure DiDRR success. Disability support organisations are disability champions and connectors but require information, skills and tools to (a) self-assess their organisational preparedness to manage in an emergency situation; (b) prioritise preparedness actions they need to take; and (c) develop a business continuity plan that takes into account duty of care to clients and disability support staff. The DiDRR in NSW project's DiDRR Guidelines provide local emergency managers in New South Wales with a starting point for hosting DiDRR community programmes and activities, incorporating disability inclusive preparedness strategies into local emergency management plans and draws upon grounded examples to illustrate success. The Disability and Disaster's Inclusive Resilience Scorecard and the complementary Roadmap to Inclusive and Resilient Cities also supports inclusive governance by providing governments and organisations with a tool to self-assess and rate how inclusive they are followed by a step-by-step roadmap on how to achieve greater inclusivity via examples of action planning and budgeting on DiDRR programmes, projects and activities.

17.5.5 *Sustainability and Scaling Up*

Models of DiDRR best practice that are grounded and refined in the field must also be (1) sustainable that is, robust and transferable enough to continue in some form after the project ends and (2) replicable and scalable in order to represent value for money and maximise impact and wider relevance. Resilience-building solutions to risk must be adapted and refined to suit the target population and context. But they also must honour principles of universal design and be transferrable. Tested and effective ways to increase transferability and scalability include:

- The inclusion of key partners and end-users—People with disabilities, DPOs, disability service provides, major DRR stakeholders with local, regional and international reach (e.g. Australian Red Cross and United Nations Office for Disaster Risk Reduction (UNDRR), and DiDRR academic specialists—in the inception of DiDRR projects and activities. This inclusive process ensures the relevance of outputs and facilitates greater end-user ownership and buy-in across all scales

(local, national, regional and international). All three projects achieved this through their considered choice of partners from the outset and formalised input via regular advisory committee meetings;
- Developing of DiDRR toolkits including things like risk and disaster preparedness information in accessible forms, DiDRR best practice guidelines and models to follow and training manuals, that can be replicated, adapted to different contexts and populations and scaled;
- Leveraging and mobilising established and wide-reaching networks of end-users (and funders where possible) to: (1) disseminate information on the needs and capabilities of people with disabilities, (2) promote the distribution and use of output products and (3) share and widely publicise grounded examples of best practice to attract broader interest, uptake and more funding to build momentum and trajectory of change in inclusion and greater justice for all; and
- Collaborating with the media to raise DiDRR awareness and reach those people with disabilities that are socially and geographically isolated and/or where mobility is limited. For example, the Disability and Disasters project broadcast a series of radio pieces and interactive call-in community 'plays' via Cambodia's Voices of People with Disabilities radio channel to engage with 4725 people with disabilities and their families in DiDRR.

17.6 Conclusion

People with disabilities do, on the whole, experience higher levels of vulnerability to hazard-related risk than the general populace. But their vulnerability is socially constructed; it is a dynamic and uneven outcome of multiple reinforcing socio-cultural, economic and political barriers that they face in their daily lives. This includes how society and DRR actors perceive people with disabilities and their abilities. Our experiences in doing DiDRR in different contexts have shown that reversing this trend—making inclusion a reality for the benefit of all—is possible but requires multiple layers of action (such as those presented in Fig. 17.1) that simultaneously address the key social, political, economic and environmental barriers (outlined in Table 17.1) that impede progress. The types of actions needed to address these barriers include: improving the knowledge base of people with disabilities, their needs and capabilities; improving access to infrastructure and early warning systems; redressing normalised discrimination and marginalisation; increasing access to social and human resources including information in accessible forms; building capacity and increasing awareness raising for disability support organisations and services, people with disabilities and DRR actors; and supporting inclusive governance. Operationalising change on multiple levels is difficult but our experiences also show that desired change in many of the above-listed points can be achieved by undertaking activities that target more than one issue. For example, capacity-building activities aimed at people with disabilities, disability support organisations and DRR actors not only increase disability awareness within emergency services but increase support and preparedness for people with disabilities. Such actions also redress stigmas, support empowerment, self-determination and leadership and facilitate collaboration, shared learning and understanding. However, a key determinant of true transformation in achieving inclusion and greater disaster justice is sustainability. The loss of momentum is common when the money and support runs out at the end of resilience-building projects. This too can be avoided with careful planning from the outset to ensure end-user buy-in and accountability from the inception of the project and to produce solutions that can be scaled up. In doing so, solutions become financially sound spring-boards for broader success.

References

Adger, W. N. (2006). Vulnerability. *Global Environmental Change, 16*(3), 268–281.

Anam, N., Khan, A. N., Bari, N., & Alam, K. J. (2002). *Unveiling Darkness: The Situation Analysis on Disaster and Disability Issues in the Coastal Belt of Bangladesh*. Dhakar. Retrieved from http://www.csid-bd.com/research/research01.pdf

ASB, & Handicap International. (2011). *Mainstreaming Disability in Disaster Risk Reduction: A Training Manual and Facilitation Guide*. Manilla. Retrieved from http://masbatepdrrmo.com/wpcontent/uploads/2015/11/Mainstreaming-Disability-in-Disaster-Risk-Reduction-A-Training-Manual-and-Facilitation-Guide.pdf

Belser, J. W. (2015). Disability and the Social Politics of 'Natural' Disaster: Toward a Jewish Feminist Ethics of Disaster Tales. *Worldviews: Environment, Culture, Religion, 19*(1), 51–68.

Boon, H. J., Pagliano, P., Brown, L., & Tsey, K. (2012). An Assessment of Policies Guiding School Emergency Disaster Management for Students with Disabilities in Australia. *Journal of Policy and Practice in Intellectual Disabilities, 9*(1), 17–26.

Brown, L. M., Haun, J. N., & Peterson, L. (2014). A Proposed Disaster Literacy Model. *Disaster Medicine and Public Health Preparedness, 8*(3), 267–275.

Busapathumrong, P. (2013). Disaster Management: Vulnerability and Resilience in Disaster Recovery in Thailand. *Journal of Social Work in Disability & Rehabilitation, 12*(1–2), 67–83. https://doi.org/10.1080/1536710X.2013.784176

Calgaro, E., & Dominey-Howes, D. (2013). Final Project Report (Milestone 7)—Increasing the Resilience of the Deaf Community in NSW to Natural Hazards. Retrieved from http://deafsocietynsw.org.au/news/entry/resilience_natural_hazards

Calgaro, E., Allen, J., Craig, N., Craig, L., & Dominey-Howes, D. (2013). Deaf Community Experience, Knowledge & Needs Assessment—Final Results Report (Milestone 2 & 3). *Increasing the Resilience of the Deaf Community in NSW to Natural Hazards—Project Reports*. Retrieved from http://deafsocietynsw.org.au/news/entry/resilience_natural_hazards

Calgaro, E., Craig, L., Johnson, K., Krongkant, P., Craig, N., Zayas, J., ... Dominey-Howes, D. (2015). Disability and Disasters: Empowering People and Building Resilience to Risk. *Global Resilience Challenge: Solution Statement*. Retrieved from http://www.globalresiliencepartnership.org/teams/disability-and-disasters/

Calgaro, E., Dominey-Howes, D., & Lloyd, K. (2014). Application of the Destination Sustainability Framework (DSF) to Explore the Drivers of Vulnerability in Thailand Following the 2004 Indian Ocean Tsunami. *Journal of Sustainable Tourism, 22*(3), 361–383. https://doi.org/10.1080/09669582.2013.826231

Calgaro, E., Lloyd, K., & Dominey-Howes, D. (2014). From Vulnerability to Transformation: A Framework for Assessing the Vulnerability and Resilience of Tourism Destinations in a World of Uncertainty. *Journal of Sustainable Tourism, 22*(3), 341–360. https://doi.org/10.1080/09669582.2013.826229

Centre for Disability Research and Policy and Natural Hazards Research Group. (2017). *Local Emergency Management Guidelines for Disability Inclusive Disaster Risk Reduction in NSW*. Sydney, NSW. Retrieved from Sydney: http://sydney.edu.au/healthsciences/cdrp/projects/Emergency%20Preparedness_brochure_August2017_WEB_ACCESS.pdf

Chou, Y., Huang, N., Lee, C., Tsai, S., Chen, L., & Chang, H. (2004). Who Is at Risk of Death in an Earthquake? *American Journal of Epidemiology, 160*(7), 688–695.

COAG. (2011). National Strategy for Disaster Resilience: Building Our Nation's Resilience to Disasters. Retrieved from https://www.ag.gov.au/EmergencyManagement/Documents/NationalStrategyforDisasterResilience.PDF

Cooperrider, D. L., & Whitney, D. (2005). *Appreciative Inquiry: A Positive Revolution in Change*. San Fransisco: Berret-Koehler Publishers Incorporated.

Craig, L., Craig, N., Calgaro, E., Dominey-Howes, D., & Johnson, K. (2019). People with Disabilities: Becoming Agents of Change in DRR. In F. I. Rivera (Ed.), *Emerging Voices in Natural Hazards Research* (pp. 323–353). Cambridge, MA: Elsevier.

Engelman, A., & Deardorff, J. (2016). Cultural Competence Training for Law Enforcement Responding to Domestic Violence Emergencies with the Deaf and Hard of Hearing: A Mixed-Methods Evaluation. *Health Promotion Practice, 17*(2), 177–185.

Engelman, A., Ivey, S. L., Tseng, W., Dahrouge, D., Brune, J., & Neuhauser, L. (2013). Responding to the Deaf in Disasters: Establishing the Need for Systematic Training for State-Level Emergency Management Agencies and Community Organizations. *BMC Health Services Research, 13*(84). https://doi.org/10.1186/1472-6963-1113-1184.

Engeström, Y. (2008). *From Teams to Knots: Activity-Theoretical Studies of Collaboration and Learning at Work.* Cambridge, UK: Cambridge University Press.

EU Humanitarian Aid and Civil Protection, ActionAid International, Handicap International, & Oxfam. (2014). *Briefing Paper: Making Disaster Risk Management Inclusive.* Johannesburg. Retrieved from http://www.actionaid.org/sites/files/actionaid/actionaid:inclusion_paper_final_170614_low.pdf

Folke, C. (2006). Resilience: The Emergence of a Perspective for Social-Ecological Systems Analyses. *Global Environmental Change, 16*(3), 253–267. Retrieved from http://www.sciencedirect.com/science/article/B6VFV-4KFV39T-1/2/21ccf91cced363dbd098af70b8eb525d

Gartrell, A., & Hoban, E. (2013). Structural Vulnerability, Disability and Access to Non-governmental Organization Services in Rural Cambodia. *Journal of Social Work in Disability and Rehabilitation, 12*(3), 194–212.

Gartrell, A., Calgaro, E., Goddard, G., & Saorath, N. (2017). *Women with Disabilities Experience of Disasters in Rural Cambodia.* Disability and Disasters: Empowering People & Building Resilience to Risk – Internal Project Report. Monash University, Melbourne & University of Sydney, Sydney.

Gaskin, C. J., Taylor, D., Kinnear, S., Mann, J., Hillman, W., & Moran, M. (2017). Factors Associated with the Climate Change Vulnerability and the Adaptive Capacity of People with Disability: A Systematic Review. *Weather Climate and Society, 9*(4), 801–814.

Good, G. A., Phibbs, S., & Williamson, K. (2016). Disoriented and Immobile: The Experiences of People with Visual Impairments during and after the Christchurch, New Zealand, 2010 and 2011 Earthquakes. *Journal of Visual Impairment & Blindness, 110*(6), 425–435.

Ha, K.-M. (2016). Inclusion of People with Disabilities, their Needs and Participation, into Disaster Management: A Comparative Perspective. *Environmental Hazards-Human and Policy Dimensions, 15*(1), 1–15.

Hunt, M. R., Chung, R., Durocher, E., & Henrys, J. H. (2015). Haitian and International Responders' and Decision-Makers' Perspectives Regarding Disability and the Response to the 2010 Haiti Earthquake. *Global Health Action, 8*(1). https://doi.org/10.3402/gha.v8.27969

IFRC. (2007a). *World Disaster Report: Focus on Discrimination*. Geneva.

IFRC. (2007b). *World Disasters Report 2007: Focus on Discrimination*. Geneva, Switzerland. Retrieved from https://www.ifrc.org/PageFiles/99876/WDR2007-English.pdf

IFRC. (2018). *World Disasters Report: Leaving No One Behind*. Geneva. Retrieved from http://www.ifrc.org/Global/Documents/Secretariat/201610/WDR%202016-FINAL_web.pdf

IPCC. (2014). Summary for Policymakers. In C. B. Field, V. R. Barros, D. J. Dokken, K. J. Mach, M. D. Mastrandrea, T. E. Bilir, et al. (Eds.), *Climate Change 2014: Impacts, Adaptation, and Vulnerability. Part A: Global and Sectoral Aspects. Contribution of Working Group II to the Fifth Assessment Report of the Intergovernmental Panel on Climate Change* (p. 32). Cambridge: Cambridge University Press.

Kelman, I., & Stough, L. (Eds.). (2015). *Disability and Disaster: Explorations and Exchanges*. New York, NY: Palgrave Macmillan US.

Kendall-Tackett, K., & Mona, L. (2005, August 18–21). *The Impact on 9/11 of People with Disabilities: Committee on Disability Issues in Psychology*. Paper Presented at the Proceedings of the American Psychological Association Conference, Washington, DC, USA.

King, J., Edwards, N., Watling, H., & Hair, S. (2019). Barriers to Disability-Inclusive Disaster Management in the Solomon Islands: Perspectives of People with Disability. *International Journal of Disaster Risk Reduction, 34*, 459–466.

Larsen, R. K., Calgaro, E., & Thomalla, F. (2011). Governing Resilience Building in Thailand's Tourism-Dependent Coastal Communities: Conceptualising Stakeholder Agency in Social–Ecological Systems. *Global Environmental Change, 21*(2), 481–491.

Lunga, W., Pathias Bongo, P., van Niekerk, D., & Musarurwa, C. (2019). Disability and Disaster Risk Reduction as an Incongruent Matrix: Lessons from Rural Zimbabwe. *Jamba (Potchefstroom, South Africa), 11*(1), 648–648.

Matin, N., Forrester, J., & Ensor, J. (2018). What Is Equitable Resilience? *World Development, 109*, 197–205. https://doi.org/10.1016/j.worlddev.2018.04.020

Murray, J. S. (2011). Disaster Preparedness for Children with Special Healthcare Needs and Disabilities. *Journal for Specialists in Pediatric Nursing, 16*(3), 226–232.

National Council on Disability. (2005). *Saving Lives: Including People with Disabilities in Emergency Planning*. Washington, DC. Retrieved from hpod.pmhclients.com/pdf/saving-lives.pdf

National Council on Disability. (2006). *The Impact of Hurricanes Katrina and Rita on People with Disabilities: A Look Back and Remaining Challenges.* Washington, DC, USA. Retrieved from http://www.ncd.gov/publications/2006/Aug072006

National Council on Disability Affairs. (2009). *National Disability Summit—Documentation Report.* Manilla, Philippines.

Neuhauser, L., Ivey, S. L., Huang, D., Engelman, A., Tseng, W., Dahrouge, D., … Kealey, M. (2013). Availability and Readability of Emergency Preparedness Materials for Deaf and Hard-of-Hearing and Older Adult Populations: Issues and Assessments. *PLoS One, 8*(2), e55614.

Nick, G. A., Savoia, E., Elqura, L., Crowther, M. S., Cohen, B., Leary, M., … Koh, H. K. (2009). Emergency Preparedness for Vulnerable Populations: People with Special Health-Care Needs. *Public Health Reports, 124*(2), 338.

Osaki, Y., & Minowa, M. (2001). Factors Associated with Earthquake Deaths in the Great Hanshin-Awaji Earthquake, 1995. *American Journal of Epidemiology, 153*(2), 153–156. https://doi.org/10.1093/aje/153.2.153

Patton, M. (2010). *Developmental Evaluation: Applying Complexity Concepts to Enhance Innovation and Use.* New York, NY: Guilford Press.

Paudel, Y. R., Dariang, M., Keeling, S. J., & Mehata, S. (2016). Addressing the Needs of People with Disability in Nepal: The Urgent Need. *Disability and Health Journal, 9*(2), 186–188.

Peek, L., & Stough, L. M. (2010). Children with Disabilities in the Context of Disaster: A Social Vulnerability Perspective. *Child Development, 81*(4), 1260–1270.

Priestley, M., & Hemingway, L. (2007). Disability and Disaster Recovery: A Tale of Two Cities? *Journal of Social Work in Disability & Rehabilitation, 5*(3–4), 23–42.

Ronoh, S. (2017). Disability through an Inclusive Lens: Disaster Risk Reduction in Schools. *Disaster Prevention and Management, 26*(1), 105–119. https://doi.org/10.1108/dpm-08-2016-0170

Ronoh, S., Gaillard, J. C., & Marlowe, J. (2017). Children with Disabilities in Disability-Inclusive Disaster Risk Reduction: Focussing on School Settings. *Policy Futures in Education, 15*(3), 380–388.

Rooney, C., & White, G. (2007). Narrative Analysis of a Disaster Preparedness and Emergency Response Survey from Persons with Mobility Impairments. *Journal of Disability Policy Studies, 17,* 206–215.

Shakespeare, T. (2013). *Disability Rights and Wrongs Revisited.* Routledge.

Sheppard, P. S., & Landry, M. D. (2016). Lessons from the 2015 Earthquake(s) in Nepal: Implication for Rehabilitation. *Disability and Rehabilitation, 38*(9), 910–913.

Smith, F., Jolley, E., & Schmidt, E. (2012). *Disability and disasters: The importance of an inclusive approach to vulnerability and social capital.* Melksham, UK. Retrieved from https://www.preventionweb.net/publications/view/34933.

Sonpal, D. (2017). *Disability, Disaster and the Law: Developing a Mandate for Disability Inclusive Law Making Process for Disaster Risk Reduction*. Taylor and Francis.

Stough, L. M., & Kang, D. (2015). The Sendai Framework for Disaster Risk Reduction and Persons with Disabilities. *International Journal of Disaster Risk Science, 6*(2), 140–149.

Subramaniam, P., & Villeneuve, M. (2019). Advancing Emergency Preparedness for People with Disability and Chronic Health Conditions: A Scoping Review. *Disability & Rehabilitation.* https://doi.org/10.1080/09638288.2019.1583781

Takahashi, A., Watanabe, K., Oshima, M., Shimada, H., & Ozawa, A. (1997). The Effect of the Disaster Cause by the Great Hanshin Earthquake on People with Intellectual Disability. *Journal of Intellectual Disability Research, 41*, 193–196.

Takayama, K. (2017). Disaster Relief and Crisis Intervention with Deaf Communities: Lessons Learned from the Japanese Deaf Community. *Journal of Social Work in Disability & Rehabilitation, 16*(3–4), 247–260.

Tannenbaum-Baruchi, C., Feder-Bubis, P., Adini, B., & Aharonson-Daniel, L. (2014). Emergency Situations and Deaf People in Israel: Communication Obstacles and Recommendations. *Disaster Health, 2*(2), 106–111.

Turner, B. L., Kasperson, R. E., Matson, P. A., McCarthy, J. J., Corell, R., Christensen, L., ... Schiller, A. (2003). A Framework for Vulnerability Analysis in Sustainability Science. *Proceedings of the National Academy of Sciences of the United States, 100*(4), 8074–8079.

Twigg, J., Kett, M., Bottomley, H., Tan, L. T., & Nasreddin, H. (2011). Disability and Public Shelter in Emergencies. *Environmental Hazards, 10*(3–4), 248–261. https://doi.org/10.1080/17477891.2011.594492

Twigg, J., Kett, M., & Lovell, E. (2018). Disability Inclusion and Disaster Risk Reduction: Overcoming Barriers to Progress. London. Retrieved from https://www.odi.org/sites/odi.org.uk/files/resourcedocuments/12324.pdf

UNESCAP. (2012). *Incheon Strategy to 'Make the Right Real' for Persons with Disabilities in Asia and the Pacific*. Bangkok. Retrieved from http://www.unescap.org/sites/default/files/Incheon%20Strategy%20%28English%29.pdf

UNESCAP. (2014a). *Asia-Pacific Meeting on Disability-Inclusive Disaster Risk Reduction: Changing Mindsets through Knowledge*. Retrieved from http://www.unescap.org/events/asia-pacific-meeting-disability-inclusive-disaster-risk-reduction-changing-mindsets-through

UNESCAP. (2014b). *Message on Disability-Inclusive Disaster Risk Reduction Delivered at the 6th AMCDRR*. Retrieved from http://www.unescap.org/announcement/message-disability-inclusive-disaster-risk-reduction-delivered-6th-amcdrr

UNESCAP. (2017). *Disability in Asia and the Pacific: The Facts—2017 Midpoint Review Edition*. Retrieved from http://www.unescap.org/sites/default/files/Disability_The_Facts_2.pdf

UNESCAP & UNISDR. (2012). *The Asia-Pacific Disaster Report 2012: Reducing Vulnerability and Exposure to Disasters*. Bangkok, Thailand.

UNISDR. (2015a). *Sendai Framework for Disaster Risk Reduction 2015–2030*. Geneva. Retrieved from http://www.preventionweb.net/files/43291_sendai-frameworkfordrren.pdf

UNISDR. (2015b). *UN World Conference on Disaster Risk Reduction, Issue Brief: Inclusive Disaster Risk Management—Governments, Communities and Groups Acting Together*. Geneva. Retrieved from http://www.wcdrr.org/uploads/Inclusive-Disaster-Risk-Management-2.pdf

Villeneuve, M. (2018). Emergency Preparedness Pathways to Disability Inclusive Disaster Risk Reduction. *Australian Journal of Emergency Management, Diversity in Disaster* (Monograph No. 3), 44–47.

Villeneuve, M., Robinson, A., Pertiwi, P. P., Kilham, S., & Llewellyn, G. (2017). The Role and Capacity of Disabled People's Organisations (DPOs) as Policy Advocates for Disability Inclusive DRR in Indonesia. In R. Djalante, F. Thomalla, M. Garschagen, & R. Shaw (Eds.), *Disaster Risk Reduction in Indonesia: Progress, Challenges, & Issues* (pp. 335–356). Cham: Springer International Publishing AG.

Weibgen, A. A. (2015). The Right to Be Rescued: Disability Justice in an Age of Disaster. *Yale Law Journal, 124*(7), 2406–2469.

White, G. W., Fox, M. H., Rooney, C., & Cahill, A. (2007). *Assessing the Impact of Hurricane Katrina on Persons with Disabilities*. Lawrence. Retrieved from http://rtcil.org/sites/rtcil.drupal.ku.edu/files/images/galleries/NIDRR_FinalKatrinaReport.pdf

Wisner, B. (2003). Disability and Disaster: Victimhood and Agency in Earthquake Risk Reduction. In C. Rodrigue & E. Rovai (Eds.), *Earthquake*. London: Routledge.

Wisner, B., Blaikie, P., Cannon, T., & Davis, I. (2004). *At Risk: Natural Hazards, Peoples, Vulnerability and Disasters* (2nd ed.). London: Routledge.

World Health Organization. (2011). *Disaster Risk Management for Health: People with Disabilities and Older People* [Press Release]. Retrieved from http://www.who.int/hac/events/drm_fact_sheet_disabilities.pdf

World Health Organization & The World Bank. (2011). *World Report on Disability*, Geneva.

CHAPTER 18

Future Pathways for Disaster Justice

Anna Lukasiewicz and Claudia Baldwin

18.1 Introduction

This book represents the beginning of a journey to explore the complexities of disaster justice in Australia. As disasters and their impacts continue to increase in the foreseeable future, issues of justice and equity will rise in prominence. Thus disaster justice provides a crucial lens for both research and policy formulation. In this chapter, we outline proposed policy approaches and a research agenda emerging from our November 2018 workshop and book chapter authors.

When we refer to natural disasters, we include rapid onset disasters, slow incremental changes due to climate change or other challenges (see Preface and Chap. 2). In Australia, our most common rapid disasters are due to flooding, cyclones and fire, however these and more slow onset disasters

A. Lukasiewicz (✉)
Fenner School of Environment and Societyand Institute for Integrated Research on Disaster Risk Science, Australian National University, Canberra, ACT, Australia
e-mail: anna.lukasiewicz@anu.edu.au

C. Baldwin
Urban Design and Town Planning, and Sustainability Research Centre, University of the Sunshine Coast, Maroochydore, QLD, Australia
e-mail: cbaldwin@usc.edu.au

© The Author(s) 2020
A. Lukasiewicz, C. Baldwin (eds.), *Natural Hazards and Disaster Justice*, https://doi.org/10.1007/978-981-15-0466-2_18

(e.g. sea level rise along the populated coast and heat stress particularly in urban areas) may be cumulative and require ongoing adaptation as further instances of inequity and injustice emerge. The sudden unpredictable and destructive capacity of bushfires explains why they are such a focus in this book—Chaps. 4, 5, 11–13, and 16 all report on fire case studies. However, urban heat waves have increased in frequency, duration and intensity internationally over the last two decades and have resulted in higher mortalities than any other disaster in Australia (Chesnais et al., 2019). Slow-onset disasters such as chronic drought are also coming to the attention of disaster management research with implications for agriculture and food security, livelihoods and community well-being. Reducing risk is complex. The various outcomes of increasing heat are concerned with similar issues around governance, integration of practice, resilience and vulnerabilities and would benefit from examination through a justice lens.

18.2 Possible Policy Approaches

A key feature of measures proposed in this book is the need for integration and collaboration—integrated thinking and approaches across the Prevention—Preparation—Response—Recovery (PPRR) spectrum, from policy to practice, involving the range of sectors and government agencies. The case studies illustrate the challenges in implementing the aim of shared responsibility due to procedural and distributive injustices that frustrate integration and collaboration. The imbalances in decision-making through exclusion of sectors such as non-governmental organisations (NGOs) and vulnerable communities in relation to disaster preparedness can result in misdirected resources that ultimately make a difference in recovery from disaster. But authors also stress that injustices are pervasive in society and stem from chronic social-economic circumstances.

Without intending to reiterate themes of the *National Strategy for Disaster Resilience* 2011, the following policy approaches suggested by the authors do reflect concrete ways of addressing coordination, communication, empowerment and capacity-building, summed up as—integration and collaboration, facilitated through a justice approach.

18.2.1 NGO and Agency Collaboration

Participation of local community leaders and relevant NGOS in all phases of disaster management such as inclusion in local emergency/disaster management committees is critical (Chap. 12). Partnerships between

NGOs and government organisations can be mutually beneficial by mainstreaming and resourcing successful NGO projects (Chap. 9). The benefits of such were evidenced in Calgaro et al.'s examples of applying resources to capacity-building across sectors and agencies in ways that better prepare agencies and the community to build resilience (Chap. 17).

18.2.2 Integration of a Risk-Based Approach in Governance, Policy and Legislation

Considering uneven impacts of disasters, whether one judges these as 'just' or not, it becomes clear that key policy levers are not in the hands of emergency management or disaster policy. Vulnerability is defined by spatial location, development decisions, socio-economic status, education, access to finance and communication and more (Handmer & Dovers, 2013). Locational disadvantage due to past development decisions, and how land use planning and legal instruments could be used were discussed in Chaps. 4–6, and 11. If reducing vulnerability is desirable, then policy attention is spread from the specifics of disaster management towards the determinants of broader social disadvantage. This invites a search for 'no-regrets' or synergistic policy options, where improving resilience to disasters is but one of the benefits of, for example, increasing social capital and community capacity. This would mean that agencies that traditionally existed in the Preparation and Response phases branch out and collaborate with more land-use planning and social service-focused agencies. This need for integrated governance across different sectors is being increasingly recognised and consistent with the principles of the Sendai Framework.

Proactive planning needs to take account of risk assessments, including climate change amplification of risks, by urban/land use planners, developers, infrastructure providers and emergency service planners. A risk justice approach considers the collective rights of the wider community including future occupants, and the overall costs and benefits to the community and broader population, including tax payers (Chap. 5). Fundamental to a risk justice approach is including informative overlays of risks and possible restrictions on construction through planning schemes, regulations, as well as a star rating or similar on land titles and building permits, revealed in property transfers and rental agreements in high risk areas, whether bushfire or flooding from storms or sea level rise. A longer term strategic line of attack involves reducing land-use and transportation inequities that affect certain sectors of society, such as youth (Chap. 11). This requires sound approaches to making trade-offs looking at opportunity costs and broader social benefits (Chaps. 4 and 8).

Given the deep sense of physical loss and emotional trauma experienced by disaster-affected communities, there is a need to consider the public interest, and for restorative justice, rather than an adversarial legal process that focuses on blame and results in winners and losers (Chaps. 6 and 7). Of course any review or inquiry into a disaster needs to understand how to minimise risk in the future, but a compensation fund can be a way that the broader community (and those who make or accept the rules) makes good on losses.

18.2.3 Participatory Processes

Fundamental to any just system of policy, planning or programme implementation are participatory processes that genuinely engage communities in decision-making. This is not only necessary for two way advice on preparation for potential disasters, but is essential to address underlying societal inequities. But it is not easy. Actions to address barriers include: empowering and resourcing contributions by Indigenous people, those with disabilities, youth, children and those without legal identity to provide insights into preparedness and response policy or programmes; improving emergency service and other agencies' knowledge base about the sector (their needs and capabilities); and ensuring risk and preparedness information is accessible and available to individuals and support organisations (Chaps. 10, 11, 14, and 15). Engaging people in threatened places about the meaning of change, and designing new ways to support social well-being that address place attachment and identify changing places, is key to building justice into resilience (Chap. 13).

Finally, policy and practice can foster mutual respect (interactive justice) and goodwill among sectors and across agencies, and community (e.g. emergency services, land use planners, community leaders and volunteers), as well as influence provision of resources.

18.3 A Research Agenda for the Future

We propose that a disaster justice research agenda establishes what disaster justice looks like for Australia in the context of its Asian neighbours. Being a relatively well-off country, Australia nevertheless has rising inequality, deep seated poverty amongst particular groups (such as our indigenous populations, single parents or the unemployed) and increasing number and costs of natural disasters. We acknowledge the privilege with its accompanying obligations, but also the severe risks in this already sunburnt country. We propose that disaster justice research focuses on:

1. Understanding features of vulnerability and resilience-building
2. Investigating existing and perceived rights, responsibilities, accountabilities, values and expectations
3. Identifying justice issues across the PPRR spectrum
4. Exploring the interlinkages between procedural, distributive and interactional justice

18.3.1 Understanding Features of Vulnerability and Resilience-Building

Procedurally just approaches to disaster management require transparency in sharing of information and in decision-making. Further research is needed on cascading climate change impacts. Current thinking about climate-fire relationships suggest that fires exacerbate global warming which in turn raises the risk of fire (Chap. 2).

To complement the significant research done in Australia on vulnerability and resilience to disasters (Australian Institute for Disaster Resilience, 2018; O'Connell et al., 2018; Parsons et al., 2016), a disaster justice focus can uncover vulnerable segments of the population that may not be considered as 'obvious' or 'deserving' justice, and the contributing factors. Chapters 9–17 discussed injustices faced by different vulnerable groups. Some vulnerable groups, such as people living in poverty, those who are stateless or live with disabilities (in Chaps. 9, 14, and 17 respectively) are obvious and visible. However, others, such as community leaders or business owners (Chaps. 12 and 15), are not groups that would traditionally be identified as vulnerable. A disaster justice perspective can bring other less obvious groups to attention, across government and community actors throughout the PPRR spectrum. One example is the treatment of senior officials and politicians in the media and post-event inquiries. With an increasing focus on responders and volunteers, pioneered through the Bushfire and Natural Hazards Cooperative Research Centre (McLennan, Whittaker, & Handmer, 2016), a disaster justice perspective examining procedural and interactional aspects would add frameworks for a more comprehensive evaluation. Likewise another sector would benefit from a distributional equity lens—the non-impacted members of the public who have burdens and costs imposed on them through taxes or insurance, paying recovery costs for others, post-disaster. This requires complex analyses of economic and social trade-offs.

While the vulnerability of Indigenous people has been acknowledged in disaster management, a justice perspective, with its focus on recognition as well as a temporal and spatial perspective, would not only acknowledge the urgency of disaster management for indigenous people given their socio-economic and cultural situation but also their role as holders of relevant knowledge and enablers of localised solutions. A disaster justice perspective would not just see a group vulnerable to disasters, but an opportunity to incorporate and practise indigenous people's distinctive sovereignty, law, governance systems and knowledge. In parts of Australia, indigenous people are re-connecting to their heritage, supported by Indigenous-led and participatory research (see Chap. 16).

Disaster justice, with a focus on recognition can also contribute to conversations around who decides what vulnerability is, and whose views on vulnerability count. For example, Kelman, Gaillard, Lewis, and Mercer (2016) discuss the disciplinary origins of vulnerability and resilience (especially ecology) and point out how the understanding of resilience as a return to pre-disaster conditions can perpetuate vulnerability, as well as injustice. However, while research on resilience has diversified (see Chap. 15), these ecological origins have shaped much of the discussion, especially when they influence IPCC reports, as Kelman et al. (2016) point out. This illustrates that the question of whose knowledges count is not strictly hypothetical. Fortunately, the 'building back better' approach is becoming more prominent and advocates an improvement on post-disaster conditions, rather than a return to what was there before. Further research on ways to improve inclusivity of marginalised groups (Hallegatte, Rentschler, & Walsh, 2018) will not only enlighten but empower.

A continued focus on the more diverse conceptualisations of vulnerability and the identification of future vulnerabilities in Australia would be helpful for disaster justice. This is especially true given the evolution of research on vulnerability—it is thus no longer used as a static description of specific groups, but rather a fluid and dynamic concept. A person might be vulnerable to one type of disaster but not to another (a flood or a bushfire) or become more or less vulnerable at different points in their life (following a separation from their partner or gaining an illness or a disability).

18.3.2 Investigating Existing and Perceived Rights, Responsibilities, Accountabilities, Values and Expectations

Another identified research area, given the discussion on vulnerability and resilience above is to consider the existence (and perceptions by different groups) of rights and responsibilities, accountabilities, values and expectations around disaster management. Some of these have been tackled in this book—for example, the tensions between private rights and the public good in Chap. 5, as well as rights and expectations of private landholders to protection from sea level rise in Chap. 6. The line between private rights and public responsibilities is continuously contested within society (Lukasiewicz, Dovers, & Eburn, 2017; McLennan & Eburn, 2015) and an investigation into how these vary along the PPRR spectrum is necessary to paint a picture of disaster justice in Australia.

While the uneven impact of disasters is an existing research and policy focus, it may be that a discussion around disaster justice can sharpen debates. The very terms 'justice' and 'injustice' involve deeply held beliefs with strong emotional and legal associations, raising issues of human rights and obligations. It can force attention to whether a particular preparatory or ameliorative action is (1) an arguable but completely optional proposition depending on personal opinion or available resources or (2) something that definitely should or even must be done on the basis of moral or even legal principle. The implications of rights in this sense asks sharp questions of the poorly understood policy goal of 'shared responsibility' (Lukasiewicz, Dovers, Robin, et al., 2017) where responsibilities may produce obligations and expectations that become equated with rights held by some in the community.

A justice lens requires that normative or moral positions about what is fair are made clear, which can be expressed as justice principles, for example, seeking equal treatment for all versus progressively weighted assistance that favours the poor. The analysis of the distributional impact, total cost, cost effectiveness and overall residual risk/risk reduction of, for example, a fire building upgrade or flood alleviation programme needs to be analysed to compare the different outcomes of adhering to either the 'equal treatment for all' or the 'favouring the poor' principles in a policy programme. This may be basic policy option analysis, but not often applied comprehensively, and not to disasters.

18.3.3 Identifying Justice Issues Across the PPRR Spectrum

Much of the vulnerability to disasters is created outside of the disaster space. Disaster justice highlights that the seeds of disaster justice are planted in everyday life. It is there, that vulnerability needs to be tackled. Disasters while to be regretted, provide opportunities for reflection and improvement as weaknesses are exposed. They effectively 'stress test' our systems of governance, institutions, livelihoods, finance and social cohesion. For example, there is now a recognition in Australia, as well as globally, that the neoliberal-prompted deregulation of the building industry and introduction of private certifiers has resulted in unacceptable and even dangerous residential situations (e.g. the Grenfell Tower in the UK; combustible cladding in Australia) where residents have been forced to leave their homes due to sudden defects (often made apparent after natural events such as storms), with little recourse against the developers. An independent expert examination of ways to improve compliance and enforcement practices in the building and constructions systems (Shergold & Weir, 2018) led to a recently announced national approach to industry reform in Australia. Any new system needs to be closely monitored and evaluated for its ability to provide fair outcomes.

If a disaster results in a break or a collapse of a system, and we genuinely accept the need to 'build back better', then the Recovery phase of PPRR becomes a point when systemic intervention is possible to transform to a 'new normal'. Addressing inequity is primarily a social (and political!) process (Kelman et al., 2016) and research into vulnerability has moved into systemic analysis and a focus on values to identify 'points of intervention' (O'Connell et al., 2018).

Reconstruction programmes, created in the immediate post-event environment, can potentially be reactive and localised. Chapters 8 and 9 on the Recovery phase of the PPRR spectrum in Pakistan and Bangladesh respectively remind us of the difficult trade-offs, with justice implications that would benefit from comprehensive analysis. A comprehensive review of reconstruction efforts, post recent disasters in Australia through a disaster justice lens, would be a timely contribution to Australian disaster management research. There have been many lessons learned and principles recommended that are yet to be comprehensively documented (Ryan, Wortley, & Shé, 2016). Therefore, opportunities to address injustices during the Recovery phase are less well understood. There may be value in developing better guidance and lessons on the carriage of both short- and long-term recovery that are inclusive and address uneven impacts and access to recovery means, informed by synthesising lessons from multiple recent recovery experiences. Equitable and just approaches to recovery and reconstruction is one amongst a number of areas where professional capacities incorporating justice perspectives warrant attention and improvement, noting that capacity-building is more often considered a need in communities rather than in agencies, even though it is needed by both.

18.3.4 Exploring the Interlinkages Between Procedural and Distributive Justice

A justice research lens on disasters forces attention on the implications of different forms of justice—broadly, distributional, procedural and interactional justice (e.g. in resource management), (see Lukasiewicz, Dovers, Robin, et al., 2017; Lukasiewicz, Bowmer, Syme, & Davidson, 2013). A strong propensity to pursue real or perceived 'unfairness' in disasters can be made (a little) more tractable by, for example, examining the procedural justice issues (involvement in planning) that may later influence distributive justice (i.e. who gets resources or who is impacted). Then, considerable experience and guidance from social and environmental justice and participatory governance could be brought to bear.

Procedural justice concerns decision-making processes and who participates, how, and whether the processes are transparent, timely and impartial enough to achieve any influence on a decision (Lukasiewicz & Baldwin, 2017). Under a justice framework one might, for example, ensure that information about disaster risk is made widely available and is tailored in an appropriate way, so that all members of a community, including those most vulnerable (e.g. people with disabilities), can advise on the best ways to mitigate impacts and facilitate recovery that does not put them in a worse position. Participatory approaches that build on social capital, ensure social learning and build capacity can contribute to a more resilient community that copes with and is informed and motivated enough to take responsibility and initiative, with support from institutions. Post-disaster, procedural justice may invite consideration of restorative processes and inquiries as opposed to or to supplement the more usual technical or adversarial inquiries (Eburn & Dovers, 2017). Evaluation of, effectiveness of and factors for successful participatory approaches is needed; this includes a participatory research approach to better inform policy.

The Sendai Framework proposes equal degrees of protection from disaster impacts and involvement in planning, response and recovery as a human right, which means that uneven access to protection or involvement violates that right. This in turn requires understanding of how disasters impact different segments of the population and those who lack opportunities for engagement. While emergency management research and agencies are paying increasing attention to this, the requirement for disaggregated data exposes significant gaps.

The law is often assumed to be a prime mechanism to address injustice. While the law is a necessary and even crucial element for delivering fair or just policy and management outcomes, deeper discussions of disaster justice reveal that statutory interpretation and case-specific legal judgements are often uncertain, limited by context and jurisdiction and may contribute to inequity and injustice. The law may be more important in setting out responsibilities and processes of decision-making than in providing redress or firm rules through judicial decisions (see Chaps. 6 and 7 that specifically deal with the law).

To finish our proposal for a research agenda, a final suggestion regarding research methods. We recommend that research be done using proactive evaluative and comparative approaches, rather than a series of case studies in order to avoid the 'death by case study syndrome' that is quite prevalent in disaster management.

References

Australian Institute for Disaster Resilience. (2018). *Australian Disaster Resilience Snapshot: What's Working? What's Needed to Take Us Forward?* Retrieved from https://knowledge.aidr.org.au/resources/australian-disaster-resilience-snapshot/

Chesnais, M., Green, A., Phillips, B., Aitken, P., Dyson, J., Trancoso, R., … Dunbar, C. 2019. *Queensland State Heatwave Risk Assessment 2019*, Queensland Fire and Emergency Services, May 2019.

Eburn, M., & Dovers, S. (2017). Reviewing High-Risk and High-Consequence Decisions: Finding a Safer Way. *Australian Journal of Emergency Management, 32*(4), 26–29.

Hallegatte, S., Rentschler, J., & Walsh, B. (2018). *Building Back Better: Achieving Resilience through Stronger, Faster, and More Inclusive Post-Disaster Reconstruction.* Washington, DC. Retrieved from https://openknowledge.worldbank.org/bitstream/handle/10986/29867/127215.pdf?sequence=4&isAllowed=y

Handmer, J., & Dovers, S. (2013). *Handbook of Disaster Policies and Institutions: Improving Emergency Management and Climate Change Adaptation* (2nd ed.). London, UK: Routledge.

Kelman, I., Gaillard, J. C., Lewis, J., & Mercer, J. (2016). Learning from the History of Disaster Vulnerability and Resilience Research and Practice for Climate Change. *Natural Hazards, 82*(Suppl. 1), 129–143.

Lukasiewicz, A., & Baldwin, C. (2017). Voice, Power, and History: Ensuring Social Justice for All Stakeholders in Water Decision-Making. *Local Environment, 22*(9), 1042–1060.

Lukasiewicz, A., Bowmer, K., Syme, G. J., & Davidson, P. (2013). Assessing Government Intentions for Australian Water Reform Using a Social Justice Framework. *Society & Natural Resources, 26*(11), 1314–1329. Retrieved from http://www.tandfonline.com/doi/abs/10.1080/08941920.2013.791903#preview

Lukasiewicz, A., Dovers, S., & Eburn, M. (2017a). Shared Responsibility: The Who, What and How. *Environmental Hazards, 16*(4), 291–313. https://doi.org/10.1080/17477891.2017.1298510

Lukasiewicz, A., Dovers, S., Robin, L., McKay, J., Schilizzi, S., & Graham, S. (Eds.). (2017b). *Natural Resources and Environmental Justice: Australian Perspectives.* Clayton, VIC: CSIRO Publishing.

McLennan, B., & Eburn, M. (2015). Exposing Hidden-Value Trade-Offs: Sharing Wildfire Management Responsibility Between Government and Citizens. *International Journal of Wildland Fire, 24*, 162–169.

McLennan, B., Whittaker, J., & Handmer, J. (2016). The Changing Landscape of Disaster Volunteering: Opportunities, Responses and Gaps in Australia. *Natural Hazards, 84*(3), 2031–2048. https://doi.org/10.1007/s11069-016-2532-5

O'Connell, D., Wise, R. M., Williams, R., Grigg, N., Meharg, S., Dunlop, M., … Crosweller, M. (2018). *Approach, Methods and Results for Co-producing a Systems Understanding of Disaster: Technical Report Supporting the Development of the Australian Vulnerability Profile.* Retrieved from https://publications.csiro.au/rpr/pub?pid=csiro:EP187363

Parsons, M., Glavac, S., Hastings, P., Marshall, G. R., McGregor, J., McNeill, J., … Stayner, R. (2016). Top-Down Assessment of Disaster Resilience: A Conceptual Framework Using Coping and Adaptive Capacities. *International Journal of Disaster Risk Reduction, 19*, 1–11. https://doi.org/10.1016/j.ijdrr.2016.07.005

Ryan, R., Wortley, L., & Shé, É. N. (2016). Evaluations of Post-Disaster Recovery: A Review of Practice Material. *ANZSOG Evidence Base, 1*(4), 1–33. Retrieved from http://hdl.handle.net/10197/8195

Shergold, P., & Weir, B. (2018). *Building Confidence: Improving the Effectiveness of Compliance and Enforcement Systems for the Building and Construction Industry Across Australia.* Canberra, ACT. Retrieved from https://www.industry.gov.au/data-and-publications/building-confidence-building-ministersforum-expert-assessment

Index[1]

A

Aboriginal people, ix, 299, 300, 302, 307–313
Accessibility, x, 127, 171, 208, 210, 212, 213, 328, 329, 331, 332
Accountable/ility, 8, 9, 16–18, 56, 87, 136, 154, 191, 280, 327, 342, 353, 355
Agency, 8, 55, 58, 64, 66, 80, 81, 109, 110, 138, 142, 143, 152, 154, 156, 171, 177, 190, 196, 207, 232, 242–245, 255, 274, 275, 280, 283, 288, 291–294, 307, 309, 310, 312, 313, 327, 337, 350–352, 357, 358
Ash Wednesday, 75, 76, 79, 80, 84, 134, 140
Australian Capital Territory (ACT), 140, 300, 310–311, 313

B

Bangladesh, viii, 169–181, 356
Black Saturday, 26, 40, 75, 76, 79–82, 94, 135, 137, 138, 140, 141, 145
Bond, 33, 240
Build back better, 154, 225, 356
Burning, prescribed, 215, 307, 308, 313
Bushfire, vii, ix, 26, 56, 73–87, 93, 134, 205–216, 239, 302, 350

C

Capacity, vi, viii, ix, 17, 31, 33, 35–37, 52, 53, 63, 78, 81, 82, 87, 94, 100, 102, 103, 108, 111, 119, 134, 135, 152, 162, 172, 226, 228, 232, 233, 244, 245, 261–264, 268, 269, 271–274, 280, 282, 283, 288, 291, 292, 294, 303, 313, 324, 326, 329, 331, 334, 336–339, 342, 350, 351, 357

---

[1] Note: Page numbers followed by 'n' refer to notes.

Cascading disasters, 4
Catastrophe/catastrophic, 8, 73–77, 79, 86, 87, 133–135, 145, 147, 170, 171, 173, 206, 207, 254, 307
Central Arnhem Land, 300, 309–311, 313
Child-Centred Disaster Risk Reduction (CCDRR), ix, 186–188, 197, 199
Children, viii, 98, 178, 179, 185–199, 207, 211, 244, 262, 264, 266–271, 287, 352
Citizen participation, 16–17
Climate
  change, vii, 13, 25–40, 53, 73, 94, 117–128, 170, 205, 239, 281, 303, 306, 308, 349
  justice, 7, 9–12, 14, 28, 120, 241
Coastal inundation, 36–40
  protection, 122
  storm surge, 37, 39, 170
Collaboration, x, 280, 300, 301, 313, 330–335, 337, 342, 350–351
Collective, vii, 81–83, 86, 93–111, 144, 244, 255, 281, 283, 288, 291, 294, 302, 303, 305, 308, 311, 312, 321, 330, 336, 339, 351
Colonial, 300, 303, 304, 309, 311, 314
Command and control, 221
Community organisations, 222, 224, 226, 227, 229, 235, 236, 247, 280
Compensation, 7, 135, 136, 139–141, 145–147, 266, 267, 284, 286, 352
Consequences, viii, 5, 13, 14, 28, 54, 77, 80, 94, 97, 98, 102, 107, 120, 122, 127, 134, 136, 143, 147, 152, 171, 187, 199, 208, 231, 233, 234, 291, 303, 309, 313
Cool burning, 307
Coping capability, 4
Country Fire Authority (CFA), 75, 78, 95, 96, 99, 102–105, 138
Criteria, 59, 60, 62, 63, 152, 292, 294
Culture, 5, 18, 189, 293, 302, 308, 311, 328, 329, 338
  cultural burning, 311
  sensitivities, 336–338
Cyclones, vii, 3, 26, 37, 170, 171, 173, 303, 349

D
Deaf, x, 326–329
Destruction, 67, 96, 170, 192, 234, 249, 250
Disability-inclusive, 333
Disability-inclusive DRR (DiDRR), x, 320, 321, 324–341
Disability support organisations, 324, 336, 339, 340, 342
Disaster relief, 107, 153, 163, 262, 272
Discrimination, 9, 10, 12, 267, 321, 341
Distribution, vii, 6, 9, 10, 12, 17–18, 27, 36, 52, 53, 59, 62–64, 68, 85, 97, 153, 155, 160, 171, 177, 198, 265–266, 268, 274, 284, 289, 292–294, 341, 354, 355, 357
  distributive justice, v, 12, 97, 100, 304, 305, 357–358
  equity, 155
  inequity, 223
Drought, vii, 4, 26, 29, 31–34, 39, 40, 75, 86, 93, 170, 350

E
Early warning, 170, 341
Earthquake, vi, viii, 3–5, 17, 154, 185–198, 263, 266, 270, 303
Economic efficiency, 155

Education, 59, 66, 82, 83, 87, 95, 101, 176, 179, 180, 186–188, 197–199, 210–214, 216, 224, 229, 231, 267, 284, 287, 291, 293, 310, 333, 336, 351
Elder, 310, 311
Emergency
 management, vi, ix, x, 51–55, 63, 66–68, 80, 221, 224–226, 229, 231, 235, 236, 248, 330, 331, 338–340, 351, 358
 preparedness, 327, 328, 331, 332
 responders, 324, 339
 services, ix, 54, 86, 103, 134, 135, 142, 190, 216, 221–224, 226, 229, 232, 233, 235, 236, 248, 326–330, 338, 339, 342, 351, 352
Employment, 16, 81, 82, 146, 172, 188, 193, 214
Empowerment, vii, ix, 17–19, 51, 180, 279, 280, 291, 327, 336–338, 342, 350
Engagement, 51, 52, 64, 65, 118, 224, 255, 280, 291, 292, 294, 301, 311–313, 329–332, 336, 339, 358
Environmental justice, 9–10, 12–14, 243, 244, 254, 255, 305, 306, 357
Episodic fires, 76
Equality/inequality, v, 11, 14, 19, 52, 68, 97, 110, 117, 152, 160–162, 180, 215, 222, 228–231, 236, 261, 262, 268, 269, 272, 280, 282, 289, 304, 314, 319, 352
Equity, v, 9, 83, 87, 97, 152, 153, 155, 157, 158, 160–164, 206, 256, 284, 291, 301, 305, 336, 349, 354
Evacuation, 14, 26, 75, 99, 102, 105, 229, 264, 265, 284, 289
Exploitation, 186, 198, 262, 270, 271, 274

F

Fatalities, 75, 98, 186
Fire
 fighting, 76, 78, 85, 223
 intensity, 79, 95, 104
 management, ix, 34, 35, 66, 299–314
First Nations, 34, 299n1, 300, 303–305, 304n5, 307, 312, 313
Flammable, vii, 74–78, 80, 87, 93, 96, 99, 104, 105, 108
Flood/flooding, vi–ix, 3, 5, 7, 15, 17, 26, 29, 33, 36–40, 55, 58, 62, 63, 66, 135, 141, 142, 146, 151–164, 170, 175, 224, 284–288, 303, 349, 351, 354, 355
Fuel, 35, 40, 74, 79, 85, 87, 95, 96, 99, 104, 110, 209, 214–216, 308

G

Geographical location, 206
Gippsland, 76
Guidelines, x, 67, 226, 230, 233, 324, 332, 333, 337, 339–341

H

Hazard
 exposure, 83, 84, 86, 97–99
 mitigation, 74, 79
 risk, 301, 331
Health
 emotional, 245
 mental, 29, 186, 196, 253, 286, 291
 spiritual, 242
Housing, vii, 15, 54, 66, 68, 79–85, 102, 151–164, 172, 208, 211, 212, 215, 263, 266, 267, 289
Humanitarian aid, viii, 171, 262, 265–266, 268, 270
Human right/s, 8, 9, 18, 68, 270, 305, 320, 355, 358

I

Identity, viii, ix, 240–246, 248, 249, 252–256, 261–275, 291, 293, 337, 352

Impact, vi, 3, 26, 51, 73, 94, 119, 135, 153, 170, 186, 206, 239, 262, 281, 320, 325, 349

Inclusion/inclusive, x, 7, 9, 64, 180, 208, 210, 211, 215, 236, 243, 256, 263, 264, 279, 280, 319–342, 350, 357

Indigenous, 7, 78, 79, 299–314, 352, 354

Information, ix, 29, 30, 35, 59, 62, 82, 85, 87, 97, 102, 103, 106, 109, 138, 155, 224, 226, 228, 229, 265, 269, 292–294, 327, 331, 336, 338, 340–342, 352, 353, 357

Infrastructure, 4, 26, 31, 34–36, 38, 39, 52, 66, 80, 83, 85, 110, 145, 191, 208–210, 213, 214, 216, 245, 255, 289, 341, 351

Injustice, v, vii, ix, 8, 9, 11, 12, 14–18, 51–56, 63–68, 172, 179, 198, 215, 244, 261–275, 301, 302, 306, 312, 319, 350, 353–356, 358

Institutional, x, 35, 51, 52, 54, 57, 67, 81, 82, 86, 87, 154, 171, 172, 279, 280, 320n1, 327, 333, 340

Insurance, 29, 35, 37, 55, 58, 63, 82, 107, 145–147, 249, 250, 284, 286, 287, 354

Intangible, 240, 245, 254

Intergovernmental Panel on Climate Change (IPCC), 4, 28, 34, 40, 283, 284, 320, 354

Intervention, 4, 9, 104, 107, 108, 170, 225–227, 288, 292, 356

K

Knowledge, vi, ix, 5, 9, 12, 18, 76, 87, 106, 157, 199, 229, 230, 232, 245, 284, 291–294, 306, 308, 313, 324, 326–328, 331, 334, 336, 337, 339, 341, 352, 354

L

Land
holders, 300, 302, 309
managers, 307
use, 27, 52, 58, 77, 80, 84, 101, 209, 308, 351, 352

Land use planning, v, 13, 56, 60, 66, 73, 74, 76, 77, 79–86, 100, 119, 122, 127, 128, 210, 216, 351

Language, 68, 146, 242, 287, 302, 311, 326, 327, 335, 338

Leaders/ship, ix, 54, 55, 58, 120, 221–236, 291, 292, 313, 327, 328, 334, 336–338, 342, 350, 352, 353

Learning, shared, 329, 330, 338, 339, 342

Liability, 136, 137, 140, 145

Link/linkage, viii, ix, 9, 52, 53, 69, 119, 172, 216, 244, 282, 287, 328, 338

Litigation, vii, 117–128, 135–141, 145, 146

Loss of home, 186

M

Macedon, 75, 76

Maladaptive adaptation/maladaptation, 125

Marginalised/marginalisation, 10, 18, 199, 227, 229, 261, 262, 264–269, 273, 274, 302, 320, 321, 338, 354

Martu, 302, 308–309
Media, v, 118, 197, 230, 247, 286, 287, 341, 353
Melbourne, ix, 75, 76, 101, 206, 208–214, 216, 304n5
Mental health, 29, 40, 186, 196, 253, 286, 291
Mobility, 77, 83, 269, 289, 341
Moral community, 7, 12–13

N

National Strategy for Disaster Resilience, 16, 18, 142, 222, 225, 226, 235, 350
Native title (rights), 309
Natural hazard, vi, 3, 10, 13, 14, 31, 54, 56, 142, 145, 170, 172, 190, 300–303, 310–314, 331
Negligent/negligence, 119, 125, 127, 137, 139, 140
Non-governmental organisations (NGOs), 64, 154, 170–173, 175, 177, 179, 181, 188, 190–192, 196, 226, 228, 235, 261, 263, 264, 266, 270, 272, 350–351

P

Pakistan, viii, 68, 151–164, 266, 356
Participation/participatory, v, ix, 16–17, 59, 101, 110, 188, 199, 227, 228, 236, 246, 255, 279–294, 303, 320, 320n1, 321, 329–333, 337, 350, 352, 354, 357
Partnership, 64, 181, 228, 293, 302, 303, 313, 327, 329, 332, 336, 339, 350
Peri-urban, ix, 56, 75, 76, 79–86, 208, 239, 246, 307
Persons/people with disabilities, x, 14, 172, 284, 292, 320–324, 330–342, 357

Place attachment/place-attachment, ix, 240–248, 251, 252, 255, 291, 352
Policy, instruments, 58–62, 67, 69
Politics/political, v–vii, 8–10, 14–17, 19, 54–63, 65, 73, 79, 81, 82, 118, 120–122, 163, 172, 178, 181, 225, 263, 269, 273, 280, 282, 284, 288, 289, 293, 302–304, 303n4, 307, 313, 321, 324, 325, 341
Post disaster, viii, ix, 15, 55, 79, 152–155, 171, 181, 187–188, 190, 191, 194, 195, 197–199, 261, 265, 267, 270, 292, 294, 324, 354, 357
Post-traumatic stress disorder (PTSD), 186, 194, 249
Poverty, viii, 11, 17, 82, 170, 186, 262, 273, 336, 352, 353
Power, vii, 10, 12, 13, 17, 27, 38, 55, 64, 78, 80, 83, 122, 128, 143, 208, 209, 226, 232, 235, 236, 243, 273, 283, 288, 292, 293, 301, 304, 313
Power, S., 32
Precautionary principle, 164
Pre-disaster, 154, 198, 263–265, 267, 270, 274, 324, 354
Preparedness, 52, 53, 63, 66, 69, 190–191, 197–199, 232, 263, 279, 281, 288, 292, 321, 326–329, 331, 332, 335, 337–342, 350, 352
Prevention, vi, 15, 53, 79, 82, 120, 199, 215–216, 226, 274
Procedural, v, vii, 6, 7, 9, 10, 12, 17–19, 52–54, 59, 62, 64, 65, 67–69, 73, 97, 100, 101, 106, 107, 118, 198, 199, 243, 289, 292–294, 319, 324, 325, 350, 353, 357–358

Property
  private, vii, 84, 86, 107, 121–123, 125
  public, 79, 125, 126, 311
Public
  good, vii, 18, 94, 95, 99–101, 105–108, 110, 111, 355
  interest, 56, 121, 126–128, 312, 352
Pyrocene, 74, 77–82, 87

R

Recognition, vii, 7, 9, 10, 12, 17–19, 28, 54, 128, 223, 232, 240, 243, 274, 301, 304–306, 309, 311, 354, 356
Reconstruction, 107, 108, 151–164, 171, 266, 279, 356, 357
Recovery, vi, viii, 4, 11, 15, 35, 52, 53, 58, 74, 86, 108, 146, 147, 170–172, 175, 177, 178, 187, 190–191, 221–236, 245, 248, 251, 261, 266, 267, 279, 283, 284, 289, 332, 350, 354, 356–358
Reflexive, 283, 305
Regional planning, 73–87
Regulatory, 60, 63, 83, 107, 108, 118, 119, 128, 252, 309
Relational fairness, 10
Representative/representation, 7, 9, 65, 97, 137–139, 224, 226, 230, 246, 248, 284, 327, 329, 331, 339
Resilience, ix, 55, 66, 73, 87, 93–111, 145, 153, 156, 157, 162, 163, 170, 171, 196, 199, 222, 223, 225, 227, 235, 240–247, 253–256, 268, 273, 279–294, 324–327, 329, 333–336, 350–355
Resource constraints, 164, 171
Respect, 64, 136, 228, 243, 300, 304, 339, 352
Response, vi, 4, 18, 30, 35, 38, 52–55, 58, 60, 63, 65, 66, 68, 69, 74, 94, 95, 99, 103, 105, 118, 122, 125, 126, 134, 135, 140–142, 144, 146, 170, 171, 175, 177, 187, 188, 194–195, 198, 199, 207, 222, 224, 226, 232, 245, 248, 251, 252, 254–256, 261–269, 272–274, 280, 283–285, 287, 288, 301, 308, 310, 313, 324, 326, 332, 333, 351, 352, 358
Responsibility/responsibilities, vi, vii, ix, 8–10, 14, 16–18, 52, 55, 63–66, 79, 81, 82, 86, 97, 98, 119, 141, 142, 147, 221, 222, 226, 227, 235, 236, 251, 252, 280, 288, 293, 300, 302, 303, 307, 309, 312, 313, 353, 355, 357, 358
Restorative justice, viii, 142–145, 352
  inquiry, 144, 147
  practice, 142–144, 147
Rights
  children's, 187, 188, 198, 199
  collective, vii, 86, 93–111, 351
  human, 8, 9, 18, 68, 270, 305, 320, 355, 358
  individual, 94, 100n1, 106, 107, 128
  legal, 118, 138
Risk
  assessment, 84, 86, 109, 351
  awareness, 335, 337
Roles, ix, 7, 16, 18, 30, 64–66, 74, 82, 93–111, 119, 120, 122, 143, 161–164, 171, 172, 175, 177, 179, 181, 187, 215–216, 225, 232, 234, 236, 245, 248, 250, 252–255, 288, 291–292, 307, 308, 328, 337, 354
Rural/urban interface, 26, 77, 79, 206, 208, 214–216

S

Scenario, 30, 31, 35, 38, 40, 56, 60, 84–85, 269
Security, 136, 142, 146, 198, 285, 350
Self-determination, 301, 303, 336–338, 342
Self-esteem, confidence, 208, 210, 211, 214
Sendai Framework, v, 16–18, 53, 98, 170, 225, 279, 280, 351, 358
Settlement
  design, 95–96
  informal, viii, 169–181, 264
  patterns, 60, 73, 76, 77, 84–86, 108
Settler, 78, 177, 181
Shared responsibility, v, 17, 55, 64, 65, 142–144, 146, 222, 225, 227, 229, 235, 236, 280, 294, 330, 332, 350, 355
Shock, 15, 156, 187, 239–256
Slow onset, 4, 281, 283, 349, 350
Slum, 15, 171–173, 175
Social
  capital, ix, 180, 209–212, 280, 283, 289, 291, 293, 351, 357
  economic, vi, 109, 350
  equity, 284, 301
  exclusion, 214, 215
  inclusion, 208, 210, 211, 215
  justice, v, 9, 10, 13, 110, 205–216, 289, 290, 303, 306
  learning, 87, 283, 292, 293, 357
  vulnerability, vi, 12, 16, 17, 261, 262, 273, 284, 285, 287
  wellbeing, 121, 210, 211
Socio-ecological systems, 78, 87
Socio-economic/political, vii–ix, 3, 15, 16, 77, 82, 83, 170, 173, 175, 176, 178, 180, 181, 210, 211, 215, 252, 254, 269, 292, 303, 351, 354
Spatial planning, vii, 73–74, 79, 80, 83, 86

Stakeholders, 6–8, 10, 12, 17, 18, 29, 35, 53, 59, 60, 97, 141, 147, 153, 154, 158, 163, 177, 246, 247, 279, 281, 284, 288, 292, 294, 301, 312, 324, 331, 334, 335, 338, 340
State Emergency Service (SES), 329
Storms, vi, vii, 3, 5, 26, 33, 37, 39, 40, 123, 170, 194, 224, 351, 356
Strategies, 16, 65, 80, 84, 187, 216, 225, 227, 228, 245, 247, 280, 281, 327, 331, 333, 340
Subsidy, 58, 157–160, 162, 163, 216
Systems approaches, 324–326, 334, 337

T

Tipping points, 77
Tools, 60, 79, 128, 136, 268, 328, 332, 334, 336–340
Trade-offs, viii, 16, 63, 152, 153, 155, 157, 160–163, 352, 354, 356
Tradition
  custodians, 299n1, 310, 311, 313
  owners, 299n1, 309, 310, 312, 313
Training, 35, 134, 170, 172, 177, 181, 309, 324, 326–329, 334, 335, 337, 341
Transform/transformation, 239, 253, 283, 293, 329, 342, 356
Transport, 27, 82, 193, 208–214, 216
Trauma, 86, 186, 194–197, 227, 246, 253, 254, 308, 352
Trigger points, 15, 319

U

Uncertainty, 32, 68, 77, 86, 121, 253, 255, 282, 292, 293
UN Convention on the Rights of Persons with Disabilities, 320
Universal design, 340
Urban planning, 93–111, 216

V

Value for money, 152, 163, 164, 340
Victoria, vii, ix, 26, 73–87, 93, 94, 101, 102, 111, 134, 135, 137, 141, 142, 145, 206, 212, 213, 215
Volunteers, 54, 55, 65, 66, 75, 78, 142, 143, 170, 188–190, 222, 224, 227, 231, 235, 236, 248, 285, 313, 326, 329, 339, 352, 353
Vulnerable/vulnerability, v, 4, 35, 53, 73, 96, 133, 152, 170, 186, 208, 228, 243, 261, 280, 302, 321, 350

W

Warning(s), 81, 87, 102, 170, 181, 185, 193, 194, 341
Weather, extreme, 26, 29, 94, 173, 175
Western Desert, 300, 302, 308, 309
Wildfire(s), vii, 56, 73, 74, 77–79, 85, 93, 94, 96, 133, 205, 206, 216, 303, 307, 308, 310, 312, 313

Y

Youth, viii, ix, 199, 206–208, 212–216, 351, 352

Printed in the United States
by Baker & Taylor Publisher Services